Energy and Climate Policies in China and India

The book explores the proactive and reactive features of China's and India's domestic and foreign policies to address two intertwined challenges: first, China and India have taken policy measures that accord with their own domestic priorities; second, both countries have had to alter the trajectory of their proactive policy measures as a result of external pressures. The book argues that China's and India's proactive and reactive policy measures to address energy insecurity and climate change have been shaped by their two-level pressures. At the domestic/unit level, both countries have had to sustain fast economic growth and eradicate poverty, in order to maximize their economic wealth. At the international/systemic level, both countries have sought to enhance their great-power status in the international system, which is characterized not only by asymmetrical interdependence, but also by global governance in general, and global energy and climate governance in particular.

FUZUO WU is an Assistant Professor at Aalborg University, Denmark. She holds a Ph.D. in International Relations from Fudan University, and was a postdoctoral researcher at Princeton University, the University of Oxford, and Yale University; a Visiting Senior Research Fellow at the Lee Kuan Yew School of Public Policy, National University of Singapore; a Research Fellow at the Center for Chinese Foreign Policy Studies at Fudan University; and a Junior Research Fellow at the Institute of South Asian Studies, Sichuan University, China. She has published a book in Chinese and several articles in peer-reviewed journals, such as the *Journal of Contemporary China, Asian Survey, Asian Perspective,* and the *Journal of Chinese Political Science.*

Energy and Climate Policies in China and India

A Two-Level Comparative Study

FUZUO WU

CAMBRIDGE
UNIVERSITY PRESS

CAMBRIDGE
UNIVERSITY PRESS

University Printing House, Cambridge CB2 8BS, United Kingdom

One Liberty Plaza, 20th Floor, New York, NY 10006, USA

477 Williamstown Road, Port Melbourne, VIC 3207, Australia

314–321, 3rd Floor, Plot 3, Splendor Forum, Jasola District Centre, New Delhi – 110025, India

79 Anson Road, #06–04/06, Singapore 079906

Cambridge University Press is part of the University of Cambridge.

It furthers the University's mission by disseminating knowledge in the pursuit of education, learning, and research at the highest international levels of excellence.

www.cambridge.org
Information on this title: www.cambridge.org/9781108420402
DOI: 10.1017/9781108333498

© Fuzuo Wu 2018

This publication is in copyright. Subject to statutory exception and to the provisions of relevant collective licensing agreements, no reproduction of any part may take place without the written permission of Cambridge University Press.

First published 2018

Printed and bound in Great Britain by Clays Ltd, Elcograf S.p.A.

A catalogue record for this publication is available from the British Library.

Library of Congress Cataloging-in-Publication Data
Names: Wu, Fuzuo (Researcher), author.
Title: Energy and climate policies in China and India : a two-level comparative study / Fuzuo Wu.
Description: New York : Cambridge University Press, 2018. | Includes bibliographical references and index.
Identifiers: LCCN 2018022099 | ISBN 9781108420402 (hardback)
Subjects: LCSH: Energy policy–China. | Energy policy–India. | Climatic changes–Government policy–China. | Climatic changes–Government policy–India. | BISAC: POLITICAL SCIENCE / Government / International.
Classification: LCC HD9502.C62 W8 2018 | DDC 333.790951–dc23
LC record available at https://lccn.loc.gov/2018022099

ISBN 978-1-108-42040-2 Hardback

Cambridge University Press has no responsibility for the persistence or accuracy of URLs for external or third-party internet websites referred to in this publication and does not guarantee that any content on such websites is, or will remain, accurate or appropriate.

For my Maker

For my Mother

Contents

List of Figures	*page* ix
List of Tables	x
Acknowledgments	xii

	Part I Introduction and Analytical Framework	1
1	Introduction	3
	Puzzling Energy and Climate Policy Behavior	5
	Argument in Brief and Propositions	5
	Literature Review	6
	Levels of Analysis	18
	Case Selection	23
	Contributions of the Book	26
	Structure of the Book	26
2	Shaping Energy and Climate Policy Behavior: Wealth, Status, and Asymmetrical Interdependence	31
	Proactive and Reactive State Actors	31
	Two-Level Pressures: Wealth, Status, and Asymmetrical Interdependence	34
	Conclusion	75
	Part II The Inside-Out	77
3	Domestic Energy Policies: Proactive	79
	Energy Security: A Top Priority	79
	Proactive Domestic Measures to Procure Energy Security	85
	Differences between China's and India's Domestic Energy Policy Approaches	100
	Conclusion	104

4	Energy Diplomacy: Proactively Preempting and Reactively Restraining	106
	Proactive and Reactive	106
	Iran	112
	Sudan	138
	Myanmar	154
	Comparison of the Three Cases	164
	Shaping Energy Diplomacy: Two-Level Pressures	166
	Conclusion	175

Part III The Outside-In 177

5	Negotiating Climate Change: Proactively Free-Riding and Reactively Burden Sharing	179
	Dual-Track Climate Diplomacy	180
	Compromises under Dual-Track Climate Diplomacy	196
	Shaping Climate Diplomacy: Two-Level Pressures	208
	Conclusion	232
6	Domestic Climate Policies: Reactive	234
	Pressures from Epistemic Communities to Address Climate Change	235
	Reactive Domestic Policy Measures to Address Climate Change	238
	Subordinating Climate Change to Energy Security	248
	Shaping Domestic Energy and Climate Policy: Two-Level Pressures	255
	Conclusion	269

Part IV Implications and Conclusion 271

7	Implications for Global Energy and Climate Governance	273
	Global Energy and Climate Governance	273
	Implications of Addressing Energy Insecurity and Climate Change	278
	Conclusion	293
8	Conclusion	295
	Principal Findings	295
	Implications for International Relations Research	298
	Understanding the Rise of China and India	301

Index 305

Figures

2.1 China's and India's two-level pressures   *page* 35
2.2 China's and India's GDP annual growth rate (%),
    2003–2013   40
2.3 Real GDP per capita, India and China with some
    developed countries, 2003–2013 (US$)   41
2.4 China's top three trading partners, 2003–2013   69
2.5 India's top three trading partners, 2003–2013   69
2.6 The USA's top three trading partners, 2003–2013   70

Tables

2.1	World top ten economies, 2015	page 41
2.2	China's and India's dependency on energy imports, 2011–2040	44
2.3	Major world oil supply disruption, 1967–2005	46
2.4	China's and India's energy mix (%), 2009–2014	48
2.5	Hard markers and soft markers of great-power status	53
2.6	China's and India's participation in nuclear non-proliferation and export control regimes	60
2.7	Top ten GDP economies, 2013	64
2.8	China, India, and US military expenditure (US$ million in 2011), 2003–2013	65
2.9	Top ten military expenditures (US$ billion), 2011	66
2.10	Exports as percentage of GDP, 2003–2013	68
2.11	Top 10 fossil fuel-rich countries, 2013	74
3.1	Selected Chinese targets for energy development in Five-Year Plans	84
4.1	China's and India's votes on the resolutions on the Iranian nuclear issue at the IAEA, 2003–2012	119
4.2	UN Security Council resolutions on the Iranian nuclear issue, 2006–2012	123
4.3	Comparison of Chinese and Indian reactiveness toward Iran, Sudan, and Myanmar	165
5.1	Current and historical CO_2 emissions, 2011	182
5.2	Joint statements with China during public negotiation sessions, December 2009–December 2015	183
5.3	Joint statements with India during public negotiation sessions, December 2009–December 2015	184
5.4	China's foreign aid related to addressing climate change in developing countries, 2000–2012	187
5.5	China's and India's non-UN climate bandwagoning diplomacy	188

List of Tables

5.6	Patents in climate mitigation-related technologies of the world, USA, EU, China, and India	226
5.7	Major compromises between China and India (developing countries) and developed countries	230
6.1	China's and India's climate change institutions, 2000s	242
6.2	China's and India's domestic climate change policies	249
6.3	The profile of energy security and climate change in China and India	250
7.1	Integration of energy and climate change by major institutions, regimes, forums, and clubs	274
7.2	Comparison of the UNFCCC/Kyoto Protocol, Sino-India climate bandwagoning, and the Paris Agreement	288

Acknowledgments

This book is developed based on the two postdoc projects I conducted under the Henry Chauncey Jr. Postdoctoral Fellowship at Yale University's International Security Studies (ISS) (2010–11) and under the Oxford–Princeton Global Leaders Fellowship Programme (GLF) at the University of Oxford and Princeton University (2012–14). Over the long course of writing and publishing this book, I have benefited enormously from many individuals and several institutions.

Joe Ng, Acquisitions Editor at Cambridge University Press, has played a key role in making it possible for this book to be published. In March 2015, Joe approached me about this book project after my seminar on Sino-Indian climate diplomacy at the National University of Singapore (NUS). In the process of finishing the manuscript, I was struggling between writing and a 350-hour teaching load per semester, as well as some pedagogical training, so I changed the self-imposed deadline to resubmit the manuscript up to three times. I really appreciate Joe's understanding, patience, and kindness. In addition, Joe and his colleagues helped me refine the title and recommended editors for editing and indexing.

Bob Keohane has guided me through every step of the process, from what I should do in response to Joe Ng's approach, how to write the book prospectus, and how to respond appropriately to comments and critiques by reviewers, to making some very insightful comments on my book prospectus and manuscript. The book has benefited tremendously from Bob's insights, and I am very proud to have such an amazing mentor.

Cambridge's six anonymous manuscript reviewers in the three-round review process provided invaluable comments, suggestions, and critiques, especially "Reader B," who was present in every round. This book's theoretical framework, case selection and implications for international relations (IR) research were developed thanks to Reader

B's insightful comments and critiques. However, I alone am responsible for any shortcomings and limitations in this book.

I give thanks to Yale ISS and Oxford–Princeton GLF for offering me postdoc fellowships which enabled me to embark on this book project. At Yale, I audited three courses: "Studies in Grand Strategy," a graduate seminar instructed by John Gaddis, Paul Kennedy, and some other scholars, as well as some retired US diplomats; "Energy, International Security, and the Global Economy," an undergraduate course instructed by Flynt Leverett; and "Environmental Diplomacy Practicum," a graduate seminar instructed by Roy S. Lee at Yale School of Forestry and Environmental Studies. I am grateful to these instructors for allowing me to audit their courses, which helped me to broaden and deepen my understanding of energy security and climate change issues. In addition, I thank John Gaddis, Wayne Hsieh, Ryan Irwin, and other participants for their useful comments on my paper on Sino-Indian energy diplomacy presented at ISS's weekly colloquium. My thanks also go to Paul Kennedy for hospitably inviting all of the ISS's predocs and postdocs to attend some formal research dinners with guest lecturers and informal dinners at his home, which provided me with the chance to communicate with other fellows and get to know more scholars. I also benefited a lot from a number of seminars on IR and international political economy (IPE) at the Yale MacMillan Center.

At Oxford, I learned a lot from a weekly colloquium on IR and GLF's annual workshop and colloquium. I thank Nagire Woods, Andrew Hurrell, and Rosemary Foot for their instruction, and Emily Jones, Thomas Hale, Arunabha Ghosh, Jiyong Jin, Pichamon Yeophantong, and other GLF fellows for their comments and suggestions on my research project.

At Princeton, a weekly colloquium on IR and IPE at the Woodrow Wilson School and weekly graduate lunch seminars on IR and politics at the Department of Politics have helped me to significantly enhance my IR theoretical foundation. Various seminars on China organized by China and the World Program (CWP) have significantly broadened my knowledge of China's relations with the world through diversified perspectives. I am grateful to Thomas Christensen and Yan Bennett for kindly giving me the same treatment as their own fellows, by not only inviting me to attend all of their research dinners, but also offering me the Bradley Research Program Fellowship.

At NUS, I am grateful to Jing Huang and the Center on Asia and Globalisation (CAG) in the Lee Kuan Yew School of Public Policy for offering a Visiting Senior Research Fellowship, where I gave four public seminars based on this book's various chapters, after one of which I was approached by Joe Ng.

At Aalborg University, I give thanks to the Department of Culture and Global Studies (CGS) for providing funding for editing and indexing, and Aalborg University Library for borrowing a few books from other libraries. Thanks to Marianne Rostgaard for allowing me to use a large office for a whole year so that I could have a separate space to pay close attention to writing the manuscript; to Susanne Aarup and Lise-Lotte Holmgreen for their help in getting me funding from the CGS; and to Peng Bo and Ashley Kim Stewart for letting me use their office for half a month during the Christmas holidays, 2015.

At Fudan University, I am deeply grateful to Du Youkang and Zhang Guihong for their great support of my career development. Also to Wang Xiaohua and Liu Fang, for their help while I was in difficulty. And to Zhang Jiegen, for helping me to get my travel documents speedily.

At Sichuan University, I thank Wen Fude and Zhang Li for their support of my career development.

Thanks to Paula Parish for editing the whole book and to Caroline Diepeveen for compiling the index.

Portions of Chapters 4 and 5 appeared in the *Journal of Contemporary China* (Taylor & Francis Group), *Asian Survey*, *Asian Perspective* (Lynne Rienner Publishers), and the *Journal of Chinese Political Science*. I am grateful to these journals and their publishers for giving permission to use the material.

Thanks to my younger sister, Fulian, and her colleague, Wang Tao, for drawing several figures. Thanks to my oldest niece, Li Tianran, for collecting all the Chinese sources used in the book. Thanks to my parents and four sisters for their constant encouragement and support.

Thanks to Shanghai Grace Church, University Church at Yale, Christ Church Cathedral, Princeton University Church, Clementi Bible Center in Singapore, and Budolfi Church in Aalborg for wonderful services on Sundays, where I have been inspired not only by excellent sermons, but also by beautiful hymns sung by the most beautiful voices.

Acknowledgments

I am so grateful to God for blessing me with all these wonderful people and institutions! Had I not been chosen by God, it would not have been possible for me to have grown up, from a careless elementary school student who barely achieved passing grades during the first three years, to be a faculty member at universities. Thanks to God's blessing, this book project attracted the attention of Cambridge's editor. I have been praying to God to bless me with the wisdom and capability to write out the manuscript that would meet Cambridge's high criteria. Daily Scripture-reading, praying, and hymn-singing have greatly encouraged me to overcome all the difficulties and challenges in the process of finishing this book. Thus, humbly, I dedicate this book to my Maker – the Almighty God!

PART I

Introduction and Analytical Framework

1 Introduction

China and India, two of the world's fastest-growing economies and most populous countries, with their ever-increasing demand for energy and their huge greenhouse gas (GHG) emissions, have already emerged as crucial players in the international system in general, and in the issue areas of energy security and climate change in particular. In other words, both China and India matter.[1] More specifically, they are the world's largest and third-largest energy consumers respectively, consuming coal extensively. They are the world's largest and third-largest net importers of oil respectively, and are projected to lead global oil demand growth over the next 20 years.[2] They are also the world's largest and third-largest emitters of GHGs on an annual basis respectively. As a result, it is almost inevitable that China's and India's efforts to address the intertwined challenges of energy insecurity[3] and climate change[4] will have significant implications for

---

[1] In the late 1990s, it was argued that China was a middle power so it was relatively unimportant, or China did not matter. See Gerald Segal, "Does China Matter?," *Foreign Affairs* 75, no. 5 (September/October 1999), 24–36. However, such a view has been seriously challenged by other scholars since the beginning of the 21st century. See Robert G. Sutter, "Why Does China Matter?," *The Washington Quarterly* 27, no. 1 (Winter 2003–4), 75–89; Barry Buzan and Rosemary Foot (eds), *Does China Matter? A Reassessment* (London: Routledge, 2004). As to the argument that India matters, see Mohammed Ayoob, "India Matters," *The Washington Quarterly* 23, no. 1 (2000), 27–39; Maya Chadda, *Why India Matters* (Boulder, CO: Lynne Rienner Publishers, 2014).

[2] International Energy Agency (IEA), *World Energy Outlook 2013* (Paris: OECD/IEA, 2013), 62.

[3] Energy insecurity may be defined as "the loss of economic welfare that may occur as a result of a change in the price or availability of energy." See D. R. Bohi and M. A. Toman, *The Economics of Energy Security* (Norwell, MA: Kluwer Academic Publishers, 1996), 1.

[4] "Climate change" is defined by the United Nations Framework Convention on Climate Change (UNFCCC) as "a change of climate which is attributed directly or indirectly to human activity that alters the composition of the global atmosphere and which is in addition to natural climate variability observed over

global energy and climate governance, given the fact that without substantial efforts on the part of China and India to enhance their energy security and limit their increased GHG emissions, any measures undertaken by other countries to address these two challenges would turn out to be much less effective. In other words, China's and India's efforts to address energy insecurity and climate change have already had and will continue to have significant ramifications for global energy and climate governance, which becomes an important factor in determining the effectiveness and evolution of global governance on the two issue-areas.[5] Other factors, such as sustainable development and human development, contribute to the equation, so that not only the economic and political, but also moral and ethical issues will invariably be brought to the fore. Simply put, both countries have a major role to play in global governance on energy security and climate change.

This introductory chapter starts with China's and India's puzzling energy and climate policy behavior and a brief statement of the main argument and propositions. It then reviews studies of the existing research on China's and India's energy security and climate policies, the literature on global energy and climate governance, as well as on China's and India's roles in global energy and climate governance. This is followed by a theoretical debate on levels of analysis in international relations, case selection, and contributions of the study, as well as a synopsis of each chapter in the book.

---

comparable time periods." See United Nations, *United Nations Framework Convention on Climate Change* (1992), 7, available at http://unfccc.int/files/essential_background/background_publications_htmlpdf/application/pdf/conveng.pdf, accessed February 19, 2018. "Energy insecurity" and "climate change" appear as two key words in the book title: Felix Dodds, Andrew Higham, and Richard Sherman (eds), *Climate Change and Energy Insecurity: The Challenge for Peace, Security and Development* (London: Earthscan, 2009).

[5] According to Robert O. Keohane, "Issue-areas are best defined as sets of issues that are in fact dealt with in common negotiations and by the same, or closely coordinated, bureaucracies, as opposed to issues that are dealt with separately and in uncoordinated fashion. Since issue-areas depend on actors' perceptions and behavior rather than on inherent qualities of the subject-matters, their boundaries change gradually over time." See Robert O. Keohane, *After Hegemony: Cooperation and Discord in the World Political Economy* (Princeton, NJ: Princeton University Press, 2005), 61.

Puzzling Energy and Climate Policy Behavior

China's and India's policy behavior to address their energy insecurity and climate change in the first decade and a half of the twenty-first century has been puzzling. In terms of their energy policies, especially their energy diplomacy, both countries have, on the one hand, employed a variety of policy measures – political, economic, military, and diplomatic – to foster close relationships with all energy-rich countries across the world, including those labeled as "pariah states," such as Iran, Sudan, and Myanmar. On the other hand, China and India have also undertaken policy measures that obviously run counter to their efforts to strengthen their energy ties with those energy-rich countries. For instance, both China and India voted against Iran at the International Atomic Energy Agency (IAEA), and China voted against Iran at the UN Security Council on the Iranian nuclear issue. Moreover, China and India even cut their oil imports from Iran against the backdrop of their increased dependence on imported oil (see more in Chapter 4). When it comes to China's and India's climate diplomacy, both countries had at one time consistently rejected taking any actions to mitigate their increased GHGs, especially CO_2 emissions, in international climate change negotiations (ICCN), which was the main force that led to the fiasco of the Copenhagen climate conference in late 2009. In stark contrast, at the Paris climate conference in late 2015, China and India agreed, along with other countries, developed and developing, not only to undertake voluntary actions to mitigate their CO_2 emissions, but also to subject their actions to legally binding, transparent procedures, which eventually led to the successful adoption of the Paris Agreement (see more in Chapter 5). Why have China and India conducted such puzzling energy and climate diplomacy? More broadly, what forces have driven China's and India's energy and climate policies in general, and their energy and climate diplomacy in particular?

Argument in Brief and Propositions

The main argument of the book is that China's and India's energy and climate policy behavior has been both proactive and reactive. On the one hand, China and India have adopted proactive energy and climate policy measures, at both national and international levels. On the other hand, both countries have had to modify or adjust their proactive policy measures in response to external pressures applied on them by state and non-state actors. China's and India's proactive and reactive energy and climate policy behaviors have been shaped by two-level pressures: at the domestic/unit level, both countries have tried to maximize their economic wealth by sustaining their fast economic growth; at the international/systemic level, both countries have tried to enhance their status as great powers in the international system, which is characterized not only by asymmetrical interdependence, but also by global governance in general and global energy and climate governance in particular.

Based on this argument, I put forward four propositions:[6]

Proposition 1: The energy policies of China and India are designed to help these countries to achieve sustained, fast economic development at the domestic level.

Proposition 2: The energy policies that China and India have pursued globally are constrained by the patterns of asymmetrical interdependence and international norms in which the two countries are enmeshed. In other words, their energy diplomacy is increasingly constrained by forces emanating from the systemic level.

Proposition 3: China's and India's negotiating stances in international climate change negotiations are increasingly shaped by asymmetrical interdependence, climate protection norms, and social opprobrium at the systemic level.

Proposition 4: Addressing climate change is not China's or India's domestic policy priority. They have only adopted domestic climate change policies when faced with increased international pressures.

Literature Review

Since the early 2000s, issues related to China's and India's energy security and climate change in general, and their relationship with global energy and climate governance in particular, have already generated considerable scholarly attention. This section explores this

[6] I owe the idea of developing propositions to Professor Robert O. Keohane, who made some very insightful and useful comments on the second version of my book prospectus. My four propositions draw directly on the two propositions recommended by Professor Keohane.

Introduction

growing literature on energy security and climate change, in addition to the literature on global governance on these two issue-areas, and the literature on the role of both countries in global energy and climate governance.

China's and India's Energy Security and Climate Change

As far as the energy security issue is concerned for both countries, there is a hot debate among scholars and experts. Much of this debate centers on the nature and impact of their national oil companies' (NOCs) "going-out" strategy, to acquire overseas equity oil and natural gas on the international energy market, as well as the impact of this on regional and international security. Some argue that seeking overseas energy has transformed both countries' foreign policy,[7] and the nature of such strategy is neo-mercantilism, aimed at locking up energy resources around the world.[8] Therefore, it has intensified competition for scarce energy resources with other energy-consuming countries, especially with the United States,[9] which might worsen

---

[7] Charles E. Ziegler, "The Energy Factor in China's Foreign Policy," *Journal of Chinese Political Science* 11, no. 1 (March 2006), 1–23; Amy Myers Jaffe and Steven W. Lewis, "Beijing's Oil Diplomacy," *Survival* 44, no. 1 (Spring 2002), 115–134; I. P. Khosla (ed.), *Energy and Diplomacy* (New Delhi: Konark Publishers PVT Ltd, 2005); Sascha Müller-Kraenner, "China's and India's Emerging Energy Foreign Policy," Discussion Paper, Bonn: Deutsches Institut für Entwicklungspolitik 15/2008, available at www.die-gdi.de/uploads/media/DP_15.2008.pdf, accessed February 19, 2018; Erica S. Downs, *China* (Washington, D.C.: Brookings Foreign Policy Studies Energy Security Series, 2006), available at www.brookings.edu/~/media/research/files/reports/2006/12/china/12china.pdf, accessed February 19, 2018; Tanvi Madan, *India* (Washington, D.C.: Brookings Foreign Policy Studies Energy Security Series, 2006), available at www.brookings.edu/~/media/research/files/reports/2006/11/india/2006india.pdf, accessed February 19, 2018.

[8] Flynt Leverett, "Resource Mercantilism and the Militarization of Resource Management: Rising Asia and the Future of American Primacy in the Persian Gulf," in Daniel Moran and James A. Russell (eds), *Energy Security and Global Politics: The Militarization of Resource Management* (New York: Routledge, 2009), 211–242; Kenneth Lieberthal and Mikkal Herberg, "China's Search for Energy Security: Implications for U.S. Policy," *NBR Analysis* 17, no. 1 (2006), 5–9.

[9] 吴磊 [Wu Lei],《能源安全与中美关系》[*Energy Security and Sino-US Relations*] (北京：中国社会科学出版社 [Beijing: China Social Sciences Press], 2009); 伍福佐 [Wu Fuzuo], "能源安全：中印面临的共同难题 [Energy Security: A Shared Problem Facing both China and India],"《南亚研究季刊》[*South Asian Studies Quarterly*] 2 (2006), 47–53; 张力 [Zhang Li], "印度的能源外交及

geopolitical competition and even lead to conflicts at the regional level, such as in the Middle East – the world's largest reservoir of proven oil and gas reserves – and in Asia,[10] as well as at the international level.[11] In addition, some argue that both countries' resource-motivated trade and investment relations with some so-called "rogue states," or "pariah states," such as Iran, Myanmar, Sudan, and other African countries, have largely compromised the efforts of the international community – especially the Western countries – to prevent nuclear proliferation, promote good governance, and protect human rights.[12] Although the International Energy Agency (IEA) does not make a similar normative judgment in its *World Energy Outlook 2007: China and India Insights*, an outlook with an exclusive focus on China's and India's energy sectors and energy-related CO_2 emissions, it still implies

其地缘政治考量 [India's 'Energy Diplomacy' and its Perspectives on Geopolitics]," 《南亚研究季刊》 [*South Asian Studies Quarterly*] 3 (2004), 34–40; 张力 [Zhang Li], "能源外交：印度的地缘战略认知与实践 [India's Energy Diplomacy: Geo-Strategic Perceptions and Practice]," 《世界经济与政治》 [*World Economics and Politics*] 1 (2005), 51–56; Philip Andrews-Speed, Xuanli Liao, and Roland Dannreuther, *The Strategic Implications of China's Energy Needs* (London: International Institute for Strategic Studies, 2002).

[10] Flynt Leverett and Jeffrey Bader, "Managing China–U.S. Energy Competition in the Middle East," *The Washington Quarterly* 29, no. 1 (Winter 2005–6), 187–201; Robert A. Manning, *The Asian Energy Factor: Myths and Dilemmas of Energy, Security and the Pacific Future* (New York, NY: Palgrave, 2000); Stein Tønnesson and Åshild Kolås, *Energy Security in Asia: China, India, Oil and Peace* (Oslo: International Peace Research Institute, 2006); Michael Wesley, *Energy Security in Asia* (London: Routledge, 2007); Charles L. Glaser, "How Oil Influences U.S. National Security," *International Security* 38, no. 2 (Fall 2013), 112–146.

[11] Michael Klare, *Resource Wars: The New Landscape of Global Conflict* (New York: Henry Holt and Company, 2002).

[12] George J. Gilboy and Eric Heginbotham, *Chinese and Indian Strategic Behavior: Growing Power and Alarm* (New York: Cambridge University Press, 2012), 231–250; Harry G. Broadman, *Africa's Silk Road: China and India's New Economic Frontier* (Washington, D.C.: The World Bank, 2007); Monica Enfield, "Africa in the Context of Oil Supply Geopolitics," in Andreas Wenger, Robert W. Orting, and Jeronim Perovic (eds), *Energy and the Transformation of International Relations: Toward a New Producer–Consumer Framework* (Oxford and New York: Oxford University Press, 2009); David Zweig and Bi Jianhai, "China's Global Hunt for Energy: The Foreign Policy of a Resource Hungry State," *Foreign Affairs* 84, no. 5 (September/October 2005), 25–38; Ian Taylor, "China's Oil Diplomacy in Africa," *International Affairs* 82, no. 5 (October 2006), 937–956.

that both countries would pose some serious challenges for the international community in terms of energy security and climate change.[13]

Others, in stark contrast, challenge the aforementioned discourse in three ways: first, pointing out the fact that both countries' NOCs have traded the majority of their acquired equity oil and gas on the international energy market rather than shipping it back to their domestic markets, and therefore, both countries' going-out strategy has increased world energy supplies;[14] second, arguing that the strategy of both countries has provided the opportunity for more cooperation between countries because of their increased interdependence;[15] and third, providing empirical evidence that China and India have already cooperated with each other in their efforts to seek overseas energy supplies.[16]

When it comes to China's and India's climate change issue, the existing literature includes research studies that use a variety of perspectives. Some scholars explore the factors behind both countries' climate policies. For instance, Yu argues that the United Nations Framework Convention on Climate Change (UNFCCC) has played an important role in shaping China's climate policy.[17] In contrast, Moore contends that it is the Chinese Communist Party's core interests that have determined its climate policy.[18] Gan explores China's climate diplomacy

---

[13] IEA, *World Energy Outlook 2007: China and India Insights* (Paris: OECD/IEA, 2007).

[14] Downs, *China*; Madan, *India*.

[15] Zha Daojiong, "China's Energy Security: Domestic and International Issues," *Survival* 48, no. 1 (Spring 2006), 179–190; 伍福佐 [Wu Fuzuo],《亚洲能源消费国间的能源竞争与合作：一种博弈的分析》[*Energy Competition and Cooperation Among Asian Energy Consuming Countries: A Game Theory Analysis*] (Shanghai: Shanghai People's Publishing House, 2010).

[16] Hong Zhao, *China and India: The Quest for Energy Resources in the Twenty-First Century* (London: Routledge, 2012); Ma Jiali, "The Energy Cooperation between China and India in the Post-Crisis Era," *Contemporary International Relations* 20, no. 2 (2010), 96–102; 张立 [Zhang Li], "浅论中印能源合作 [Brief Comments on Sino-Indian Energy Cooperation],"《国际问题研究》[*International Studies*] 1 (2008), 26–29; 龚伟 [Gong Wei], "印度能源外交与中印合作 [India's Energy Diplomacy and Sino-Indian Cooperation],"《南亚研究季刊》[*South Asian Studies Quarterly*] 1 (2011), 29–34.

[17] Yu Hongyuan, *Global Warming and China's Environmental Diplomacy* (New York: Nova Science Publishers, Inc., 2008).

[18] Scott Moore, "Strategic Imperative? Reading China's Climate Policy in Terms of Core Interests," *Global Change, Peace, and Security* 22, no. 4 (June 2011), 147–157.

through the perspective of its capacity building,[19] and Zhang investigates climate change and China's national security.[20] In addition, Zhang also compares China's climate change-related cooperation with both Japan and the United States.[21] Bo examines China's climate diplomacy through its willingness and capacity to cooperate in global climate governance.[22] Ma explores how China has internalized the international climate institutions.[23] Both Rajamani and Vihma point out that India's climate policy has been shaped by its priorities, such as economic development and poverty eradication, as well as by international pressure.[24] Rajamani explores China's and India's climate policy in ICCN using a moral perspective, by arguing that both countries' negotiating stance is legitimate according to the existing climate change regime, but it is not sagacious because climate change will negatively impact their own poorest people and other poorer nations in the developing world.[25] Siddiqi argues that there is more cooperation than competition between China and India in areas of both energy and

---

[19] 甘均先 [Gan Junxian], 《中国气候外交能力建设研究》 [*A Study of China's Climate Diplomacy and its Capacity Building*] (Beijing: China Social Sciences Press, 2013).

[20] 张海滨 [Zhang Haibin], 《气候变化和中国国家安全》 [*Climate Change and China's National Security*] (Beijing: Current Affairs Press, 2010).

[21] 张海滨 [Zhang Haibin], "应对气候变化：中日合作与中美合作比较研究 [Addressing Climate Change: A Comparative Study of Sino-Japan and Sino-US Cooperation]," 《世界经济与政治》 [*World Economics and Politics*] 1 (2009), 38–48.

[22] 薄燕 [Bo Yan], "合作意愿与合作能力 – 一种分析中国参与气候变化全球治理的新框架 [Cooperative Will and Cooperative Capacity – A New Framework for Analyzing China's Participation in Global Climate Governance]," 《世界经济与政治》 [*World Economics and Politics*] 1 (2013), 135–155.

[23] 马建英 [Ma Jianying], "国际气候制度在中国的内化 [The Internalization of International Climate Institutions in China]," 《世界经济与政治》 [*World Economics and Politics*] 6 (2011), 91–121.

[24] Lavanya Rajamani, "India and Climate Change: What India Wants, Needs, and Needs to Do," *India Review* 8, no. 3 (August 2009), 340–374; Antto Vihma, "India and the Global Climate Governance: Between Principles and Pragmatism," *The Journal of Environment Development* 20, no. 10 (January 2011), 1–26.

[25] Lavanya Rajamani, "China and India on Climate Change and Development: A Stance that Is Legitimate but not Sagacious?," in Steven Bernstein, Jutta Brunnee, David G. Duff, and Andrew J. Green (eds), *A Globally Integrated Climate Policy for Canada* (Toronto: University of Toronto Press, 2007), 104–127.

climate change.[26] Rajan investigates India's stance in ICCN in the context of North–South relations.[27] Michaelowa and Michaelowa examine India's negotiating behavior in ICCN against the backdrop of its rise in the international system.[28] Wu examines Sino-Indian cooperation in ICCN and its implications for the international climate change regime, as well as China's negotiating strategy,[29] and argues that China's climate diplomacy has been shaped by its desire to maximize its wealth and status in the international system, characterized by its asymmetrical interdependence with Western countries, especially the United States and the European Union (EU).[30]

Global Energy and Climate Governance

The early twenty-first century has seen a rapidly growing body of scholarship on global energy and climate governance. With regard to global energy governance, Ann Florini, Benjamin K. Sovacool, and colleagues have carried out a series of research studies on this topic, focusing on questions of who governs energy,[31] mapping global energy governance,[32] existing gaps in global energy governance and how to bridge those gaps.[33] Sovacool and Florini also investigate the IEA's

[26] Toufiq Siddiqi, "China and India: More Cooperation than Competition in Energy and Climate Change," *Journal of International Affairs* 64, no. 2 (2011), 73–90.
[27] Mukund Govind Rajan, *Global Environmental Politics: India and the North–South Politics of Global Environmental Issues* (New Delhi: Oxford University Press, 1997).
[28] Katharina Michaelowa and Axel Michaelowa, "India as an Emerging Power in International Climate Negotiations," *Climate Policy* 12, no. 5 (2012), 575–590.
[29] Fuzuo Wu, "China's Pragmatic Tactic in International Climate Change Negotiations: Reserving Principles with Compromise," *Asian Survey* 53, no. 4 (2013), 778–800; Fuzuo Wu, "Sino-Indian Climate Cooperation: Implications for the International Climate Change Regime," *Journal of Contemporary China* 21, no. 77 (2012), 827–843.
[30] Fuzuo Wu, "Shaping China's Climate Diplomacy: Wealth, Status and Asymmetrical Interdependence," *Journal of Chinese Political Science* 21, no. 2, (June 2016), 199–215.
[31] Ann Florini and Benjamin K. Sovacool, "Who Governs Energy? The Challenges Facing Global Energy Governance," *Energy Policy* 37, no. 12 (2009), 5239–5248.
[32] Navroz K. Dubash and Ann Florini, "Mapping Global Energy Governance," *Global Policy* 2, S1 (September 2011), 6–18.
[33] Ann Florini and Benjamin K. Sovacool, "Bridging the Gaps in Global Energy Governance," *Global Governance* 17, no. 1 (2011), 57–74.

role in global energy governance[34] and the complications of global energy governance.[35] Dries Lesage, Thijs Van de Graaf, and Kirsten Westphal explore global energy governance in a multipolar system, and argue that "multipolarity is a major political factor that has to be seriously taken into account when constructing a global energy regime capable of delivering a sustainable energy future."[36] Andreas Goldthau and Jan Martin Witte argue that global energy governance is unlikely to return to the past, as in the 1970s, when oil-producing countries intentionally cut oil supplies to consuming countries due to political reasons, because "more transparency would reduce uncertainty in international energy markets and thus adjustment costs for both consumers and producers,"[37] bringing about a new rule in global energy governance.[38] Christian Downie investigates the group of rising countries known as the BRICs (Brazil, Russia, India, and China) in global energy governance, and argues that "the BRICs are unlikely to have either the capacity or the willingness to drive global energy governance reform."[39]

Existing literature also explores some specific institutions or clubs in global energy governance. For instance, the IEA, established in 1974 to respond to the 1973 oil shock, serving to coordinate the energy policies of its 28 member countries from the Organization for Economic Cooperation and Development (OECD), has become a major research focus. Examples include Florini, who explores the IEA's role in global governance; Van de Graaf, who analyzes systematically the various pressures faced by the IEA, as well as the resources it has at its

---

[34] Ann Florini, "The International Energy Agency in Global Energy Governance," *Global Policy* 2, S1 (September 2011), 40–50.
[35] Benjamin K. Sovacool and Ann Florini, "Examining the Complications of Global Energy Governance," *Journal of Energy & Natural Resources Law* 30, no. 3 (2012), 235–263.
[36] Dries Lesage, Thijs Van de Graaf, and Kirsten Westphal, *Global Energy Governance in a Multipolar World* (Burlington, VT: Ashgate, 2010).
[37] Andreas Goldthau and Jan Martin Witte, "Back to the Future or Forward to the Past? Strengthening Markets and Rules for Effective Global Energy Governance," *International Affairs* 85, no. 2 (2009), 379.
[38] Andreas Goldthau and Jan Martin Witte (eds), *Global Energy Governance: The New Rules of the Game* (Washington, D.C.: Brookings Institution Press, 2010).
[39] Christian Downie, "Global Energy Governance: Do the BRICs Have the Energy to Drive Reform?" *International Affairs* 91, no. 4 (2015), 780.

disposal to cope with those stresses;[40] and Leverett, who investigates the rising powers, the IEA, and the global energy governance architecture.[41] Kohl explores the IEA's role in facilitating cooperation among energy-consuming countries in the global energy order.[42] All these scholars acknowledge that although the IEA remains the single most important organization for energy-importing countries, due to its rather limited mandate and membership, the IEA's legitimacy and efficiency in global energy governance has been increasingly challenged. Lesage, Van de Graaf, and Westphal explore the role the G8 has played in global energy governance since the 2005 Gleneagles summit.[43] Downie explores the Group of Twenty's (G20) role in global energy governance and identifies the means that would enable G20 to drive more than piecemeal change.[44]

In addition, there is a consensus in the existing literature that the fundamental feature of global energy governance is fragmentation. For instance, Van de Graaf employs the creation of the International Renewable Energy Agency (IRENA) as a case study to explore the fragmentation of global energy governance.[45] He further examines it by investigating in detail the politics and institutions in global energy governance.[46] Given

---

[40] Thijs Van de Graaf, "Obsolete or Resurgent? The International Energy Agency in a Changing Global Landscape," *Energy Policy* 48 (September 2012), 233–241.
[41] Flynt Leverett, "Consuming Energy: Rising Powers, the International Energy Agency, and the Global Energy Architecture," in Alan S. Alexandroff and Andrew F. Cooper (eds), *Rising States, Rising Institutions: Challenges for Global Governance* (Washington, D.C.: Brookings Institution Press, 2010), 240–265.
[42] W. L. Kohl, "Consumer Country Energy Cooperation: the International Energy Agency and the Global Energy Order," in Goldthau and Witte, *Global Energy Governance*, 195–220.
[43] Dries Lesage, Thijs Van de Graaf, and Kirsten Westphal, "The G8's Role in Global Energy Governance since the 2005 Gleneagles Summit," *Global Governance* 15, no. 2, (2009), 259–277.
[44] Christian Downie, "Global Energy Governance in the G-20: States, Coalitions, and Crises," *Global Governance* 21, no. 3 (2015), 475–492.
[45] Thijs Van de Graaf, "Fragmentation in Global Energy Governance: Explaining the Creation of IRENA," *Global Environmental Politics* 13, no. 3 (August 2013), 14–33.
[46] Thijs Van de Graaf, *The Politics and Institutions of Global Energy Governance* (London: Palgrave Macmillan, 2013).

such a fragmented situation, Hirst and Froggatt investigate the reform of existing global energy governance.[47]

Similarly, the existing literature has dealt with global climate governance through a variety of perspectives. Some explore its fundamental features. For instance, Robert Keohane and David Victor focus on characterizing the structure and composition of emerging institutions and argue that there is a "regime complex" that governs global climate change, and that "in the case of climate change, the structural and interest diversity inherent in contemporary world politics tends to generate the formation of a regime complex rather than a comprehensive, integrated regime."[48] Victor explores why states have not created a robust international regime to govern climate change by investigating the gridlock under the UNFCCC, and puts forward some practical policy recommendations to overcome this.[49] Karlsson-Vinkhuyzen and McGee argue that global climate governance is in an era of fragmentation, which is centered on the UNFCCC, whose legitimacy has been under challenge with the increased emergence of some minilateral forums of climate change, such as the Asia-Pacific Partnership on Clean Development and Climate Change (APP), the Major Economies Forum on Energy and Climate Change (MEF) and meetings, and the G8.[50] In the same vein, Cadman edited volume examines global climate governance through the dimension of institutional legitimacy.[51] Like global energy governance, global climate governance is also in a state of fragmentation, and the existing literature has explored this aspect at great length. For instance, Biermann, Pattberg,

---

[47] Neil Hirst and Antony Froggatt, "The Reform of Global Energy Governance," Grantham Institute for Climate Change, Discussion paper No. 3 (December 2012), available at www.chathamhouse.org/sites/files/chathamhouse/public/Research/Energy,%20Environment%20and%20Development/1212granthamreport_energygovernance.pdf, accessed February 19, 2018.

[48] Robert O. Keohane and David G. Victor, "The Regime Complex for Climate Change," *Perspectives on Politics* 9, no. 1 (March 2011), 7–23.

[49] David G. Victor, *Global Warming Gridlock: Creating More Effective Strategies for Protecting the Planet* (New York: Cambridge University Press, 2011).

[50] Sylvia I. Karlsson-Vinkhuyzen and Jeffrey McGee, "Legitimacy in an Era of Fragmentation: The Case of Global Climate Governance," *Global Environmental Politics* 13, no. 3 (August 2013), 56–78.

[51] Timothy Cadman (ed.), *Climate Change and Global Policy Regimes: Towards Institutional Legitimacy* (Basingstoke: Palgrave Macmillan, 2013).

van Asselt, and Zelli use global climate governance as a case study to explore the fragmentation of global governance architectures.[52]

Some studies focus on specific actors in global climate governance. For instance, Bakker and Francioni's co-edited volume explores the EU and the US leadership roles in global climate governance.[53] Peter Newell investigates the role of non-state actors in global climate change politics.[54]

Others employ multidisciplinary and multi-theoretical approaches to investigate global climate governance. For instance, Held, Fane-Hervey, and Theros explore global climate governance through science, economics, politics, and ethics.[55] Methmann, Rothe, and Stephan employ some interpretative approaches to explore global climate governance, specifically focusing on constructing and deconstructing GHGs.[56] Dryzek and Stevenson employ a deliberative systems approach to explore how global climate governance can be democratized, that is, "bringing climate governance increasingly under the control of the global demos or demoi," so as to map and evaluate the existing global governance of climate change, as well as to prescribe reforms to improve the legitimacy and effectiveness of this governance.[57]

Still others explore the global climate governance after the Kyoto Protocol. For instance, Matthew J. Hoffmann explains how and why some new governance experiments have emerged, drawing upon a database of such initiatives to ascertain how these fit together and how they influence what is defined as environmental governance.[58]

---

[52] Frank Biermann, Philipp Pattberg, Harro van Asselt, and Fariborz Zelli, "The Fragmentation of Global Governance Architectures: A Framework for Analysis," *Global Environmental Politics* 9, no. 4 (November 2009), 14–40.

[53] Christine Bakker and Francesco Francioni (eds), *The EU, the US and Global Climate Governance* (Farnham: Ashgate Publishing Ltd, 2014).

[54] Peter Newell, *Climate for Change: Non-State Actors and the Global Politics of the Greenhouse* (Cambridge: Cambridge University Press, 2000).

[55] David Held, Angus Fane-Hervey, and Marika Theros (eds), *The Governance of Climate Change: Science, Economics, Politics and Ethics* (Chichester: Polity Press, 2011).

[56] Chris Methmann, Delf Rothe, and Benjamin Stephan (eds), *Interpretive Approaches to Global Climate Governance: (De)Constructing the Greenhouse* (London: Routledge, 2013).

[57] John S. Dryzek and Hayley Stevenson, *Democratizing Global Climate Governance* (Cambridge: Cambridge University Press, 2014).

[58] Matthew J. Hoffmann, *Climate Governance at the Crossroads: Experimenting with a Global Response after Kyoto* (New York: Oxford University Press, 2011).

Asim Zia explores post-Kyoto climate governance by exploring the politics of scale, ideology, and knowledge.[59]

China and India in Global Energy and Climate Governance

So far, there is limited research on the role of China and India in global energy and climate governance. For instance, Vihma explores India's relationship with global climate governance through its trade-offs between maintaining principles and seeking its pragmatic interests.[60] Dubash argues that the role of India in global energy governance is shifting from a rule-taker to a rule-maker.[61] Bo Kong argues that "energy governance in China has experienced considerable capacity decay in the era of reform and globalization," which is due largely to "severe state fragmentation at the central level, increasing autonomy of substate actors at the local level and the rise of state-owned flagship energy corporations at the industry level."[62] Zhuang Guiyang explores global climate governance in the post-Kyoto era and China's strategic options, and argues that the development of a low-carbon economy should be one of China's strategic options.[63] Bo Yan and Chen Zhimin examine the roles played by China and the EU in global climate governance.[64] Stevenson explores India and global climate governance through the perspective of normative congruence-building in India and its impact on India's climate policy and diplomacy.[65] Carl J. Dahlman devotes half of his monograph to investigating how China and India

---

[59] Asim Zia, *Post-Kyoto Climate Governance: Confronting the Politics of Scale, Ideology, and Knowledge* (New York: Routledge, 2013).
[60] Vihma, "India and the Global Climate Governance," 69–94.
[61] Navroz K. Dubash, "From Norm Taker to Norm Maker? Indian Energy Governance in Global Context," *Global Policy* 2, no. S1 (September 2011), 66–79.
[62] Bo Kong, "Governing China's Energy in the Context of Global Governance," *Global Policy* 2, no. S1 (September 2011), 51–65.
[63] 庄贵阳 [Zhuang Guiyang], "后京都时代国际气候治理与中国的战略选择 [Post-Kyoto International Climate Governance and China's Strategic Options]," 《世界经济与政治》 [*World Economics and Politics*] 8 (2008), 6–15.
[64] 薄燕和陈志敏 [Bo Yan and Chen Zhimin], "全球气候治理中的中国和欧盟 [China and the EU in Global Climate Governance]," 《现代国际关系》 [*Contemporary International Relations*] 2 (2009), 44–50.
[65] Hayley Stevenson, "India and International Norms of Climate Governance: A Constructivist Analysis of Normative Congruence Building," *Review of International Studies* 37, no. 3 (July 2011), 997–1019.

are influencing the global environment due to their increased CO_2 emissions.[66] Wu explores China and India's climate diplomacy and its implications for global climate governance.[67]

In summary, the existing literature has provided some rich and profound insights and multiple perspectives for studying China's and India's energy security and climate change against the backdrop of global energy and climate governance. However, there are at least three limitations in the existing literature. First of all, there is a noticeable tendency in the literature on global energy and climate governance to treat global governance on energy and climate separately, that is, "global energy governance" and "global climate governance," which might have been caused by the fact that they "have been conducted in different arenas" although they are "physically closely related."[68] However, due to the fact that "fossil fuel dominance of the energy sector is the single most important reason for climate change risks,"[69] it is important to explore how the international community has tried to address climate change through governing the energy sector at the global level. In other words, we need to explore what efforts have been devoted to integrating energy security and climate change into global governance.

Second, China and India, the world's largest and third-largest energy consumers and GHG emitters respectively, have not been given full consideration in the literature on global energy and climate governance. More specifically, despite their fundamental importance, the implications of both countries' efforts to address energy security and climate change have been a fairly neglected feature in studies on global energy and climate governance. As noted above, scholarly

[66] Carl J. Dahlman, *The World Under Pressure: How China and India Are Influencing the Global Economy and Environment* (Stanford, CA: Stanford University Press, 2012).
[67] Fuzuo Wu, "Sino-Indian Climate Diplomacy: Implications for Global Climate Governance," paper presented at the International Studies Association (ISA) Global South Caucus Conference, Singapore, January 8–10, 2015 and at ISA's 56th Annual Convention, New Orleans, USA, February 18–21, 2015.
[68] Lesage, Van de Graaf, and Westphal, *Global Energy Governance in a Multipolar World*, 1.
[69] International Energy Agency, "World Energy Outlook 2012 Factsheet," 2012, 1, available at www.worldenergyoutlook.org/media/weowebsite/2012/factsheets.pdf, accessed February 19, 2018.

attention has mainly been paid to the aspect of both countries' foreign policy on these two issue-areas.

Moreover, what still remains under-researched in existing literature is to explore, in a theoretical, systematic, and synthetic manner, both countries' policies on how to address energy insecurity and climate change, as well as their implications for global energy and climate governance. This book fills these gaps. Simply put, this book is intended to extend our theoretical and empirical understanding of China's and India's behavior in such specific issue-areas as energy security and climate change.

Levels of Analysis

This book employs factors arising at both the domestic and international levels, to create a two-level pressures analytical framework to explain China's and India's policy behavior on energy security and climate change. To be sure, there is no consensus among IR scholars about the levels of analysis, so before creating the two-level pressures analytical framework, it is necessary to explore the debate within IR. The focal point of this theoretical debate is which is the most appropriate level – system/international, unit/state or individual – to explain a particular state's foreign policy behaviors.

Historically, this debate can be traced back to the late 1950s and early 1960s. More specifically, in 1959, Kenneth Waltz's *Man, the State and War* initiated the approach of three levels of analysis to explain the causes of states' behavior, that is, "first image" (individual level), "second image" (domestic/unit level), and "third image" (international/systemic level).[70] In 1960, David Singer put forward "The Level-of-Analysis Problem in International Relations," in which he reduced the causal levels from Waltz's three to two, that is, the international system and the national state.[71] Since Waltz's and Singer's initiatives, the levels-of-analysis debate has been a continuing process. In the late 1970s, Peter Gourevitch initiated "The Second Image Reversed" analytical framework, arguing that forces such as war and trade at the international system level shape states' domestic

---

[70] Kenneth N. Waltz, *Man, the State and War: A Theoretical Analysis* (New York: Columbia University Press, 1959).
[71] J. David Singer, "The Level-of-Analysis Problem in International Relations," *World Politics* 14, no. 1 (1961), 77–92.

political development.[72] Based on this theoretical framework, in the mid-1980s, Robert Keohane tested how the international political economy of modern capitalism – especially the transmission of inflation and recession and the deterioration of the terms of trade, stemming from the erosion of US hegemony and the rise of exports from less developed countries, thanks to the expansion of capitalism on a world scale – had shaped Western European countries' domestic economic policy choices in the 1970s.[73] In the mid-1990s, Robert Keohane and Helen Milner's jointly edited volume, *Internationalization and Domestic Politics*, "firmly within the 'second image reversed' tradition," theoretically established and empirically tested the hypothesis that "we can no longer understand politics within countries – what we still conventionally call 'domestic' politics – without comprehending the nature of the linkages between national economies and the world economy, and changes in such linkages."[74] There have been other followers of "the second image reversed" analytical framework, hence Andrew Moravcsik reviewed this "tradition" at great length in a volume dedicated to expansion of the application of the two-level game theoretical framework developed by Robert Putnam, as discussed below.[75]

Generally speaking, a majority of theorists in three mainstream IR theories – neorealism, neoliberal institutionalism, and constructivism – give a certain primacy to the systemic level over other levels of analysis. For instance, according to Kenneth Waltz, states' behavior can be studied from the "inside-out" (namely, unit level) or from the "outside-in" (namely, systemic level).[76] The former approach locates the source of states' behavior within the states themselves, such as states'

---

[72] Peter Gourevitch, "The Second Image Reversed: The International Sources of Domestic Politics," *International Organization* 32, no. 4 (Autumn 1978), 881–912.
[73] Robert O. Keohane, "The World Political Economy and the Crisis of Embedded Liberalism," in John H. Goldthorpe (ed.), *Order and Conflict in Contemporary Capitalism* (Oxford: Clarendon Press, 1984), 15–38.
[74] Robert O. Keohane and Helen V. Milner (eds), *Internationalization and Domestic Politics* (Cambridge: Cambridge University Press, 1996), 1.
[75] Andrew Moravcsik, "Introduction: Integrating International and Domestic Theories of International Bargaining," in Peter Evans, Harold Jacobson, and Robert Putnam (eds), *Double-Aged Diplomacy: International Bargaining and Domestic Politics* (Berkeley, CA: University of California Press, 1993), 11.
[76] Kenneth N. Waltz, *Theory of International Politics* (Reading, MA: Addison-Wesley, 1979), 63.

political or economic systems, and the attributes of their leaders, as well as their domestic interest groups.[77] In contrast, the "outside-in" approach attributes the source of states' behavior to the systemic level. As to the relative importance of the elements located at the unit level and the systemic level, Waltz stressed that states' behavior has stemmed from the forces at the systemic level, and domestic differences between states are considered to be relatively less important, for the reason that "external pressures stemmed from the international level are assumed to be strong and straightforward enough to make similarly situated states behave alike, regardless of their internal characteristics."[78] Thus, according to Waltz, in order to understand why a state behaves in a specific way, it is necessary to investigate the characteristics of the international system, since the forces derived from the international system have played an important role in shaping states' behavior.[79]

Like Waltz, Stephen Krasner points out that "In foreign affairs the nature of the international system, its inherent anarchy, places many restraints on the freedom of action of any given state."[80] In the same vein, Robert Keohane regards the systemic level as the starting point to explain states' behavior: "I focus on the effects of system characteristics because I believe that the behavior of states, as well as other actors, is strongly affected by the constraints and incentives provided by the international environment."[81] Moreover, Keohane argues that an international-level analysis is "neither an alternative to studying domestic politics, nor a mere supplement to it," but "a precondition for effective comparative analysis," and without the knowledge of forces at the international level, "we lack an analytic basis for identifying the role played by domestic interests."[82] Furthermore, although in his edited volume, *Realism and its Critics*, Keohane recognizes that more attention should be paid to the interactions between the

---

[77] Gideon Rose, "Neoclassical Realism and Theories of Foreign Policy," *World Politics* 51, no. 1 (1998), 149.
[78] Ibid. [79] Waltz, *Theory of International Politics*.
[80] Stephen D. Krasner, *Defending National Interest: Raw Material Investments and U.S. Foreign Policy* (Princeton, NJ: Princeton University Press, 1978), 17.
[81] Keohane, *After Hegemony*, 26.
[82] Keohane, "The World Political Economy and the Crisis of Embedded Liberalism," 16.

international level and domestic level, or "internal–external interactions,"[83] he still emphasizes the importance of the systemic or structural level because "we must understand the context of action before we can understand the action itself," and "it provides an irreplaceable component for a thorough analysis of action, by states or nonstate actors, in world politics."[84] By the same token, Alexander Wendt has focused on the structure and effects of international systems, or taken a "systems theory" approach to IR, by attributing states' behavior to a distribution of ideas and identities at the systemic level.[85]

However, these systemic-level determinant theories have been challenged empirically and theoretically. Empirically, neorealism and neoliberal institutionalism systemic theories failed to anticipate the end of the Cold War, which exposes both theories' limitations.[86] Even Keohane himself recognizes the explanatory limitation of systemic-level analysis in his empirical studies on US foreign oil policy in the 1940s, as well as its hegemonic leadership in the 1950s, by noting that these policies had been largely shaped by US domestic factors, such as its private power and corporate interests and its domestic politics, stating that "policy depends not merely on systemic conditions but on domestic politics as well."[87] Theoretically, since the late 1980s, when Putnam's felicitous and influential metaphor of the "two-level game,"[88] and the 1990s, when neoclassical realism came into being,[89] both of which combine factors originating at both the unit and systemic levels to explain states' foreign policy behavior, there has been "a growing consensus among scholars [that] embraces models integrating the unit level and the systemic level of analysis."[90] Accordingly,

---

[83] Robert O. Keohane (ed.), *Realism and its Critics* (New York: Columbia University Press, 1986), 191.
[84] Ibid., 193.
[85] Alexander Wendt, *Social Theory of International Politics* (Cambridge: Cambridge University Press, 1999), 5, 11.
[86] John Lewis Gaddis, "International Relations Theory and the End of the Cold War," *International Security* 17, no. 3 (Winter 1992–93), 5–58.
[87] Robert O. Keohane, *International Institutions and State Power: Essays in International Relations Theory* (Boulder, CO: Westview Press, 1989), 13.
[88] Robert D. Putnam, "Diplomacy and Domestic Politics: The Logic of Two-Level Games," *International Organization* 42, no. 3 (Summer 1988), 427–460.
[89] Rose, "Neoclassical Realism and Theories of Foreign Policy," 144–172.
[90] Harald Muller and Thomas Risse-Kappen, "From the Outside In and From the Inside Out: International Relations, Domestic Politics, and Foreign Policy," in David Skidmore and Valerie M. Hudson (eds), *The Limits of State Autonomy:*

more and more literature that combines international and domestic levels has emerged.[91]

Following this trend, this book integrates factors at both unit and systemic levels to create a "two-level pressures" model to explain China's and India's energy security and climate change policy behavior. Thus, the two-level analysis brings this book close to Putnam's two-level games and neoclassical realism. Nevertheless, the two-level analysis in this book differs from these two existing two-level analytical frameworks. Specifically, neoclassical realists treat the factors at the systemic level as independent variables, and those at the unit level as intervening variables in shaping states' foreign policy behavior. In Rose's words, "the scope and ambition of a country's foreign policy is driven first and foremost by its place in the international system and specifically by its relative material power capabilities ... the impact of such power capabilities on foreign policy is indirect and complex, because systemic pressures must be translated through intervening variables at the unit level."[92] In Schweller's words, variables at the domestic level "act as transmission belts that channel, mediate, and (re)direct policy outputs in response to external forces (primarily changes in relative power)."[93]

In contrast, I argue that China's and India's foreign policy in the two issue-areas under consideration, namely energy security and climate change, have stemmed, first and foremost, from factors at the unit level, while the factors that stimulate both countries to modify their energy diplomacy and climate diplomacy have stemmed from the international or systemic level. Moreover, China's and India's domestic climate policy behavior can only be explained by factors arising at the systemic level, rather than the unit level, within both countries, that

---

Societal Groups and Foreign Policy Formulation (Boulder, CO: Westview Press, 1993), 47.

[91] A full discussion of the literature in this regard is beyond the realm of this book. Besides, there is already a joint research study that reviews how scholars in IR and international political economy have combined factors stemming from the systemic level and state level to explain international relations. See Stephen Chaudoin, Helen V. Milner, and Xun Pang, "International Systems and Domestic Politics: Linking Complex Interactions with Empirical Models in International Relations," *International Organization* 69, no. 2 (Spring 2015), 275–309.

[92] Rose, "Neoclassical Realism and Theories of Foreign Policy," 146.

[93] Randall L. Schweller, "Unanswered Threats: A Neoclassical Realist Theory of Underbalancing," *International Security* 29, no. 2 (Fall 2004), 164.

is, "the second image reversed."[94] Thus, one of the main innovations of this book is that it proves that in the two issue-areas, that is, energy security and climate change, China's and India's foreign policy behavior has been driven largely by forces at the unit level, while both countries' domestic policy on climate change has been driven by factors at the systemic level. In contrast with Putnam and his later followers' emphasis on domestic statesmen and domestic politics, and some subnational actors such as interest groups,[95] this book emphasizes states' domestic economic development. Simply put, this research combines the "inside-out" or unit level with the "outside-in" or systemic level to explain China's and India's policy behavior in the two issue-areas, that is, energy security and climate change.

Case Selection

This book focuses on China and India not only because their energy and climate change policy behaviors are puzzling, but also for reasons of methodology and research design. First, the main objective of this book is to test the explanatory power of the two-level pressures model in shaping the proactive and reactive feature of developing countries' policy behavior, so it is useful to select two of the relatively dominant emerging powers as a means of simplifying the analysis, with a focus on two specific issue-areas, that is, energy security and climate change. If we can demonstrate that China and India act in an identical way when it comes to energy and climate change – proactive on the former and reactive on the latter – then the next step is to ask whether or not such an observation is in line with what we observe in other issue-areas that entail global public goods or bads (e.g. trade, health), and what these two cases can tell us for the larger population of cases exhibiting similar characteristics (e.g. latecomers, asymmetric power relations). I discuss in the conclusion how we might take this next step.

Second, because China and India in the first 15 years of the twenty-first century were quite similar in a number of respects, this study compares their policy outcomes that were under broadly comparable conditions.

---

[94] Gourevitch, "The Second Image Reversed."
[95] Putnam and colleagues later further developed his two-level games framework in an edited volume. See Peter Evans, Harold Jacobson, and Robert Putnam (eds)., *Double-Aged Diplomacy: International Bargaining and Domestic Politics* (Berkeley, CA: University of California Press, 1993).

Most importantly, China and India, with their largest poor populations in the world (see Chapter 2), were in the process of poverty reduction, industrialization, and modernization through their rapid economic growth. As a result, both countries' energy consumption and CO_2 emissions have increased so dramatically that they have become the world's largest and third-largest energy consumers and CO_2 emitters respectively. Thus, domestically, China and India faced comparable challenges. Internationally, both countries faced the same realities in the international system, that is, asymmetrical distribution of power on the one hand and global governance on the other. Against this backdrop, it is reasonable to argue that both China's and India's proactive and reactive policy behaviors on the issue-areas of energy security and climate change have been shaped by pressures stemming from domestic and international levels.

However, there are several important differences in the domestic and international contexts in which China and India operate, which can lead to variations in how the two countries react to common two-level pressures. Domestically, China's and India's political systems are totally different, that is, China is characterized by a one-party authoritarian regime, while India is a democratic country. This difference accounts for the variation in their domestic energy policy behavior (see Chapter 3). Moreover, in terms of the size of economy, China's gross domestic product (GDP) is four times that of India. When it comes to CO_2 emissions, in absolute emission terms, China is above 28 percent of global emissions, while India is only around 6 percent; China's per capita emissions are about four times higher than those of India. These differences make China face much stronger international pressures than India, which accounts for the variation in the reactive level of both countries' foreign and domestic climate policies (see Chapters 5 and 6). At the global level, China is a permanent member of the UN Security Council, whereas India has not been able to obtain such membership. This difference has also led to stronger international pressures on China than India in resolving international crises, which in turn leads to the variation in the reactive level of China's and India's policy behavior (see Chapter 4). In short, this book employs the method that approximates to controlled comparison.[96]

---

[96] For details on this method, see Alexander George and Andrew Bennett, *Case Studies and Theory Development in the Social Sciences* (Cambridge, MA: MIT Press, 2005), 151–160.

In addition, this research builds deliberately and explicitly on rational choice theory.[97] "This approach to the study of politics begins with the premise that we can usefully study political actors, given their exogenously defined preferences, as if they were instrumentally rational."[98] Although there are some limitations to this approach, "it provides a baseline premised on a relatively uncomplicated situation characterized by purely self-interested and rational behavior."[99] According to this approach, China and India are assumed to be two state actors that rationally maximize their national interests – material (wealth) and ideational (status) – in the international system.

The sources on which I have drawn are the abundant primary and secondary material data related to China's and India's energy consumption and CO_2 emissions, including the official documents and speeches made available by both Chinese and Indian governments, UNFCCC and the UN Environment Programme (UNEP). Moreover, this study has benefited enormously from my two postdoc fellowships at Yale, Oxford, and Princeton, where I attended a number of workshops, seminars, and conferences related to topics on energy, climate change, and IR, which provided me with valuable opportunities to converse with and interview a number of Chinese and Indian academics, and retired and incumbent officials and diplomats in relevant areas, in addition to some IR scholars. This book has also benefited from interviews and communications with some of the Chinese diplomats who resided at the Center for Chinese Foreign Policy Studies at the Institute of International Studies at Fudan University. Some of the interviews and communications will be cited in the pages that follow.

---

[97] In this book, rationality is essentially the ontological basis rather than an analytical tool. I owe this point to an anonymous reviewer. For detailed discussion of rationality in IR, see Miles Kahler, "Rationality in International Relations," *International Organization* 52, no. 4 (Autumn 1998), 919–941; Peter J. Katzenstein, Robert O. Keohane, and Stephen D. Krasner, "International Organization and the Study of World Politics," *International Organization* 52, no. 4 (Autumn 1998), 645–685.
[98] J. Samuel Barkin, "Realist Constructivism," *International Studies Review* 5, no. 3 (September 2003), 329.
[99] Keohane, *After Hegemony*, 70.

Contributions of the Book

This book makes four contributions to scholarship. First, as discussed above, although China and India have had a significant impact on global governance in general and on global energy and climate change governance in particular, there are few comprehensive analyses on these issues. This book fills a gap in the literature by examining, in considerable depth, the domestic and foreign policies China and India have adopted to address energy insecurity and climate change, and the implications for global energy and climate governance, and by providing new details on this important topic.

Second, it offers a better analytical framework, that is, the "two-level pressures" model, combining the constraints at the systemic level with the material and ideational priorities at the unit level, to explain China's and India's domestic and foreign policy in the two issue-areas – energy security and climate change.

Third, the two-level pressures analytical framework has broader applications, beyond China's and India's energy security and climate change issues: it can usefully inform China's and India's international behavior in other issue-areas against the backdrop of their rise in the international system, and adds scholarship to rising powers' behavior in the international system.

Furthermore, this research adds empirical evidence to the "second image reversed" analytical framework: that is, a state's domestic policy preferences have been shaped by international factors. Specifically, this book proves that China's and India's domestic climate policies have been shaped by international pressures stemming from both countries' climate diplomacy and the epistemic communities.

Structure of the Book

In this introduction, I have provided a general overview of the argument, theory, case selection, and contributions of this book as well as the limitations within the existing literature related to China and India and global energy and climate governance.

Chapter 2 develops a theoretical framework to analyze China's and India's domestic and foreign policies to address energy insecurity and climate change during the first 15 years of the twenty-first century, in order to provide an analytical tool for the book. The central argument

Introduction

of this chapter is that China's and India's policy behaviors during this period have been both proactive and reactive, shaped by their two-level pressures. At the domestic level, China and India are faced with pressures to prolong their economic development trajectory so as to maximize their wealth, indicated by both GDP and per capita GDP, or to develop their economies, reduce poverty, and raise the living standards of their huge populations: hence, both countries have been proactively employing policy measures to sustain their economic growth. At the international level, China and India have sought to enhance their status in the international system, so both countries have had to modify or adjust, to a certain extent, their proactive policy measures as a result of external pressures stemming from asymmetrical interdependence and even international norms, rules, and principles embedded in global governance.

Chapter 3 explores China's and India's domestic policy measures to address their energy insecurity. The main point of this chapter is that China's and India's policy measures to ensure energy security are proactive, which stems from the fact that energy security has been regarded as one of the key national security issues that has some strategic imperatives for both countries' economic development. Accordingly, to procure energy security has been China's and India's top policy priority. Both countries have taken measures to develop renewable energy, expand nuclear energy, enhance their energy efficiency, and reduce their energy intensity, as well as to establish strategic petroleum reserves of their own. This chapter tests that proposition 1 is true: that is, that the energy policies of China and India are designed to help these countries to achieve sustained, fast economic development at the domestic level.

Chapter 4 explores China's and India's foreign policy targeted at procuring overseas energy supplies, namely Sino-Indian energy diplomacy. It first illuminates the proactive and reactive feature of such diplomacy. On the one hand, China and India have adopted proactive policy measures, either economic, political, diplomatic, or even military, to boost their energy ties with all energy-exporting countries across the world, including the so-called "pariah states" in the international system. On the other hand, China and India have had to adjust and modify those proactive policy measures toward "pariah states" as a result of external pressures. To test this argument, this chapter conducts three case studies, exploring China's and India's energy diplomacy toward Iran, Sudan, and Myanmar, and compares similarities and differences between their policies. In addition, this chapter analyzes how China's and India's energy diplomacy has been shaped by two-level pressures. At the domestic level, China and India have tried to facilitate their NOCs going out to seek overseas equity oil and natural gas, to maximize their NOCs' commercial profits, which is a way for China and India to maximize their domestic wealth, not only by hedging their energy supply security, but also by maximizing their revenue through the taxes paid by their NOCs, and increasing employment for their huge labor markets. At the international level, in order to enhance their status in the international system, China and India have become sensitive to pressures on their energy foreign policy behavior stemming from social opprobrium and asymmetrical interdependence. Thus, Chapter 4 tests that proposition 2 is correct: that is, that the energy policies that China and India have pursued globally are constrained by the patterns of asymmetrical interdependence and international norms in which the two countries are enmeshed.

Chapter 5 explores China's and India's foreign policy behavior in the international climate change negotiations, aimed at testing proposition 3: that China's and India's negotiating stances in international climate change negotiations are increasingly shaped by asymmetrical interdependence, climate protection norms, and social opprobrium at the systemic level. In accordance with the analytical framework developed in Chapter 2, this chapter first explores the proactive and reactive features of Sino-Indian climate change diplomacy. More specifically, this chapter points out that China and India have been conducting dual-track climate change diplomacy, that is, UN track and non-UN track. Under the UN track, China and India have been actively participating in the post-Kyoto Protocol negotiations under the UNFCCC. Under these negotiations, China and India had continued to maintain the traditional negotiating positions they have upheld since the inception of climate change negotiations in the early 1990s, that is, it was the developed countries that should take the lead in mitigating GHG emissions, while transferring technologies and providing financial assistance to developing countries in order to help the latter adapt to the negative impact of climate change. Faced with increased pressures on them to undertake the obligation to mitigate their increased GHG emissions, China and India have proactively adopted the strategy of building new coalitions to facilitate their increasingly weakened bargaining power against developed countries and their negotiating allies and blocs, due to the fragmentation of their traditional negotiation coalition – the Group of 77 plus China (G-77/China). In parallel, China and India have also been bandwagoning some non-UN climate change arrangements initiated by the United States and the EU. The outcome of China's and India's dual-track climate diplomacy is that both countries have made some significant compromises and shifted away from free-riding to burden-sharing.

Two-level pressures have shaped China's and India's climate diplomacy. At the domestic level, there is a hot debate within both countries on whether or not they should undertake mitigation actions to address climate change, by reducing their GHG emissions. In addition, the tangible profits both countries have accrued from their participation in the Clean Development Mechanism (CDM) have stimulated them to make compromises in ICCN. At the international level, China's and India's efforts to acquire great-power status make them react to pressures emanating from the international climate protection norms, social isolation, and social opprobrium. Moreover, China's and India's asymmetrical interdependence on developed countries, especially the United States and the EU, for transferring climate mitigation-related technologies, as well as other developing countries' dependence on developed countries not only for technology transfers, but also financial assistance to adapt to climate change, has prompted both countries to make compromises.

Chapter 6 explores China's and India's domestic policy measures to address climate change by testing proposition 4. The main argument of this chapter is that addressing climate change is not a national priority for China and India, so both countries have only undertaken domestic climate policies when faced with increased external pressures that stem from their dual-track climate diplomacy, as well as from epistemic communities. As a result, China's and India's climate policies are reactive.

China's and India's domestic proactive energy security policies explored in Chapter 3 and reactive climate policies have been shaped by their two-level pressures. At the domestic level, both countries have adopted a low-carbon development strategy to maximize their energy security and economic growth, while climate mitigation is only a co-benefit of their transition to a low-carbon economy. At the international level, on the one hand, both China and India have tried to seek great-power status through becoming leaders in the global renewable energy industry; on the other hand, both countries have been asymmetrically dependent on US and EU markets and on the transfer of advanced clean energy technology for their development of renewable energy.

Chapter 7 explores the broader implications of China and India addressing energy insecurity and climate change for global energy and climate governance. It first explores the characteristics of current global energy and climate governance. Then it examines three implications of China's and India's domestic energy-climate policy and international energy-climate diplomacy for global energy and climate governance. First, we will see that both countries' national and international policies to address energy insecurity and climate change have broadened energy and climate governance beyond the realms of the existing global energy and climate institutions or regimes. Second, China and India may achieve energy security and emission reductions independently of the performance of the existing energy and climate regimes based on their own national and international policy rules, principles, norms, and procedures, which significantly calls into question the effectiveness, legitimacy, and appropriateness of the existing global energy and climate governance. A further implication is that restructuring the existing global energy and climate institutions or regimes so as to fully integrate both China and India is the only way to achieve the ultimate goals of global energy and climate governance, namely globalized energy security and a stabilized climate.

Chapter 8 concludes the book by summarizing the six principal findings and considering the implications of this book's two-level pressures analytical framework for broader international relations research, as well as our understanding of China's and India's policy behavior in the context of their rise in the international system.

2 | Shaping Energy and Climate Policy Behavior: Wealth, Status, and Asymmetrical Interdependence

This chapter develops a theoretical framework for analyzing China's and India's policy behavior in the issue-areas of energy security and climate change during the first 15 years of the twenty-first century in order to provide an analytical tool for the book. The main argument is that China's and India's energy and climate policy behavior has been both proactive and reactive, which has been shaped by two-level pressures. To develop this argument, this chapter is arranged as follows. The first section defines the terms "proactive" and "reactive." The second section develops the two-level pressures analytical framework characterized by wealth, status, and asymmetrical interdependence to explain Sino-Indian proactive and reactive policy behavior. This section includes three subsections which examine, first, wealth, status, and asymmetrical interdependence – three terms in IR theory – and then explore how China and India have been seeking wealth and status under asymmetrical interdependence. A brief conclusion is drawn in the final section of this chapter.

Proactive and Reactive State Actors

In this study, China and India are assumed to be self-interested actors that rationally seek wealth and status by intentionally calculating the costs and benefits of alternative courses of action.[1] In other words, China and India, two state actors, are assumed to be rational egoists – a basic assumption shared by both neorealism and neoliberal Institutionalism.[2] More specifically, both states have rationality which, according to Robert Gilpin, "is not historically or culturally bound but that individuals ... attempt to achieve their interests and goals by

---

[1] Detlef Sprinz and Tapani Vaahtoranta, "The Interest-Based Explanation of International Environmental Policy," *International Organization* 48, no. 1 (Winter 1994), 78.
[2] Keohane, *International Institutions and State Power*, 40.

the most efficient means possible."[3] Similarly, Robert Keohane defines rationality as meaning that states have consistent, ordered preferences, and that they calculate the costs and benefits of alternative courses of action in order to maximize their utility in view of those preferences.[4] In other words, states typically make rational choices based on their calculation of their own national costs and benefits. Put differently, state actors know how and when to employ an appropriate means to pursue their own specific interests or objectives. According to Hans Morgenthau, the primary objective of foreign policy or diplomacy is "the promotion of the national interest by peaceful means."[5] To do so, one of the fourfold tasks of diplomacy is that it "must employ the means suited to the pursuit of its objective."[6] This principle implies that states' policy measures should remain flexible or appropriate to promote their national interests. In Morgenthau's words, "foreign policy ought to be rational in view of its own moral and practical purposes."[7] Under such circumstances, states can employ proactive measures to seek their interests while adjusting or modifying these proactive measures in response to external influences or pressures, either explicit or implicit, based on their own cost and benefit calculations. In other words, a state's policy behavior can be both proactive and reactive.[8]

The proactive feature of a state's foreign policy can be defined as a state independently taking its own policy initiatives to seek its own national interests and objectives based on its policy priorities. To define the reactive feature of a state's foreign policy, I draw on an academic debate in the late 1980s and the 1990s that considered whether or not Japan was a reactive state. This debate was initiated by Kent Calder, who argued that Japan was typically a "reactive" state from the

---

[3] Robert Gilpin, *War and Change in World Politics* (New York: Cambridge University Press, 1981), xii.
[4] Keohane, *After Hegemony*, 27.
[5] Hans J. Morgenthau, *Politics Among Nations: The Struggle for Power and Peace* (New York: Alfred A. Knopf, 1948), 419.
[6] Ibid.
[7] Hans J. Morgenthau, *Politics Among Nations: The Struggle for Power and Peace* (7th edn), rev. Kenneth Thompson and W. David Clinton (New York: McGraw-Hill, 1985), 10.
[8] I have tested this argument by using China's WMD-related export control policy after 2004. See Fuzuo Wu, "China's Responses to External Pressures on its WMD-Related Exports after 2004: Reactive and Proactive," *Journal of Contemporary China* 24, no. 93 (2015), 511–530.

perspective of international relations theory, due to the fact that Japan could not independently make its own foreign aid policy. In spite of "its enormous economic size" (being the largest creditor in the world since 1985), "its substantial population," and "its pre-1945 history of pro-activism in the international system,"[9] Japan had to modify its foreign aid policy toward several countries due to US pressure. Thus, according to Calder, the concept of "reactive state" does not deny "the strategic intent of much Japanese policy making nor its successful implementation in many instances." Instead, it means that "the impetus to policy change is typically supplied by outside pressure, and that reaction prevails over strategy in the relatively narrow range of cases where the two come into conflict."[10]

According to the systemic approach in international relations theory,[11] a state's behavior is essentially seen as a reaction to the actions of other states, and even a state's purposive and aggressive behavior can be simply regarded as an active form of reaction to the international environment, such as recalling ambassadors, imposing economic sanctions, forming alliances and engaging in arms races and so on.[12] However, based on Calder's definition, we can identify two fundamental characteristics of a reactive foreign policy: first, the state fails to undertake major independent foreign policy initiatives although it has the power and national incentives to do so; and second, it responds to outside pressures for change in an erratic, unsystematic, and often incomplete manner.[13] Thus, a "proactive and reactive foreign policy" can be defined as a being where a state typically adopts some independent foreign policy measures to secure its national interests, but occasionally needs to modify or adjust those measures in response to external influences or pressures, which can be: persuasion, accompanied by some obvious material threats or promises; social opprobrium applied by state and/or non-state actors; and international norms that govern state actors' behavior in the international system.

---

[9] Kent E. Calder, "Japanese Foreign Economic Policy Formation: Explaining the Reactive State," *World Politics* 40, no. 4 (July 1988), 519.
[10] Ibid., 518.   [11] Waltz, *Theory of International Politics*.
[12] Akitoshi Miyashita, "Gaiatsu and Japan's Foreign Aid: Rethinking the Reactive-Proactive Debate," *International Studies Quarterly* 43, no. 4 (December 1999), 725.
[13] Calder, "Japanese Foreign Economic Policy Formation," 519.

To expand upon this definition of proactive and reactive policy behavior, it is useful to consider the distinction between rational/logical actions and nonlogical actions drawn by Vilfredo Pareto: "logical or rational actions as those in which means–ends relationships for the performer and for an experienced outside observer are the same. Nonlogical behavior occurs when there is divergence between the two."[14] Nonlogical behavior is not necessarily irrational, but is still based on rational actors' rational choice. Reactive policy behavior may be seen as states' nonlogical actions based on their rational choice: a flexible policy tool to reach their ultimate goals.

Two-Level Pressures: Wealth, Status, and Asymmetrical Interdependence

We will now consider the two-level pressures analytical framework that integrates elements stemming from the unit and systemic levels, which will serve us best in understanding such complex issue-areas as energy security and climate change. More specifically, at the domestic/unit level, China's and India's self-induced pressure is to maximize their wealth through economic growth. To reach this goal, both countries have had to ensure their energy security by procuring sufficient energy supplies to fuel their economic growth and to help their NOCs increase their commercial profits so as to expand employment for their huge labor forces and increase governmental revenues. At the international/systemic level, both China and India face asymmetrical interdependence related to power and energy resources. Moreover, at the international level, China and India have been seeking great-power status, so they need to accumulate both hard and soft status markers. Over the first 15 years of the twenty-first century, these pressures have evolved gradually, as the realization of China's and India's domestic preferences have been increasingly constrained by forces stemming from the systemic/international level. Simply put, the "two-level pressures" model describes how the execution of China's and India's domestic preferences has been constrained by systemic or international factors (see Figure 2.1).

---

[14] Cited in Krasner, *Defending National Interest*, 15–16.

Shaping Energy and Climate Policy Behavior 35

```
Domestic/unit-level pressures          International/system-level pressures
┌─────────────────────────────┐        ┌─────────────────────────────┐
│ Wealth = Economic growth    │        │ Asymmetrical interdependence│
│                             │        │   • Power                   │
│   • Energy security         │        │   • Energy resources        │
│   • Procuring energy supplies│       │ Status                      │
│   • Helping NOCs increase   │        │   • Hard markers            │
│     their commercial benefits│       │   • Soft markers            │
└─────────────────────────────┘        └─────────────────────────────┘
              │      ╲          ╱           │
              │       ╲        ╱            │
              ▼        ╲      ╱             ▼
┌─────────────────────────────┐        ┌─────────────────────────────┐
│China's and India's proactive│◄──────►│China's and India's reactive │
│energy and climate policy    │        │energy and climate policy    │
│behavior                     │        │behavior                     │
└─────────────────────────────┘        └─────────────────────────────┘
```

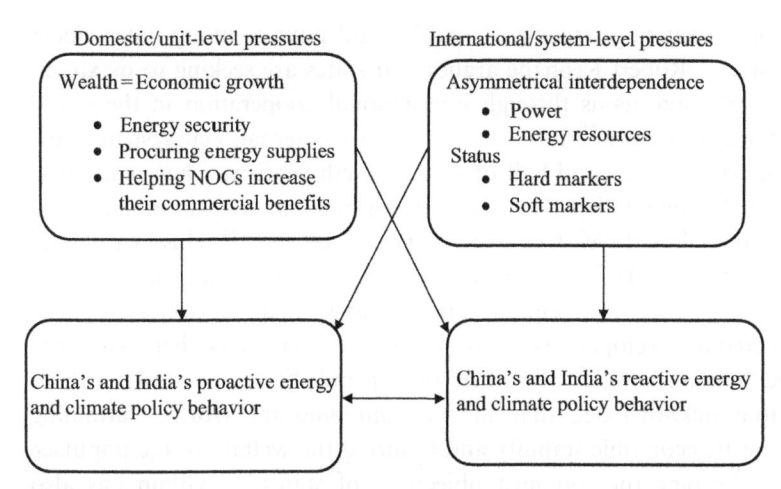

Figure 2.1 China's and India's two-level pressures
Source: Author
Note: → represents hypothesized causal relation; ↔ represents the shift of the trajectory of China's and India's energy and climate change policy behavior.

Figure 2.1 shows the linkage between these two-level pressures, that is, domestic/unit-level pressures make both China and India take some proactive energy and climate policy measures, while international/systemic-level pressures lead both countries to adjust or shift those policy measures. Thus, the linkage between these two-level pressures is established by the shift of the trajectory of China's and India's energy and climate policy.

Wealth

There is a debate among IR theorists about the ultimate motivations or goals that have shaped states' behavior in the international system. Hans Morgenthau argues that states are struggling for power,[15] while Kenneth Waltz and John Mearsheimer contend that states are seeking survival.[16] Robert Gilpin argues that "International relations continue

---

[15] Morgenthau, *Politics Among Nations*.
[16] Waltz, *Theory of International Politics*; John J. Mearsheimer, *The Tragedy of Great Power Politics* (New York: W.W. Norton and Company, 2001).

to be a recurring struggle for wealth and power among independent actors."[17] Robert Keohane argues that states are seeking to maximize their absolute gains through international cooperation in the world political economy.[18] In the contemporary international system, especially since the second half of the twentieth century, when many new states became independent, the death rate among states has been extremely low. In Waltz's words, "Few states die."[19] This fact implies that the survival of most states is not at risk in the existing international system. Under such circumstances, states' main efforts have been devoted to developing their economies, so as to increase their prosperity and maximize wealth for their own people.[20] Even realists such as Gilpin acknowledge that in the contemporary world, "attaining domestic economic stability and ensuring the welfare of the populace have become the foremost objectives of states."[21] Gilpin has also stated that "The importance of economic factors in global politics has grown continuously with the expansion of a highly interdependent world market economy."[22] Simply put, seeking wealth and trying different means to maximize it has become a top priority for states in the international system, whose external environment is characterized by the absence of imminent security threats and the low probability of war between major powers, providing them with strong incentives to make economic development their top priority.[23]

If wealth rather than survival is the most important goal for states in the current international system, then we need to investigate how to define and measure states' wealth. In his seminal work, *After Hegemony*, Keohane summarizes three typical definitions for wealth: "anything (capital, land, or labor) that can generate future income; it is composed of physical assets and human capital (including embodied knowledge)"

---

[17] Gilpin, *War and Change in World Politics*, 7.
[18] Keohane, *After Hegemony*; Robert O. Keohane, "Institutional Theory and the Realist Challenge after the Cold War," in David A. Baldwin (ed.), *Neorealism and Neoliberalism: The Contemporary Debate* (New York: Columbia University Press, 1993), 269–300.
[19] Waltz, *Theory of International Politics*, 95.
[20] Keohane, *After Hegemony*, x.
[21] Gilpin, *War and Change in World Politics*, 19.   [22] Ibid., 96.
[23] Thomas G. Moor and Dixia Yang, "Empowered and Restrained: Chinese Foreign Policy in the Age of Economic Interdependence," in David Lampton (ed.), *The Making of Chinese Foreign and Security Policy* (Stanford, CA: Stanford University Press, 2001), 225–226.

(Gilpin); "the annual produce of the land and the labour of a society" (Adam Smith); and "the means of material want satisfaction" (Karl Polanyi).[24] However, according to Keohane, all these definitions have their shortcomings. Gilpin's definition limited wealth to investment goods, but excluded assets that merely provide value in consumption.[25] Although Adam Smith's definition overcame Gilpin's shortcoming, it created another one, since it only referred to a flow of income rather than a stock of assets – an "ordinary contemporary usage of wealth."[26] According to Keohane, the problem with Polanyi's definition lies in the fact that it excluded the immaterial satisfaction – the ultimate purpose of material consumption.[27] To overcome these shortcomings in the existing definitions, Keohane defines wealth as "the 'means of want satisfaction,' or anything that yields utility, whether in the form of investment or consumption ... Thus 'the pursuit of wealth' in the world political economy refers to the pursuit of marketable means of want satisfaction, whether these are to be used for investment or consumed by their possessors."[28]

With regard to the measurement of states' wealth, according to Keohane and Nye, "National economic welfare will usually be the dominant political goal," and they regard a rising gross national product (GNP) as a critical political indicator.[29] In the same vein, Mearsheimer regards GNP, which represents a state's entire output over one year, as the most commonly used indicator of a state's wealth, even though, at the same time, he acknowledges that it is a "poor indicator" of his so-called "latent power."[30] Empirically, the most commonly used indicator of a state's wealth is GDP, "an aggregate measure of production equal to the sum of the gross values added of all resident, institutional units engaged in production (plus any taxes, and minus any subsidies, on products not included in the value of their outputs)."[31] This indicator has been used by the World Bank as one of

---

[24] Keohane, *After Hegemony*, 19. [25] Ibid. [26] Ibid. [27] Ibid.
[28] Ibid.
[29] Robert O. Keohane and Joseph S. Nye, *Power and Interdependence* (3rd edition) (Beijing: Peking University Press, 2004), 34.
[30] Mearsheimer, *The Tragedy of Great Power Politics*, 62–63.
[31] OECD, "Glossary of Statistical Terms," Gross Domestic Product, available at http://stats.oecd.org/glossary/detail.asp?ID=1163, accessed February 22, 2018.

the major indicators of states' economic size and development levels.[32] This research uses the GDP and its related real GDP per capita as two fundamental indicators to measure states' wealth.

To ensure sustained GDP growth, namely economic growth, a state needs energy security, which "essentially implies ensuring uninterrupted supplies of energy at reasonable prices."[33] To elaborate, a state's energy security encompasses at least three dimensions. First, a state's supply of energy must be adequate in volume, and there is a supply level below which national security would be jeopardized. Second, the supply of energy must be uninterrupted and continuous, as any supply shortfalls could have a serious impact on a state's economic and political development. In addition, energy must be available at "reasonable prices."[34] In other words, sufficient, reliable, and affordable supplies of energy have been a fundamental necessity for any state's economic activities.[35] Unstable supplies and high energy prices will upset the general functioning of a state's economy. Simply put, without sufficient energy supplies, a state's ambition to procure economic development will be highly unlikely to succeed. Accordingly, ensuring sufficient energy supplies both at national and international levels has been one of the policy priorities for most energy-consuming states. In this context, this research equates states' wealth-seeking behavior with their energy-seeking behavior. Hence, the objective of states' energy-seeking behavior is to ensure security of supply as well as to increase competitive economic behavior in the international energy markets.[36]

States will typically undertake domestic and foreign policy measures to reach these objectives. At the domestic level, accumulating large stockpiles of oil has been a common practice among almost all the major oil-importing countries since the late 1970s, after those countries' economies suffered from two oil crises which happened in the

---

[32] World Bank, *World Development Indicators 2014*, available at https://openknowledge.worldbank.org/bitstream/handle/10986/18237/9781464801631.pdf?sequence=1, accessed February 22, 2018.
[33] Mason Willrich, *Energy and World Politics* (New York: Free Press, 1975), 67.
[34] Bhupendra Kumar Singh, *India's Energy Security: The Changing Dynamics* (New Delhi: Pentagon Energy Press, 2010), 7.
[35] World Energy Council, *Energy for Tomorrow's World: The Realities, the Real Options and the Agenda for Achievement* (New York: St. Martin's Press, 1993), 39–74.
[36] Krasner, *Defending National Interest*, 14.

early and late 1970s.[37] In addition, states have usually promoted domestic development through tariffs, specific tax incentives, and guaranteed supporting programs. States have also directly invested in developing renewable energy resources. At the international level, some states have signed either bilateral or multilateral agreements for energy supplies, in addition to encouraging their domestic energy companies to invest in energy-rich countries. Simply put, states can take both domestic and foreign policy measures to ensure energy supply security.

Obviously, all these policy measures targeted at energy security, domestic and foreign, have been motivated by individual state's domestic attributes, especially domestic energy reserves, energy consumption mix, and energy supply and demand. Under such circumstances, a state's energy policy behavior, especially at the international level, can be characterized as an "inside-out" process, or "outcomes are much more nearly unit determined than they are system influenced."[38] Thus, this research attributes the causal factors of a state's energy policy behavior, both domestic and foreign, to the state's own attributes at the unit/state level. In stark contrast, a state's climate policy, domestic and foreign, has usually been regarded by states as hindering their domestic economic growth, so the state's climate policy has been largely stimulated by external influences, such as the international climate change negotiations under the UNFCCC. Thus, this study treats a state's climate policy as a process of "outside-in" – that is, the factors for a state's climate policy are derived from the systemic level.

China and India Seeking Wealth

China and India have been systematically and persistently seeking to develop their economies: in China, since 1978, when the Deng Xiaoping-led reform and opening-up started, and in India, since 1991, when reform and economic liberalization were initiated by the Rao government. Since the start of the twenty-first century, China and India have entered a new phase in their economic development

---

[37] Keohane, *After Hegemony*, 217–237.
[38] Waltz, *Theory of International Politics*, 47.

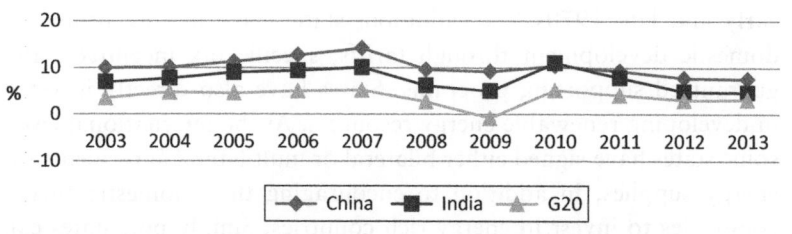

Figure 2.2 China's and India's GDP annual growth rate (%), 2003–2013
Source: International Monetary Fund, Gross Domestic Product (GDP) Volume, available at www.principalglobalindicators.org/Pages/Default.aspx, accessed February 22, 2018; 中华人民共和国国家统计局 [National Bureau of Statistics of the People's Republic of China], 《全国年度统计公报》 [National Annual Statistical Bulletin], various years, available at www.stats.gov.cn/tjsj/tjgb/ndtjgb, accessed February 22, 2018

and in the evolution of both states as two of the fastest-rising world economies. In the period of 2003–13, for instance, China's and India's GDP average development rates were 10.2 percent and 7.5 percent respectively, which is significant, especially when compared with the G20's average growth rate of 3.7 percent (see Figure 2.2). With their rapid economic growth rate, China and India have become two of the largest economies in the world. By 2015, China's and India's nominal GDP in current prices were ranked second and seventh respectively in the world, while their GDP in purchasing power parity (PPP) were ranked first and third in the world (see Table 2.1). In other words, in terms of PPP, China's GDP surpassed that of the United States, becoming the largest economy, while India surpassed Japan, becoming the third-largest economy. Moreover, as a consequence of their rapid growth in GDP, their real GDP per capita has grown significantly, which is another important indicator of the wealth that China and India have obtained (see Figure 2.3). For instance, in 2013, China's and India's GDP per capita reached $6,807.40 and $1,497.50 respectively. This attainment has already put China and India in the World Bank's "upper middle income country" and "lower middle income country" brackets respectively. At the same time, however, if we compare their figures to the GDP per capita of over $25,000 for South Korea, over $38,000 for Japan, and over $53,000 for the United States, it is clear that China and India still have a great deal of room for further growth.

Table 2.1 *World top ten economies, 2015*

Ranking	Top ten GDP nominal (US$ billion) Country	GDP	Top ten GDP based on PPP (US$ billion) Country	GDP
1	United States	17,949	China	19,392
2	China	10,983	United States	17,947
3	Japan	4,123	India	7,962
4	Germany	3,358	Japan	4,830
5	UK	2,849	Germany	3,841
6	France	2,422	Russia	3,718
7	India	2,091	Brazil	3,192
8	Italy	1,816	Indonesia	2,842
9	Brazil	1,773	UK	2,679
10	Canada	1,552	France	2,647

Source: International Monetary Fund, *World Economic Outlook*, April 2016, available at http://www.imf.org/en/Data, accessed February 22, 2018

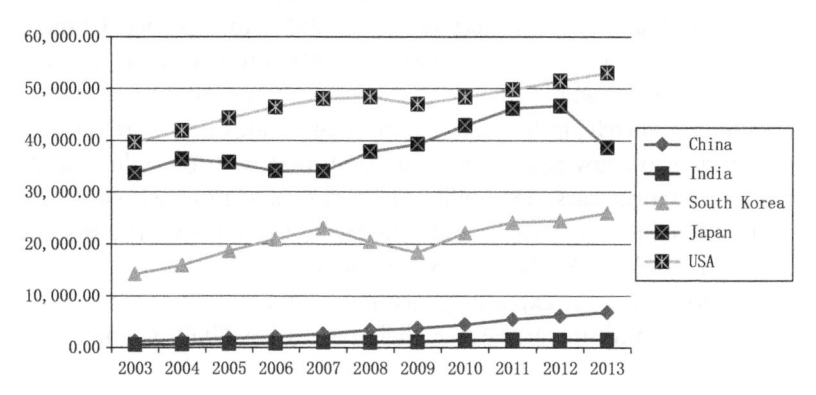

Figure 2.3 Real GDP per capita, India and China with some developed countries, 2003–2013 (US$)
Source: World Bank, "GDP per capita (current US$)," available at http://data.worldbank.org/indicator/NY.GDP.PCAP.CD?page=2, accessed February 22, 2018

Such rapid economic growth in China and India has led to a spectacular rise in living standards for the great masses of Chinese and Indian people. According to the World Bank, both countries' "extreme poverty" rate, defined as the rate of people living on $1.25 per day, has fallen from 60 percent in China and 50 percent in India in 1990, to barely 9 percent in 2010 in China and 32 percent in 2009 in India.[39] While this rate still falls short of the living standards of developed countries, China's and India's achievements have been widely hailed as part of the greatest poverty eradication in history. At the same time, however, according to the World Bank, due to China and India having the largest populations in the world, both countries have two of the largest populations still living under extreme poverty. By 2012, for instance, China's and India's extreme poverty population, respectively, accounted for 12 percent and 33 percent of the world total.[40] Not surprisingly, even the IEA admits that "Achieving economic development is of paramount importance for India, where average GDP per capita is less than 40% of the global average."[41]

Higher up on the income scale, China's and India's middle classes are growing rapidly and accessing more and better consumer goods. A typical example is the dramatic increase of car ownership in China. According to research, from 2000 to 2014, the number of cars purchased per month has grown sixfold, increasing from about 250,000 to over 1.5 million cars sold every month. In 2013, for instance, Chinese people bought more than 20 million cars, 400 times as many as were purchased in the whole of 1990, when fewer than 50,000 cars were bought.[42] Similarly, in India, the number of various transport vehicles is expected to grow significantly to 2040, that is, "260 million additional passenger cars, 185 million new two- and three-wheelers and nearly 30 million new trucks and vans are added to the vehicle stock," according to the IEA's *World Energy Outlook 2015*.[43]

Due to their ever-expanding economic development, their large populations, and an increase in their per capita GDP, China's and

---

[39] World Bank, PovcalNet, available at http://iresearch.worldbank.org/PovcalNet/index.htm?2, accessed February 22, 2018.

[40] Population Reference Bureau, "2014 World Population Data Sheet," available at www.prb.org/pdf14/2014-world-population-data-sheet_eng.pdf, accessed February 22, 2018.

[41] IEA, *World Energy Outlook 2016* (Paris: OECD/IEA, 2016), 324.

[42] Joseph O'Mahoney and Zheng Wang, "China's 1989 Choice: The Paradox of Seeking Wealth and Democracy," *The Wilson Quarterly* 38, no. 3 (Fall 2014), available at http://wilsonquarterly.com/quarterly/summer-2014-1989-and-the-making-of-our-modern-world/chinas-1989-choice-paradox-seeking-wealth-and-democracy, accessed February 22, 2018.

[43] IEA, *World Energy Outlook 2015* (Paris: OECD/IEA, 2015), 467.

India's demand for energy in general and oil in particular has been growing so dramatically that they have already become the largest and third-largest energy consumers in the world respectively. According to the IEA's *World Energy Outlook 2014*, "China is the dominant force behind global demand growth for the next decade, accounting for more than one-third of the increase. But after 2025, India takes over as Chinese growth slows down noticeably."[44] Thus, both countries' domestically endowed but limited energy resources are no longer able to meet their significantly increased demand, so both countries have already had to resort to overseas energy resources to fill the huge gap between their supply and demand. So far, China and India have become net importers of all fossil fuels, that is, oil, natural gas, and coal (see Table 2.2).

From Table 2.2, we can see that imported oil accounted for 56 percent and 75 percent, respectively, of China's and India's total oil demand in 2011, and the share is projected to increase to 82 percent and 77 percent in China in 2035 and 2040 respectively, and to 92 percent in both 2035 and 2040 in India. Moreover, both countries' oil imports have been highly dependent on the Middle East: more than half of China's oil imports and more than 60 percent of India's oil imports come from this region. According to the IEA, Sino-Indian dependence on the Middle East for oil imports will continue to rise.[45] Although the share of imported oil accounts for only 10 percent in China's total energy consumption, its transportation and some industrial sectors are totally dependent on oil. According to the IEA's 2013 projection, China was expected to surpass the United States, to become the world's largest oil importer by the early 2020s.[46] In practice, however, according to the US Energy Information Administration (EIA), China had surpassed the United States as the largest oil importer by 2014.[47] In terms of India's oil imports, according to the IEA, by 2014, India had already become the third-largest importer of crude oil in the world, behind China and the United States.[48]

---

[44] IEA, *World Energy Outlook 2014*, 53.   [45] Ibid.
[46] IEA, *World Energy Outlook 2013*, 23.
[47] According to the EIA, in 2014, China's daily oil imports were 6.1 million barrels per day, while that of the United States was 5.1 million barrels per day; see EIA, "China" (May 14, 2015), available at www.eia.gov/beta/international/analysis_includes/countries_long/China/china.pdf, accessed February 22, 2018.
[48] IEA, *World Energy Outlook 2015*, 468.

Table 2.2 *China's and India's dependency on energy imports, 2011–2040*

		2011	2012	2035	2040	% of major importing sources
Oil	China	56%	54%	82%	77%	Saudi Arabia 20, Angola 12, Iran 11, Russia 7, Oman 7, Iraq 5, Sudan 5, Venezuela 5, Kazakhstan 5, Kuwait 4, UAE 3, Brazil 3, Congo 2, others 11
	India	75%	74%	92%	92%	Africa 22, Saudi Arabia 18, Iran 11, other Middle East 34, Western hemisphere 10, others 5
Gas	China	22%	27%	40%	39%	Australia 30, Qatar 19, Indonesia 17, Malaysia 13, Yemen 7, Nigeria 6, Trinidad 2, Russia 2, Egypt 2, others 2
	India	32%	32%	42%	46%	Qatar 65, others 35
Coal	China	6% (2010)	8%	3%	8%	Indonesia and Australia 50, others 50
	India	15% (2010)	25%	34%	39%	Indonesia, Australia and others*

Sources: IEA, *World Energy Outlook 2012* (Paris: OECD/IEA, 2012), 68, 76, 120, 169; IEA, *World Energy Outlook 2014* (Paris: OECD/IEA, 2014), 81; US Energy Information Administration, "China," available at www.eia.gov/countries/cab.cfm?fips=CH, and "India," available at www.eia.gov/countries/cab.cfm?fips=IN, accessed February 18, 2018
* *Note:* No percentage date available.

When it comes to natural gas, according to the IEA, India and China started importing liquefied natural gas (LNG) in 2004 and in 2006 respectively.[49] In 2011, imported natural gas accounted for 22 percent of China's total gas demand and 32 percent of India's, and it is projected to reach 40 percent and 42 percent respectively in 2035. By then, China will be the world's second-largest importer of natural gas, after the EU.[50] Much of their imported natural gas will come from Middle Eastern countries, such as Iran and Qatar, in addition to Russia.

[49] IEA, *World Energy Outlook 2012*, 68.
[50] IEA, *World Energy Outlook 2007*, 171.

With regard to coal, although China's and India's reserves are relatively rich, accounting for 13.3 percent and 7.0 percent respectively of the global total,[51] their coal mining and production cannot keep pace with their increased energy demand. As a consequence, both countries have already been net importers of coal and are projected to become two of the largest coal importers in the world, with the majority of their coal imports coming from Australia and Indonesia.[52] According to the IEA, India is expected to be the largest coal importer by the early 2020s.[53]

China's and India's present and future high dependence on overseas energy supplies poses some real and potential security threats to their efforts to maintain their sustained economic development to maximize their wealth. First of all, given their high dependency on foreign sources for energy, especially their high dependence on oil imports from the Middle East, it is plausible to say that both countries' energy supply security, especially oil supply security, faces some potential threat of disruption, given the fact that the Middle East has long been an unstable region, accounting for eight of the nine major world oil supply disruptions since the 1960s (see Table 2.3). More specifically, the Six Day War in 1967 led to the loss of 2 million barrels of oil per day (mb/d); the Arab–Israel War and Arab oil embargo in late 1973 and early 1974 led to the loss of 4.3 mb/d, which in turn led to the first oil crisis experienced by the Western energy-consuming countries. The Iranian Revolution of November 1978 to April 1979 witnessed a production loss of 5.6 mb/d and this led to the second oil crisis; the Iran–Iraq war in 1980–81 led to the loss of 4.1 mb/d; Iraq's invasion of Kuwait led to the loss of 4.3 mb/d; between June and July 2001, Iraq oil export suspension led to the loss of 2.1 mb/d; and the Iraq war in 2003 led to the loss of 2.3 mb/d. At present, this region's situation is characterized by the perennial Israeli-Palestinian conflict, Islamic extremism and terrorism, Iran's controversial nuclear program, and the fragility of oil-based autocracies, all of which threaten the long-term stability of oil and gas supplies from this region. Therefore, Flynt Leverett points out that "For policy-makers and strategic

---

[51] BP, *BP Statistical Review of World Energy* (June 2013), 30, available at www.bp.com/content/dam/bp/pdf/statisticalreview/statistical_review_of_world_energy_2013.pdf, accessed November 15, 2014.
[52] IEA, *World Energy Outlook 2012*, 68.
[53] IEA, *World Energy Outlook 2013*, 23.

Table 2.3 *Major world oil supply disruption, 1967–2005*

Year	Incident	Gross initial loss of supplies (mb/d)
Sept. 2005	Hurricanes Katrina and Rita	1.5
Mar.–Dec. 2003	War in Iraq	2.3
Dec. 2002–Mar. 2003	Venezuela strike	2.6
Jun.–Jul. 2001	Iraq oil export suspension	2.1
Aug. 1990–Jan. 1991	Iraqi invasion of Kuwait	4.3
Oct. 1980–Jan. 1981	Outbreak of Iran–Iraq War	4.1
Nov. 1978–Apr. 1979	Iran Revolution	5.6
Oct. 1973–Mar. 1974	Arab-Israeli War and Arab oil embargo	4.3
Jun.–Aug. 1967	Six Day War	2.0

Source: IEA, *World Energy Outlook 2007*, 186

planners in Beijing and New Delhi, this growing reliance on imported hydrocarbons – and, especially, hydrocarbons from the Persian Gulf – is at the heart of the energy-security challenge facing their countries."[54] Thus, seeking energy security has become an important national security goal for China and India.

In addition, due mainly to their high dependence on overseas oil supplies, especially those from the Middle East, China and India have increasingly suffered from high and volatile oil import bills. In 2012, for instance, oil import bills for both countries reached historical highs – $270 billion for China and $140 billion for India.[55] Thus, oil accounts for the bulk of spending in China and India. According to the IEA, by 2040 China's and India's bills for energy imports will triple because of China's increased imports of oil and gas, and India's increased imports of oil, gas, and coal, with India spending more on coal than on gas.[56] Thus, the IEA points out that "specific vulnerability to a possible physical shortage in supply and to high and volatile

[54] Flynt Leverett, "Resource Mercantilism and the Militarization of Resource Management: Rising Asia and the Future of American Primacy in the Persian Gulf," in Daniel Moran and James A. Russell (eds), *Energy Security and Global Politics: The Militarization of Resource Management* (New York: Routledge, 2009), 215.
[55] IEA, *World Energy Outlook 2014*, 83.   [56] Ibid., 83, 93.

oil-import bills becomes increasingly concentrated in the major emerging oil-importing economies of Asia: China and India in particular."[57]

Aside from the energy supply security challenge, China and India have been faced with some negative environmental impacts, resulting directly from their fossil fuel – especially coal-dominated – energy consumption. As Table 2.4 shows, China's and India's fossil fuel consumption mix is similar. Between 2009 and 2014, for instance, coal, oil, and natural gas accounted for around 69 percent, 17 percent, and 5 percent respectively in China on average, and roughly 53 percent, 30 percent, and 9 percent respectively in India, of their total energy consumption. As a result, both countries' CO_2 (the leading GHG) emissions have increased so dramatically that they have already become the world's largest and third-largest CO_2 emitters respectively. According to the IEA, the energy sector accounts for around two-thirds of all GHG emissions globally.[58] Both countries' CO_2 emissions increased by 9 percent and 8 percent respectively in 2011, and by 3.1 percent and 6.8 percent respectively in 2012, a significant portion of which was due to increased coal demand in both countries.[59] Given the dominant share of fossil fuels, especially coal, in their energy mix,[60] China's and India's CO_2 emissions will continue to grow. According to the IEA, "China remains the world's largest emitter by a wide margin until 2040 in the New Policies Scenario," while "India overtakes the United States in terms of annual CO_2 emissions just before 2040," even though India's cumulative emissions and per-capita emissions remain low.[61] Increased CO_2 emissions will lead to further climate change, which has already caused an increase in the frequency and severity of

---

[57] Ibid., 82.   [58] Ibid., 87.
[59] IEA, *World Energy Outlook 2012*, 243, 244; IEA, *World Energy Outlook 2014*, 87.
[60] According to the IEA, by 2040, fossil fuels will continue to account for the dominant share of world primary energy demand under three different scenarios, that is, 74 percent under new policies, 79 percent under current policies, and 58 percent under 450 Scenario. In China and India, coal alone will account for 45 percent and 48 percent respectively. See IEA, *World Energy Outlook 2016*, 64, 22, 232. According to the IEA, "450 Scenario sets out an energy pathway consistent with the goal of limiting the global increase in temperature to 2 °C by limiting concentration of greenhouse gases in the atmosphere to around 450 parts per million of CO_2," available at www.iea.org/publications/scenariosandprojections, accessed February 24, 2018.
[61] IEA, *World Energy Outlook 2014*, 89.

Table 2.4 *China's and India's energy mix (%), 2009–2014*

		Oil	Natural gas	Coal	Nuclear	Hydropower	Renewable
2014	China	17.5	5.6	66.0	1.0	8.1	1.8
	India	28.3	7.1	56.5	1.2	4.6	2.2
2013	China	17.4	5.3	67.7	1.0	7.2	1.6
	India	29.4	7.8	54.4	1.3	5.0	2.1
2012	China	17.3	4.9	68.8	1.0	7.1	1.2
	India	30.3	9.3	52.7	1.3	4.6	1.9
2011	China	17.2	4.5	70.3	1.0	6.0	0.9
	India	30.4	10.7	50.3	1.4	5.6	1.7
2010	China	17.7	4.0	70.4	1.0	6.6	0.5
	India	30.5	11.1	50.1	1.1	4.9	1.5
2009	China	16.8	3.5	72.6	1.0	6.0	0.3
	India	31.5	9.7	51.7	1.0	5.0	1.3
Average	China	17.3	4.6	69.3	1.0	6.8	1.0
	India	30.1	9.3	52.6	1.2	5.0	1.8

Source: Calculated by author. BP, *BP Statistical Review of World Energy* (June 2015), 11, 25, 33, 35, 36, 38, 40, available at www.bp.com/content/dam/bp/pdf/energy-economics/statistical-review-2015/bp-statistical-review-of-world-energy-2015-full-report.pdf, accessed February 23, 2018

extreme weather events in recent years in both countries, which in turn has had some negative impacts on their energy supply and demand.[62]

Simply put, while China and India seek to maximize their wealth through their sustained, rapid economic development, both countries are faced with the intertwined challenges, that is, energy insecurity and climate change.

Status

States, as with human individuals with growing wealth, usually seek higher status among other state actors in the international system. In Yong Deng's words, "Growing wealth generates an expectation of greater respect."[63]

[62] Ibid., 248.
[63] Yong Deng, *China's Struggle for Status: The Realignment of International Relations* (New York: Cambridge University Press, 2008), 2.

Apart from seeking to maximize their wealth, the nonmaterial desire of states for a higher status in the international system through building up their social identity has also played an important role in shaping their policy behavior. In other words, status is another preference for states, aside from wealth,[64] or status is an element of a state's national interests.[65] However, "A state's status is fundamentally different from its material position in the international system, because status is a product of social construction."[66] Status is defined by T. V. Paul, Deborah Welch Larson, and William C. Wohlforth as "collective beliefs about a given state's ranking on valued attributes (wealth, coercive capabilities, culture, demographic position, socio-political organization, and diplomatic clout)."[67] According to Henri Tajfel, social identity is "that part of an individual's self-concept which derives from his knowledge of his membership of a social group (or groups) together with the value and emotional significance attached to that membership."[68] Such an in-group identity has significantly shaped individual's behavior by comparison and contrast between in-groups and out-groups. Individuals usually self-categorize their own identity in a social group in which there are certain rules or norms for appropriate behavior of in-group members. Once an individual's behavior is perceived by other in-group members to be inconsistent with its self-categorized identity, other actors would apply opprobrium to this individual. Under such circumstances, this individual has to adjust its behavior so as to minimize the cost of opprobrium stemming from its inappropriate behavior that runs counter to its in-group identity.[69] This point was made by John Harsanyi in the 1970s, that is, "Apart from economic payoffs, social status (social rank) seems to be the most

---

[64] William C. Wohlforth, "Status Dilemmas and Interstate Conflict," in T. V. Paul, Deborah Welch Larson, and William C. Wohlforth (eds), *Status in World Politics* (New York: Cambridge University Press, 2014), 117.
[65] Anne L. Clunan, "Why Status Matters in World Politics," in Paul, Larson, and Wohlforth (eds), *Status in World Politics*, 289.
[66] Ibid., 273.
[67] T. V. Paul, Deborah Welch Larson, and William C. Wohlforth, "Status and World Order," in Paul, Larson, and Wohlforth (eds), *Status in World Politics*, 3–29.
[68] Henri Tajfel, *Human Group and Social Categories: Studies in Social Psychology* (Cambridge: Cambridge University Press, 1981), 251.
[69] Alastair Iain Johnston, *Social States: China in International Institutions, 1980–2000* (Princeton, NJ: Princeton University Press, 2008), 76.

important incentive and motivating force of social behaviour."[70] Thus, we can say that "Identities are not the opposite of rational self-interest; in fact, identities in part determine which interests are important."[71]

Moreover, states, like individuals, live in an international system in which there is a set of rules and norms that govern, or at least influence, states' interactions. Alastair Iain Johnston argues that there exist normative structures at the international level that have constraining effects on states' behavior.[72] According to Johnston, the desire of states to maximize status and wealth are not necessarily compatible, so they have had to make trade-offs: in order to obtain status, states should first accumulate some markers of status through their policy behavior in the international system – among others, norm-abiding behavior has become a marker of status.[73] Under such circumstances, states that seek higher status in the international system have had to be sensitive to the evaluation of their behavior by other actors, either state or non-state. Specifically, states are sensitive to social opprobrium that might greatly damage their efforts to build up their international image. In other words, status-seeking states have "the desire to minimize opprobrium costs."[74] In the same vein, Oran Young points out that "Policy makers, like private individuals, are sensitive to the social opprobrium that accompanies violations of widely accepted behavioural prescriptions. They are, in short, motivated by a desire to avoid the sense of shame or social disgrace that commonly befalls those who break widely accepted rules."[75] Simply put, the desire of states to maximize their status has played an important role in eliciting their pro-social behavior in the international system because status is "intersubjective," namely "a state only has a particular status when others recognize it as such."[76] In other words, other states' recognition is an important factor in the construction of a state's status. This point was also made by Kenneth Waltz, in his reference to the historical event during the

---

[70] John Harsanyi, *Essays on Ethics, Social Behavior, and Scientific Explanation* (Dordrecht: D. Reidel, 1976), 204.
[71] David C. Kang, *China Rising: Peace, Power, and Order* (New York: Columbia University Press, 2007), 79–80.
[72] Johnston, *Social States*, xiv. [73] Ibid., 75. [74] Ibid., xxvi.
[75] Oran Young, "The Effectiveness of International Institutions: Hard Cases and Critical Variables," in James Rosenau and Ernest-Otto Czempiel (eds), *Governance without Government: Order and Change in World Politics* (Cambridge: Cambridge University Press, 1992), 176–177.
[76] Clunan, "Why Status Matters in World Politics," 289.

Cold War when the United States conferred superpower status on China.[77] Accordingly, China gained its "superpower status" in the 1970s, not because of the material attributes it possessed, but because the United States – the existing superpower – treated it as such.[78]

China and India Seeking Status

For China and India, economic wealth has provided them with a fundamental resource to seek great-power status in the international system, because great-power status would enable them to win voluntary deference from other state leaders and peoples for their possession of "certain special rights and duties, namely the right to play a part in determining issues that affect the peace and security of the international system as a whole and the responsibility of modifying their policies in the light of the managerial responsibilities they bear."[79] Although China and India have the largest populations in the world, neither country has been able to ascend to great-power status. In other words, population is not the major criteria for measuring a great power, a fact which receives consensus among IR theorists. For instance, according to Kenneth Waltz, "states are not placed in the top rank because they excel in one way or another. Their rank depends on their score on all of the following items: size of population and territory, resource endowment, economic capability, military strength, political stability and competence."[80] Hedley Bull also pointed out that "A population of 100 million or more today is not sufficient to confer

---

[77] Kenneth Waltz noted that, "Thus Nixon, when he was President, slipped easily from talking of China's becoming a superpower to conferring superpower status on her. In one of the statements that smoothed the route to Peking, he accomplished this in two paragraphs [...]. And the headlines of various news stories before, during, and after his visit confirmed China's new rank. This was the greatest act of creation since Adam and Eve, and a true illustration of the superpower status of the United States. A country becomes a superpower if we treat it like one. We create other states in our image." See Waltz, *Theory of International Politics*, 130.

[78] Paul, Larson, and Wohlforth, "Status and World Order," 9.

[79] Hedley Bull, *The Anarchical Society: A Study of Order in World Politics* (New York: Columbia University Press, 1977), 202.

[80] Waltz, *Theory of International Politics*, 131.

superpower status upon a nation."[81] John Mearsheimer specifically echoes this point: "Both China and India ... had much larger populations than either the Soviet Union or the United States during the Cold War, but neither China nor India achieved great-power status because they were nowhere near as wealthy as the superpowers."[82] With a great deal of wealth accumulated through their economic development over the past three decades, China and India have tried to obtain great-power status through their accumulation of both hard and soft status markers in the international system.[83] In other words, China and India are status maximizers.[84]

As Table 2.5 shows, there are a number of basic markers that mark states' higher status in the international system.[85] More specifically, permanent membership of the UN Security Council (UNSC) has been the most important marker of a state's great-power status.[86] This lies in the fact that it "was built on the premise of great power management of the system, and the permanent five members reflect the distribution

[81] Hedley Bull, "Population and the Present World Structure," in William Alonso (ed.), *Population in an Interacting World* (Cambridge, MA: Harvard University Press, 1987), 79.
[82] Mearsheimer, *The Tragedy of Great Power Politics*, 61.
[83] This categorization is modeled on Joseph Nye's "hard power" and "soft power." See Joseph S. Nye, Jr., *Soft Power: The Means To Success in World Politics* (New York: Public Affairs, 2004).
[84] Randall L. Schweller and Xiaoyu Pu argue that rising states do not necessarily choose to maximize their status in the international system due to the fact that their priority is to sustain their economic growth, so they are inward-looking and unwilling to undertake the responsibilities associated with great-power status. See Randall L. Schweller and Xiaoyu Pu, "After Unipolarity: China's Visions of International Order in an Era of U.S. Decline," *International Security* 36, no. 1 (Summer 2011), 41–72. At the same time, these two scholars largely agree that rising powers have been seeking higher status. See Xiaoyu Pu and Randall L. Schweller, "Status Signaling, Multiple Audience, and China's Blue-Water Naval Ambition," in Paul, Larson, and Wohlforth (eds), *Status in World Politics*, 141–162.
[85] There are some other markers, such as club membership. See Thomas J. Volgy, Renato Corbetta, J. Patrick Rhamey, Jr., Ryan G. Baird, and Keith A. Grant, "Status Considerations in International Politics and the Rise of Regional Powers," in Paul, Larson, and Wohlforth (eds), *Status in World Politics*, 58–84.
[86] Vincent Pouliot, "Setting Status in Stone: The Negotiation of International Institutional Privileges," in Paul, Larson, and Wohlforth (eds), *Status in World Politics*, 192–215.

Table 2.5 *Hard markers and soft markers of great-power status*[87]

	Hard markers	Soft markers
International peace and security	Permanent membership of UN Security Council	Pro-social behavior or norm-abiding behavior
Military	• Nuclear weapons • Power-projection navy with aircraft carriers	• Non-proliferation and export control • Good governance • Free trade • Contribution to provision of global public goods such as global energy security and a stabilized climate
International economy	• Voting power in the World Bank and the International Monetary Fund • Multilateral economic arrangements	
International sports	Olympic Games	

Source: Author

of power at the end of World War II."[88] In addition, the UNSC holds the authority in the maintenance of international peace and security, which includes: 1) determining the existence of a threat to the peace or an act of aggression; 2) calling upon the parties to a dispute to settle it by peaceful means and recommending methods of adjustment or terms of settlement; 3) imposing sanctions or even authorizing the use of force to maintain or restore international peace and security.[89]

China has been one of the five permanent members on the UNSC (or P-5, namely Britain, China, France, Russia, and the United States) since the 1970s, while India has not obtained that status. According to Stephen Cohen, "A permanent seat on the UN Security Council (with veto power) would have been the ultimate certification of India's status

---

[87] Paul, Larson, and Wohlforth list some other status markers, including membership of elite clubs such as G8, leadership positions in international organizations, formal state visits, summit meetings, and inclusion in informal problem-solving groups. See Paul, Larson, and Wohlforth, "Status and World Order," 10–11.

[88] Deborah Welch Larson and Alexei Shevchenko, "Status Seekers: Chinese and Russian Responses to U.S. Primacy," *International Security* 34, no. 4 (Spring 2010), 70.

[89] United Nations, "The Security Council," available at www.un.org/en/sc, accessed February 23, 2018.

as a major power."[90] It is no surprise that India has been striving to obtain permanent membership. To reach this goal, India has tried to win non-permanent membership on the UNSC as a stepping stone.[91] Since 1992, India has secured such membership seven times. The latest non-permanent membership was granted in October 2010. In order to secure this two-year rotating seat on the UNSC, in the run-up to the vote, India's then External Affairs Minister, S. M. Krishna, met leaders of a record 56 countries on the sidelines of the UN General Assembly's annual session in September 2010. After winning this non-permanent membership of the UNSC, India's envoy to the UN, Hardeep Singh Puri, said, "We have worked hard ... we have pushed for every single vote."[92] Since 2005, India has been working intensely with Brazil, Germany, and Japan, or the so-called G-4, a group composed of all the states that are seeking permanent membership of the UNSC, to expand permanent membership. In 2006, the Indian government nominated Shashi Tharoor for election to the post of UN Secretary General, which can be identified as the first real manifestation of its great-power status aspirations.[93]

A second important status marker is in the military realm, that is, nuclear weapons and a power-projection navy with aircraft carriers. In terms of the role of nuclear weapons in marking a state's status, Scott Sagan pointed out that nuclear weapons are not only "tools of national security," but "can also serve as international normative symbols of modernity and identity."[94] In other words, nuclear weapons acquisition has been regarded as serving important symbolic functions – both shaping and reflecting a state's identity, which is embedded in "deeper norms and shared beliefs about what actions

[90] Stephen Philip Cohen, *India: Emerging Power* (Washington, D.C.: Brookings Institution Press, 2001), 33.
[91] 张贵洪 [Zhang Guihong] and 王磊 [Wang Lei], "印度政治大国梦与金砖国家合作 [India's Political Great Power Dream and BRICs Cooperation]," 《复旦学报 (社会科学版)》 [*Fudan Journal* (Social Sciences Edition)] 55, no. 6 (2013), 167–174.
[92] "India Elected to UNSC as Non-permanent Member," *The Times of India*, October 12, 2010, available at http://timesofindia.indiatimes.com/india/India-elected-to-UNSC-as-non-permanent-member/articleshow/6737610.cms, accessed February 23, 2018.
[93] P. R. Kumaraswamy, "India's Energy Dilemma with Iran," *South Asia: Journal of South Asian Studies* 36, no. 2 (2013), 288.
[94] Scott D. Sagan, "Why Do States Build Nuclear Weapons? Three Models in Search of a Bomb," *International Security* 21, no. 3 (Winter 1996–1997), 55.

are legitimate and appropriate in international relations."[95] All P-5 members are nuclear weapons states.

In the military arena, India had already developed nuclear weapons to further its great-power ambition, which has significantly challenged the Nuclear Non-Proliferation Treaty (NPT). This treaty divides the states in the international system into nuclear weapons states and non-nuclear weapons states, based on whether or not the states had nuclear weapons by January 1, 1967.[96] Thus, the NPT divides the world into the "haves" and the "have-nots." Moreover, this status quo has been extended indefinitely since the mid-1990s.[97] Although this treaty has been in force since 1970, India has never signed up to it. According to T. V. Paul, "India's opposition to the NPT has been primarily shaped by the fact that the NPT cements the status of the five declared nuclear weapons states as the major powers in the international system, thereby denying India the opportunity to rise to their ranks purely because it was a latecomer."[98] India's defiance of this treaty culminated in its nuclear tests in both 1974 and 1998, after which India officially declared to the world that it was a nuclear weapons state.[99] According to Baldev Raj Nayar and T. V. Paul, the Indian nuclear test in 1998 was a "declaration that the present status hierarchy in the international system was no longer acceptable and needed to be modified by accommodating India."[100] In other words, India's nuclear

---

[95] Ibid., 73.
[96] According to the NPT's Article 9 (3), "a nuclear-weapon State is one which has manufactured and exploded a nuclear weapon or other nuclear explosive device prior to 1 January 1967." See United Nations, "The Treaty on The Non-Proliferation of Nuclear Weapons," available at www.un.org/en/conf/npt/2005/npttreaty.html, accessed February 23, 2018.
[97] According to Decision 3 of the 1995 Review and Extension Conference of the Parties to the Treaty on the Non-Proliferation of Nuclear Weapons held in New York on April 17–May 12, 1995, the NPT has been extended indefinitely. See United Nations Office for Disarmament Affairs, "The 1995 Review and Extension Conference of the Parties to the Treaty on the Non-Proliferation of Nuclear Weapons," April 17–May 12, 1995, available at www.un.org/disarmament/wmd/nuclear/npt1995, accessed February 23, 2018.
[98] T. V. Paul, "The Systemic Bases of India's Challenges to the Global Nuclear Order," *The Nonproliferation Review* 6, no. 1 (Fall 1998), 1–11.
[99] Gaurav Kampani, "New Delhi's Long Nuclear Journey: How Secrecy and Institutional Roadblocks Delayed India's Weaponization," *International Security* 38, no. 4 (Spring 2014), 79–114.
[100] Baldev Raj Nayar and T. V. Paul, *India in the World Order: Searching for Major-Power Status* (Cambridge: Cambridge University Press, 2003), 231.

option has been largely driven by its status-seeking desires. However, India's nuclear weapons state status had remained illegitimate due to its violation of the NPT regime. But since the United States signed the civil nuclear cooperation agreement with India in July 2005, its status as a legitimate nuclear weapons state has now been recognized by the United States and the international community.[101]

Apart from nuclear weapons, a strong navy, especially one equipped with aircraft carriers that enable power projection, has also been regarded as a hard marker of a state's prestige. According to Robert Ross, one of the essential elements that has supported and maintained the United States' supremacy in the post-Cold War international system has been its power-projection navy fleet with carrier-deployed aircraft, since such a navy has enabled US power to be projected into distant theatres. Specifically, the US victory over Iraq in 1991, its rapid destruction of the Iraqi government in 2003, as well as its air war against Afghanistan's Taliban government in 2001 could be largely attributed to its carrier-based capabilities.[102]

China has embarked on efforts to build a blue-water navy with aircraft carriers with a maritime power projection capability to boost its great-power status. Ross argues that China's efforts to build a blue-water navy has been mainly motivated by Chinese nationalism, which deems it a humiliation that "the navies of not only the United States but also of India and Japan can sail the South China Sea, while China's navy lacks such a capability."[103] Moreover, China was greatly "embarrassed by the contrast between the Australian and U.S. leadership of the 2004 Indonesian tsunami maritime rescue mission and China's peripheral role,"[104] and it also feels humiliated by the fact that China cannot protect its own overseas citizens from violence, so in order to win international respect for China's national strength, China must build a blue-water navy equipped with aircraft carriers.[105] According to Michael A. Glosny, Phillip C. Saunders, and Robert S. Ross, China's effort to do so is understandable, since "After thirty years of impressive economic growth and the corresponding development of national pride, the Chinese people desire the international prestige and status

---

[101] Dinshaw Mistry, *The US-Indian Nuclear Agreement: Diplomacy and Domestic Politics* (New Delhi: Cambridge University Press, 2014).
[102] Robert S. Ross, "China's Naval Nationalism: Sources, Prospects, and the U.S. Response," *International Security* 34, no. 2 (Fall 2009), 55.
[103] Ibid., 66.   [104] Ibid.   [105] Ibid., 66–67.

commensurate with China's domestic and international successes. An aircraft carrier would be one reflection of Chinese success."[106]

When it comes to status markers in the international economy, a state's voting share in the World Bank and the International Monetary Fund – the Bretton Woods institutions established at the end of World War II under the leadership of the United States – has been the most important indicator. It is well-known that the characteristics of these two regimes "are often hierarchical in their structure and functions and in that manner embody the status hierarchy."[107] According to Ngaire Woods, although these two organizations have universal membership, their voting quotas have long been limited to a small number of big states, so there have been calls for reform of both organizations' voting structures since the 1990s, because "The quotas on which votes are based are seen by member states as important symbols of both status and commitment," in addition to each individual member's relative economic weight and prestige.[108]

China and India have been leaders in advocating an increase in the share of their voting quotas in the World Bank and the IMF, based on the increased wealth they have accumulated through their rapid economic growth.[109] For instance, in 2010, urged on by the emerging economies, including China and India, G20 finance ministers and central bank governors agreed on a doubling of IMF members' quotas – financial stakes that determine voting power in the institution – that would shift 6 percent of voting shares from developed countries to emerging markets and developing countries.[110] According to this reform plan, China's voting shares would rise from the sixth-largest to the third-largest, only behind the United States and Japan, while

---

[106] Michael A. Glosny, Phillip C. Saunders, and Robert S. Ross, "Debating China's Naval Nationalism," *International Security* 35, no. 2 (Fall 2010), 174.
[107] Larson and Shevchenko, "Status Seekers," 70.
[108] Ngaire Woods, "The Challenge of Good Governance for the IMF and the World Bank Themselves," *World Development* 28, no. 5 (2000), 823, 829.
[109] Alan Heston, "What Can Be Learned about the Economies of China and India from Purchasing Power Comparisons?," in Barry Eichengreen, Poonam Gupta, and Rajiv Kumar (eds), *Emerging Giants: China and India in the World Economy* (New York: Oxford University Press, 2010), 8.
[110] International Monetary Fund, "G-20 Ministers Agree 'Historic' Reforms in IMF Governance," October 23, 2010, available at www.imf.org/external/pubs/ft/survey/so/2010/new102310a.htm, accessed February 23, 2018.

India's voting shares would be elevated from the rank of 23rd to 8th.[111] However, this reform plan had been stalled for several years until 2016, because the United States Congress declined to ratify IMF reforms and give emerging markets the voting rights they had been asking for. Against this backdrop, China and India, together with other emerging economies, especially Brazil, Russia, and South Africa (the so-called BRICS), launched a $100 billion development bank, dubbed the New Development Bank (NDB) in July 2014.[112] Moreover, in October 2014, China officially launched its new development bank – the Asian Infrastructure Investment Bank (AIIB), with India as one of its founding members.[113] Regardless of what these new banks would be like, they reflected China's and India's desire to gain a higher status in the international financial regimes, in order to match their increased economic importance in the global economy.

A further marker of a state's status is in the arena of international sports, that is, most states consider hosting the Olympic Games to be a symbol of their major-power status. For instance, Russian President Vladimir Putin regarded Russia being awarded the 2014 Winter Olympics as a "judgment of our country," while Brazilian President Luiz Inácio Lula da Silva tearfully proclaimed Rio de Janeiro's selection to host the 2016 Olympic Games as Brazil beginning to "receive the respect we deserve."[114] Not surprisingly, China has also resorted to hosting the Olympic Games to boost its great-power status. More specifically, although China lost its bid to Australia for the 2000 Olympics, it bid once again for the 2008 Games. To ensure its success, with the approval of the Chinese State Council, a Beijing 2008 Olympic Games Bid Committee was established as early as September 1999, to work on the organization of the Beijing application. Before the International Olympic Committee's inspection, the Chinese government mobilized a large number of people to clean up and repaint many parts of Beijing. As a result of these officially sponsored efforts,

---

[111] International Monetary Fund, "Illustration of Proposed Quota and Voting Shares," 2010, available at www.imf.org/external/np/sec/pr/2010/pdfs/pr10418_table.pdf, accessed February 23, 2018.

[112] Mamta Badkar, "The BRICS Are Trying to Create a Competitor to the IMF: Here's Why it Won't Have a Big Impact," *Business Standard*, July 25, 2014, available at www.businessinsider.com/brics-bank-doesnt-threaten-imf-2014-7, accessed February 23, 2018.

[113] See AIIB's official website, available at www.aiib.org, accessed February 23, 2018.

[114] Larson and Shevchenko, "Status Seekers," 70.

in September 2001, Beijing won the bid for the 2008 Olympic Games.[115] A commentator remarked that the Beijing Olympic Games showed that China, as a modernizing nation, was yearning for great-power status.[116]

Apart from their accumulation of hard status markers, China and India have also attempted to increase some of the soft status markers embedded in global governance. Specifically, China and India have tried to convince an international audience that they are norm-abiders. This is illustrated by their policies toward nuclear non-proliferation and export control regimes (see Table 2.6). China has not only signed the NPT, but has also become a member of some of the nuclear-related export control regimes, such as the Zangger Committee and Nuclear Suppliers Group (NSG), stimulated by the fact that membership of these multilateral export control regimes usually signals that states comply with the rules and regulations when they export nuclear and nuclear-related goods, items, and missiles. In other words, membership of these regimes means that a state is being recognized as a "good citizen" by the non-proliferation community. Although China has been admitted into the Zangger Committee and the NSG, its membership request was declined by the Missile Technology Control Regime (MTCR) in 2004, due mainly to its record of exporting missile behavior.[117] As regards India's non-proliferation policy, although India has been in persistent and vocal opposition to the nuclear non-proliferation and missile-control regimes, and it only has membership of the IAEA, "India has not supplied nuclear materials to any potential nuclear states, nor has it engaged in selective proliferation, as most of the current nuclear weapons powers have done."[118] Thus, India's de facto compliance with the non-proliferation and export control regimes has

---

[115] Pere Berkowitz, George Gjermano, Lee Gomez, and Gary Schafer, "Brand China: Using the 2008 Olympic Games to Enhance China's Image," *Place Branding and Public Diplomacy*, no. 3 (2007), 164–178.

[116] Xin Xu, "Modernizing China in the Olympic Spotlight: China's National Identity and the 2008 Beijing Olympiad," *The Sociological Review* 54, no. 2 (December 2006), 104.

[117] China's norm-abiding behavior in non-proliferation and export control goes beyond the nuclear arena. For instance, China has complied with the rule of on-site inspection under the Chemical Weapons Convention. On China's non-proliferation and export control policy, see Fuzuo Wu, "Pragmatic Compliance: China's Policy toward Multilateral Export Control Regimes," in Allen Carlson and Ren Xiao (eds), *New Frontiers in China's Foreign Relations* (New York: Lexington Books, 2011), 151–170.

[118] Nayar and Paul, *India in the World Order*, 15.

Table 2.6 *China's and India's participation in nuclear non-proliferation and export control regimes*

NPT and export control regimes	China	India
Nuclear Non-Proliferation Treaty (NPT)	1992	n/a
International Atomic Energy Agency (IAEA)	1984	1957
Nuclear Suppliers Group (NSG)	2004	n/a
Zangger Committee	1997	n/a
Missile Technology Control Regime (MTCR)	n/a	n/a
Comprehensive Test Ban Treaty (CTBT)	1996	n/a

Source: Author, based on the lists of member states on the websites of these regimes[119]
Note: n/a represents no membership.

been used by the United States as the main rationale for it to violate the NPT to sign a civil nuclear cooperation agreement with India.

Although China and India have tried to accumulate both hard and soft status markers to achieve their long-desired great-power status, this does not necessarily guarantee that they have been recognized by the international community as great powers, since "status requires acceptance from others."[120] In other words, whether or not China and India are great powers, especially responsible great powers, is not just dependent upon their own self-identification, but on other actors' acceptance. Simply put, their great-power status will not be legitimate until such status is recognized by other actors in the international system. Under such circumstances, China and India have had to be sensitive to social opprobrium from other actors in the international system. More specifically, both states are sensitive to being portrayed as isolated, obstructive players in the international system, since such accusations clash with their ambitions to seek great-power status, "an identity that is supported by other actors in the system."[121] China's and India's participation in the institutions discussed above means

[119] The lists of these regimes' websites are as follows: NPT, www.un.org/disarmament/wmd/nuclear/npt; IAEA, www.iaea.org/node/16963; NSG, www.nuclearsuppliersgroup.org/en/participants1; Zangger Committee, http://zanggercommittee.org/members.html; MTCR, http://mtcr.info/partners; CTBT, www.ctbto.org/the-treaty/status-of-signature-and-ratification, all accessed February 23, 2018.
[120] Larson and Shevchenko, "Status Seekers," 82.
[121] Alastair Iain Johnston and Paul Evans, "China's Engagement with Multilateral Security Institutions," in Alastair Iain Johnston and Robert S. Ross (eds),

that their behavior and policies are under scrutiny, not only from a particular audience within those institutions, but also from the whole international community. Thus, image or reputational concerns play an important role in shaping the trajectory of both countries' foreign policies and practices in certain issue-areas. Put differently, China's and India's policy behavior has been stimulated by their desire to maximize a diffused image as a responsible great power whose identity as such requires that their behavior should be in compliance with the norms and rules regulating interstate behavior embedded in those institutions of which they are members, and to make a contribution to global governance by providing certain public goods, including global energy security and a stabilized climate, which are monitored by a large number of state and non-state actors in the international system.

Asymmetrical Interdependence

Apart from domestic pressures to seek wealth, and international pressures involved in the process of seeking great-power status, states' policy behavior has also been shaped by another systemic pressure, namely asymmetrical interdependence. As mentioned above, although mainstream IR scholars disagree about the ultimate goals states have been seeking, all agree that states have been living in an international system characterized by anarchy – not necessarily in a situation of chaos, but with no world central government governing states' behavior. In practice, the international system is characterized not only by the distribution of power and a hierarchy of prestige, but also by "a set of rights and rules that govern or at least influence the interactions among states."[122] Moreover, we have witnessed over time, particularly during the second half of the twentieth century, and more especially during the 1970s, the increased expansion of cross-border trade, capital, and human resources, meaning that states have become increasingly interdependent, "situations characterized by reciprocal effects among countries or among actors in different countries."[123] Such interdependence has had a significant impact on states' behavior.

Engaging China: The Management of an Emerging Power (New York: Routledge), 1999, 252.
[122] Gilpin, *War and Change in World Politics*, 34.
[123] Keohane and Nye, *Power and Interdependence*, 7.

Specifically, interdependence has severely constrained the freedom of action of governments, and has even affected their internal organization.[124] On certain issue-areas, the interdependent system even "affects the content of the policies which they (states) formulate."[125] According to Keohane and Nye, interdependence does not necessarily guarantee mutual benefits, nor "evenly balanced mutual dependence" and "pure dependence." In a real interdependent world, interdependence entails constraints or costs, the example they cite to test their argument being that an oil-importing country is likely to be more dependent on a continued flow of oil than a country importing some luxury goods.[126] They point out that "evenly balanced mutual dependence" and "pure dependence" are two extremes, which is usually rare in the real world; instead, asymmetrical interdependence is the most normal situation under which "Less dependent actors can often use the interdependent relationship as a source of power in bargaining over an issue and perhaps to affect other issues."[127] In other words, asymmetrical interdependence provides less dependent states with a source of power to get the more dependent states to "do something they otherwise would not do."[128] The more dependent states, as rational actors, when faced with the power or influence of the less dependent states, usually "resist the temptation to disrupt or break this interdependent relationship because by so doing would incur the high costs to their economic growth."[129]

In the existing international system, asymmetrical interdependence stems mainly from the distribution of military and economic power among states. States that are more militarily and economically powerful usually have the capability to provide side payments and/or impose sanctions to get the less powerful and more dependent states to do things they would otherwise not do. Under such circumstances, more dependent states have had to alter their original policy behavior to accommodate pressures stemming from asymmetrical interdependent relationships with the more powerful states in the international system.

A second source of asymmetrical interdependence is derived from the fact that natural resources, especially energy resources, are unevenly distributed among countries. That countries with rich energy resources

---

[124] Ibid.   [125] Gourevitch, "The Second Image Reversed," 894.
[126] Keohane and Nye, *Power and Interdependence*, 8.   [127] Ibid., 9.
[128] Ibid., 10.   [129] Ibid., 35.

can employ their energy supply as a source of power to influence those energy-poor and importing countries' policy choices lies in the fact that uninterrupted and sufficient energy supplies are prerequisites for any economic activities and development. In Kenneth Waltz's words, "Those who have what others want or badly need are in favored positions."[130]

China and India Facing Asymmetrical Interdependence

China and India have sought wealth and status in an international political economic system that is characterized by asymmetrical interdependence which arises from the asymmetrical distribution of power and energy resources.

With regard to the asymmetrical distribution of power in the international system, it is well-known that, since the end of the Cold War, not only economic, but also military power have been unevenly concentrated in the United States, with some scholars regarding the post-Cold War system as being characterized by unipolarity.[131] Although Paul Kennedy and Robert Keohane began to worry about the United States' decline in the international system in the 1980s,[132] three decades on, it remains the dominant power in the international system,[133] in spite of its relative economic decline, overextension of its military commitments, and diminishing soft power since the early twenty-first century. The primacy of the United States in the first 15 years of the twenty-first century can best be illustrated by the following indicators.

First, the United States had accounted for the largest share of world GDP up until 2013. In Table 2.7, we can see that the United States was the largest economy in 2013, accounting for about 22.4 percent of world GDP, while China and India account for 12.3 percent and 2.4 percent respectively. Although GDP, the absolute size of a country's

---

[130] Waltz, *Theory of International Politics*, 147.
[131] G. John Ikenberry, Michael Mastanduno, and William C. Wohlforth, *International Relations Theory and the Consequences of Unipolarity* (New York: Cambridge University Press, 2011).
[132] Paul Kennedy, *The Rise and Fall of Great Powers* (New York: Vintage Books, 1989); Keohane, *After Hegemony*.
[133] Harsh V. Pant (ed.), *Indian Foreign Policy in a Unipolar World* (New York: Routledge, 2009), 8.

Table 2.7 *Top ten GDP economies, 2013*

Rank	Country	GDP (US$ millions)	% world GDP
1	United States	16,800,000	22.4
2	China	9,240,270	12.3
3	Japan	4,901,530	6.5
4	Germany	3,634,823	4.8
5	France	2,734,949	3.6
6	United Kingdom	2,521,381	3.4
7	Brazil	2,245,673	2.9
8	Russian Federation	2,096,777	2.8
9	Italy	1,876,797	2.4
10	India	1,826,769	2.4

Source: World Bank, "Gross domestic product 2013," available at http://databank.worldbank.org/data/download/GDP.pdf, accessed February 23, 2014

economy, is an important indicator of a country's wealth, as mentioned above, the more salient indicator to reflect a country's economic wealth is per capita GDP income. US per capita GDP income is one of the highest in the world, and is several times higher than those of China and India (see Figure 2.3). Thus, although China's phenomenal economic growth has successfully posed a major challenge to US economic dominance, as Michael Beckley points out, the United States is still a much wealthier and more innovative nation than China, and its advantages "will persist well into this century."[134]

Second, the United States still maintains the largest military expenditure. Table 2.8 shows that, in the period 2003–13, the United States devoted the largest share of its GDP to its military, roughly 4.2 percent of its GDP annually, compared with China's 2 percent and India's 2.6 percent annually. In this table, we can also see that the gap between China's and the USA's military expenditure has narrowed dramatically over the past decade, but the United States is still the largest military spender. For instance, in 2011, its total military spending of $711 billion was more than the total amount of the next nine-largest military spenders combined, accounting for 40 percent of the world total, nearly five times that of China, the second-largest military spender

[134] Michael Beckley, "China's Century? Why America's Edge Will Endure," *International Security* 36, no. 3 (Winter 2011/12), 77.

Table 2.8 *China, India, and US military expenditure (US$ million in 2011), 2003–2013*

	China (% share of GDP)	India (% share of GDP)	USA (% share of GDP)
2003	57390 (2.1)	29165 (2.8)	507781 (3.7)
2004	63560 (2.1)	33879 (2.8)	553441 (3.9)
2005	71496 (2.1)	36054 (2.8)	579831 (4.0)
2006	83928 (2.1)	36225 (2.5)	588837 (3.9)
2007	96782 (2.1)	36664 (2.3)	604292 (4.0)
2008	106640 (2.0)	41585 (2.6)	649003 (4.3)
2009	128734 (2.2)	48963 (2.9)	701048 (4.8)
2010	136239 (2.1)	49159 (2.7)	720282 (4.8)
2011	147268 (2.0)	49634 (2.6)	711338 (4.7)
2012	159620 (2.0)	49459 (2.5)	671097 (4.4)
2013	171381 (2.0)	49091 (2.5)	618681 (3.8)

Source: Stockholm International Peace Research Institute (SIPRI), "SIPRI Military Expenditure Database," available at www.sipri.org/databases/milex, accessed February 23, 2018.

(see Table 2.9).[135] Not surprisingly, Joseph Nye declared, "military power is largely unipolar, and the United States is likely to retain primacy for quite some time."[136]

Moreover, the United States has maintained its supremacy in nuclear capabilities. According to US researchers, since the end of the Cold War, the strategic nuclear balance – mutual assured destruction (MAD) – has shifted dramatically and the United States has stood "on the cusp of nuclear primacy," a "situation in which the one country with primacy can destroy its adversary's nuclear retaliatory capabilities in a disarming strike."[137] US nuclear primacy may last for a decade or longer.[138] In addition, according to more recent research, since the George W. Bush administration, the United States has been

---

[135] Stockholm International Peace Research Institute (SIPRI), "SIPRI Military Expenditure Database," available at www.sipri.org/databases/milex, accessed February 23, 2018.
[136] Joseph S. Nye Jr., "The Twenty-First Century Will Not Be a 'Post-American' World," *International Studies Quarterly* 56, no. 1 (March 2012), 215.
[137] Keir A. Lieber and Daryl G. Press, "The End of MAD? The Nuclear Dimension of US Primacy," *International Security* 30, no. 4 (Spring 2006), 8.
[138] Ibid.

Table 2.9 *Top ten military expenditures (US$ billion), 2011*

Rank	Country	Military expenditure	% World military expenditure
1	United States	711	40.9
2	China	143	8.2
3	Russia	71.9	4.1
4	UK	62.7	3.6
5	France	62.5	3.5
6	Japan	59.3	3.4
7	India	48.9	2.8
8	Saudi Arabia	48.5	2.7
9	Germany	46.7	2.6
10	Brazil	35.4	2.0

Source: SIPRI, "Background Paper on SIPRI military Expenditure Data, 2011", available at www.mapw.org.au/files/downloads/SIPRI%20factsheet%20on%20military%20expenditure%202011.pdf, accessed March 4, 2017

pursuing nuclear strategic superiority through the development of a "new triad," apart from its development of missile defenses and counterforce capabilities, all of which could "reduce or eliminate China's ability to launch a retaliatory strike."[139] Furthermore, until the first decade of the twenty-first century, the United States has been the only country that owns global projection capabilities, as indicated by its aircraft carriers or long-range bombers. Despite embarking on a major effort to build a blue navy, according to a report from the Independent Task Force sponsored by the April 2007 Council on Foreign Relations, even by 2030, there is "no evidence to support the notion that China will become a peer military competitor of the United States."[140] Accordingly, Christopher Layne argues that although US power has been waning, "It is beyond dispute, however,

---

[139] Fiona S. Cunningham and M. Taylor Fravel, "Assuring Assured Retaliation: China's Nuclear Posture and U.S.–China Strategic Stability," *International Security* 40, no. 2 (Fall 2015), 7.

[140] Council on Foreign Relations, *U.S.–China Relations: An Affirmative Agenda, a Responsible Course* (Washington, D.C.: Council on Foreign Relations, 2007), 54, available at www.cfr.org/content/publications/attachments/ChinaTaskForce.pdf, accessed February 23, 2018.

that the United States still enjoys a commanding preponderance of power over its nearest rivals."[141]

It is the United States' primacy in the international system – a structural reality – that has put some restraints, to various degrees, on China's and India's efforts to seek wealth and status in the international system. The asymmetrical interdependence of China and India on the United States stems from the fact that both China's and India's efforts to seek wealth and status in the international system have largely relied upon US support.

Economically, China's and India's economic growth has largely depended on exports. Table 2.10 shows that the United States, China, and India are unevenly dependent on exports. From 2003 to 2008, the share of exports in China's economy increased significantly, reaching 39 percent of its GDP in 2006. Although its dependence on exports has dropped significantly since 2009, they still accounted for 26 percent of its GDP in 2013. On average, China's dependence on exports in this period was more than 29 percent. India's dependence on exports is lower than China's, but it has steadily risen since 2003: by 2013, the share of exports in India's GDP had reached 25 percent. In stark contrast, the United States' dependence on exports is much lower than both China's and India's, with its average dependence on exports between 2003 and 2013 being just over 11 percent.

A more salient factor, and one that has greatly influenced China and India, is the fact that the United States has been the primary market for both countries' exports. In other words, the importance of the United States for Chinese and Indian exports is far greater than vice versa. In 2013, for example, the United States accounted for 17.0 percent and 12.1 percent of China's and India's total exports respectively, whereas China accounted for 7.3 percent of US total exports, while India has not even entered into the list of the United States' top ten trading partners. Thus, the United States' dependence on both China's and India's markets is low (see Figures 2.4, 2.5 and 2.6). This clearly implies that the cost of terminating their bilateral trade with the United States is higher for both China and India than it is for the United States. Put differently, China and India would suffer more

---

[141] Christopher Layne, "The Waning of U.S. Hegemony: Myth or Reality? A Review Essay," *International Security* 34, no. 1 (Summer 2009), 150.

Table 2.10 Exports as percentage of GDP, 2003–2013

	2003	2004	2005	2006	2007	2008	2009	2010	2011	2012	2013	average
China	30	34	37	39	38	35	27	29	29	27	26	29.3
India	15	18	19	21	20	24	20	22	24	24	25	21.1
USA	9	10	10	11	12	13	11	12	14	14	14	11.8

Source: World Bank, "Data: Exports of goods and services (% of GDP)," available at http://data.worldbank.org/indicator/NE.EXP.GNFS.ZS/countries?page=2, accessed February 23, 2018

Shaping Energy and Climate Policy Behavior 69

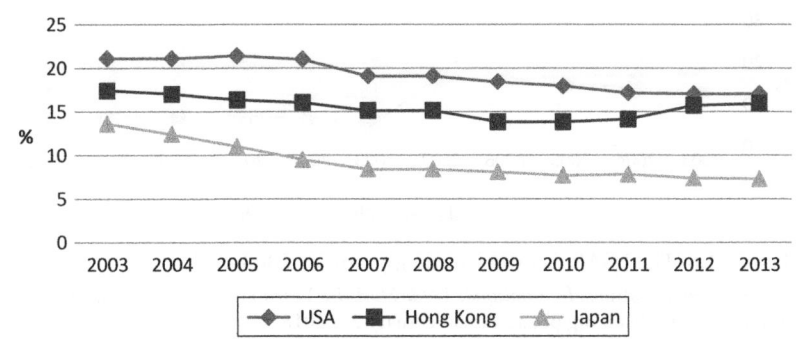

Figure 2.4 China's top three trading partners, 2003–2013
Source: United Nations, *International Trade Statistics Yearbook*, various years, available at http://comtrade.un.org/pb, accessed February 23, 2018

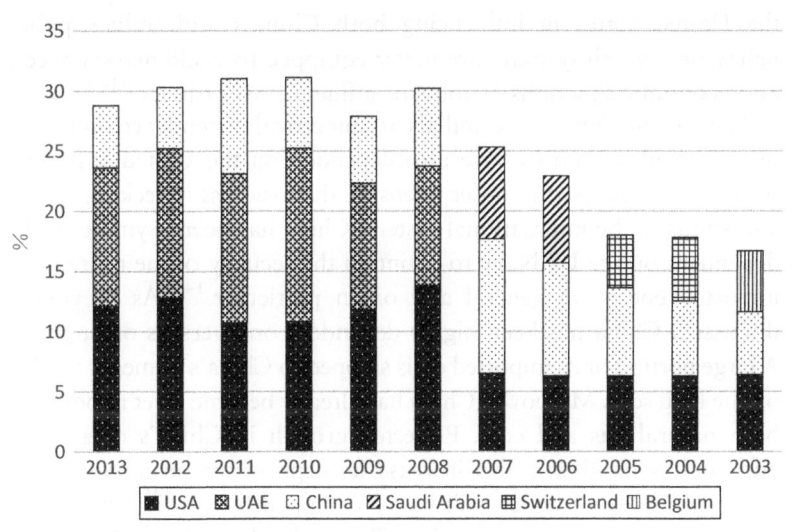

Figure 2.5 India's top three trading partners, 2003–2013
Source: United Nations, *International Trade Statistics Yearbook*, various years, available at http://comtrade.un.org/pb, accessed February 23, 2018

acutely from a disruption of trade with the United States, since both states' dependence on exports is high. The asymmetry of China's and India's export dependence on the United States is a long-term trend. Not surprisingly, Chinese officials expressed their concerns,

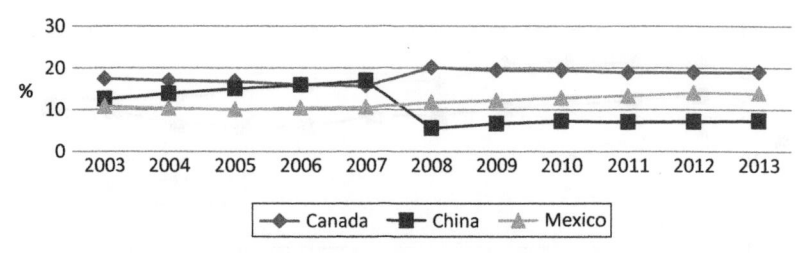

Figure 2.6 The USA's top three trading partners, 2003–2013
Source: United Nations, *International Trade Statistics Yearbook*, various years, available at http://comtrade.un.org/pb, accessed February 23, 2018

in early 2008, about their inability to diversify away from the US market.[142] Against this backdrop, China's and India's asymmetrical dependence on the US market has become a source of power for the United States in influencing both China's and India's policy behavior. "Wealthy states are better equipped to wield market access and economic sanctions as tools of influence over others."[143]

Apart from China's and India's asymmetrical dependence on the US market, both countries have also depended on the United States for some specific support in order to ensure their success in seeking wealth and status in the international system. China has been asymmetrically dependent on the US Navy to maintain the security of the route of its imported energy in general and oil in particular.[144] As previously discussed, China has been highly dependent on overseas oil supplies. A large portion of its imported oil is shipped to China's domestic market on the high seas. Moreover, China has already become a net importer of both natural gas and coal. Projected growth in China's overall gas and coal demand will also increase its dependence on imported gas and coal, much of which will be also shipped by sea.[145] This means that China's economic growth will largely depend on the energy resources shipped through the sea lanes of communication (SLOC).

[142] Daniel W. Drezner, "Bad Debts: Assessing China's Financial Influence in Great Power Politics," *International Security* 34, no. 2 (Fall 2009), 33.
[143] Beckley, "China's Century?," 56.
[144] Andrew Kennedy, "China and the Free Rider Problem: Exploring the Case of Energy Security," *Political Science Quarterly* 130, no. 1 (Spring 2015), 27–50.
[145] China's dependence on imports for natural gas will also rise to 40 percent by 2030. See Table 2.4.

It is common knowledge that the SLOC's security has been maintained by the US Navy. Although China has already embarked on efforts to build a blue-water navy, such a navy will not be capable of undertaking this task. Against this backdrop, China will continue to depend on the US Navy for the security of its imported energy that is transported by sea. As Aaron Friedberg points out, "Given China's heavy reliance on maritime transport and the prevailing balance of power at sea, the primary danger to supplies in transit now comes from the U.S. Navy,"[146] so "Beijing has little choice but to maintain the best possible relationship with the United States."[147] Some recent research points out even more bluntly that, given its navy's maritime dominance, "the United States has substantial coercive capability against both major oil importers and major oil exporters that rely on maritime transport,"[148] and "China remains vulnerable to coercion in the midstream from U.S. naval supremacy."[149] Simply put, the supremacy of the US Navy in maintaining the security of SLOC has become another source of power through which the United States is able to influence China's policy behavior.

India has also depended on the United States asymmetrically to fulfill its great-power status ambitions. More specifically, India relied totally on the United States to obtain internationally recognized status as a nuclear weapons state. According to Atteridge et al., "US support was crucial for the realization of India's nuclear ambitions (with the conclusion of a US–India Civil Nuclear Agreement in 2008)."[150] As mentioned above, India self-ascribed its status as a nuclear weapons state after its second nuclear test in 1998. However, this self-ascribed status was regarded as illegitimate for its violation of the norms of NPT. Under US leadership, the international community imposed economic sanctions against India. So, India had been faced with

---

[146] Aaron L. Friedberg, "'Going Out': China's Pursuit of Natural Resources and Implications for the PRC's Grand Strategy," *NBR Analysis* (Spring 2006), 25.
[147] Ibid., 14.
[148] Llewelyn Hughes and Austin Long, "Is There an Oil Weapon? Security Implications of Changes in the Structure of the International Oil Market," *International Security* 39, no. 3 (Winter 2014/15), 187.
[149] Ibid.
[150] Aaron Atteridge, Manish Kumar Shrivastava, Neha Pahuja, and Himani Upadhyay, "Climate Policy in India: What Shapes International, National and State Policy?" *Ambio* 41, no. S1, (2012), 71.

"the difficulties of persuading the United States to accept its de facto nuclear status" since the 1990s.[151] India's illegitimate status as a de facto nuclear weapons state remained unchanged until 2005, when the George W. Bush administration took the initiative to sign a civil nuclear cooperation agreement with India,[152] through which the United States abandoned its strict nuclear non-proliferation policy and recognized India's status as a nuclear weapons state.[153] After three years of negotiations, the United States not only altered its domestic non-proliferation legislation, but also forced other members of NSG to give India an exemption, and urged India to sign an agreement with the IAEA. The agreement was eventually signed into law in October 2008. Since then, India's status as a nuclear weapons state has been recognized by the United States and the international community. Thus, India has become a legitimate nuclear weapons state without having its signature on the NPT and the Comprehensive Test Ban Treaty (CTBT), and without providing a commitment to the proposed Fissile Materials Cut-Off Treaty (FMCT).[154] According to some scholars, the US-Indian civilian nuclear cooperation agreement "has virtually rewritten the rules of the global nuclear regime by underlining India's credentials as a responsible nuclear state that should be integrated into the global nuclear order. The nuclear agreement creates a major exception to the U.S. prohibition of nuclear assistance to any country that does not accept international monitoring of all its nuclear facilities."[155] Simply put, the US-Indian civil nuclear cooperation agreement "openly

---

[151] Gaurav Kampani, "New Delhi's Long Nuclear Journey: How Secrecy and Institutional Roadblocks Delayed India's Weaponization," *International Security* 38, no. 4 (Spring 2014), 111.
[152] C. Raja Mohan, *Impossible Allies: Nuclear India, United States and the Global Order* (New Delhi: India Research Press, 2006), 35, 57–58, 77–78.
[153] George Perkovich, "Global Implications of the U.S.–India Deal," *Daedalus* 139, no. 1 (Winter 2010), 20–31; and T. V. Paul, "The US–India Nuclear Accord: Implications for the Nonproliferation Regime," *International Journal* 62, no. 4 (Autumn 2007), 845–861.
[154] For details of the process of the US and Indian civilian nuclear agreement, see Mistry, *The US-Indian Nuclear Agreement*.
[155] C. Raja Mohan, "India and the Emerging Non-Proliferation Order: The Second Nuclear Age," in Harsh V. Pant (ed.), *Indian Foreign Policy in a Unipolar World* (New York: Routledge, 2009), 12.

acknowledged India as a legitimate nuclear power," and therefore ended India's 30-year quest for such recognition.[156]

Apart from its legitimate nuclear weapons status, India will also be relying on the United States to obtain its membership of the multilateral export control regimes, that is, NSG, MTCR, the Australia Group, and the Wassenaar Arrangement, so as to enhance its status as a responsible nuclear weapons state, but also to gain material benefits from its membership. India's efforts paid off to a certain extent. In February 2011, for example, the United States agreed to work toward the inclusion of India in these regimes, and to "devise a road map for simultaneous movement" on India's full membership as the "next logical step towards global recognition of India as a nuclear weapons state and an emerging military power."[157]

Furthermore, it is only the United States that can help India obtain permanent membership on the UN Security Council. In November 2011, during his visit to India, US President Barack Obama pledged to support India's quest for such status. Not surprisingly, C. Raja Mohan points out that "New Delhi has acknowledged that U.S. support is necessary for India's rise to be successful."[158] In the same vein, T. V. Paul and Mahesh Shankar argue that the success of India's efforts to seek great-power status has largely depended on the United States' status accommodation.[159]

Apart from China's and India's asymmetrical interdependence on the United States stemming from the distribution of economic and military power, the two countries have also had to face an asymmetrical interdependence stemming from the uneven distribution of energy resources. As discussed above, China and India have relied highly on imported fossil fuels, which are unevenly distributed in certain regions and areas. China's and India's fossil fuel resources are so limited that they are unlikely to meet either country's increased energy consumption. Table 2.11 shows that oil

---

[156] Ashton B. Carter, "America's New Strategic Partner?," *Foreign Affairs* 85, no. 4 (July/August 2006), 33.

[157] K. P. Nayar, "US Push for Nuclear Club Entry," *The Telegraph*, February 17, 2011. In fact, on January 19, 2018, India was admitted by the Australia Group as its 43rd member.

[158] C. Raja Mohan, "India and the Balance of Power," *Foreign Affairs* 85, no. 4 (July/August 2006), 32.

[159] T. V. Paul and Mahesh Shankar, "Status Accommodation through Institutional Means: India's Rise and the Global Order," in Paul, Larson, and Wohlforth (eds), *Status in World Politics*, 165–191.

Table 2.11 *Top 10 fossil fuel-rich countries, 2013*

Oil		Natural gas		Coal	
Country	% Share of total	Country	% Share of total	Country	% Share of total
Venezuela	17.7	Iran	18.2	United States	26.6
Saudi Arabia	15.8	Russia	16.8	Russia	17.6
Canada	10.3	Qatar	13.3	China	12.8
Iran	9.3	Turkmenistan	9.4	Australia	8.6
Iraq	8.9	United States	5.0	India	6.8
Kuwait	6.0	Saudi Arabia	4.4	Germany	4.5
United Arab Emirates	5.8	United Arab Emirates	3.3	Ukraine	3.8
Russia	5.5	Venezuela	3.0	Kazakhstan	3.8
Libya	2.9	Nigeria	2.7	South Africa	3.4
United States	2.6	Algeria	2.4	Indonesia	3.1

Source: BP, *BP Statistical Review of World Energy* (June 2014), 6, 20, 30, available at www.bp.com/content/dam/bp-country/de_de/PDFs/brochures/BP-statistical-review-of-world-energy-2014-full-report.pdf, accessed February 23, 2018

is richly concentrated in some Middle Eastern countries, especially Saudi Arabia, Iran, Iraq, and United Arab Emirates, as well as in Russia, while natural gas is mainly located in Iran, Qatar, and Russia. In contrast, coal, the major source of CO_2 emissions contributing to global warming, is richly distributed in the United States, China, and India. In Table 2.2, we can see that both countries' oil imports have become highly dependent on Middle Eastern countries. For instance, China's and India's dependence on Saudi Arabia stood at 19 percent and 20 percent respectively in 2013. In addition to Saudi Arabia, Iran has also been an important oil supplier for both countries. In 2013, for instance, Iranian oil supplies accounted for 8 percent and 6 percent of China's and India's total oil imports. Although both countries are rich in coal, they have already become coal importers (see Table 2.2). China's and India's fossil fuel dependence on overseas suppliers, especially their dependence on several major suppliers, leads both countries to be not only sensitive, but also vulnerable to potential supply disruptions. Thus, China and India have tried to boost their energy ties with their major energy suppliers. As a result, both countries' policies have turned out to be exposed to pressures applied by some of their largest suppliers. This topic will be explored in detail in Chapter 4.

In summary, China and India have been faced with two-level pressures. At the domestic level, both countries have tried to maximize their economic wealth by sustaining their rapid economic growth and poverty eradication programs. At the international level, both countries have been trying to enhance their status in the international system, which is governed by a number of norms, rules, and principles. Therefore, China and India have had to comply with these norms, rules, and principles in their policy behavior, in order to gain recognition of their enhanced status from other state actors. Moreover, at the international level, China and India have also been faced with the pressure of asymmetrical interdependence, arising not only from their asymmetrical interdependence on the US market for their exports, on the United States for maintaining global public goods, especially the security of the SLOC for China, and on the United States for its support in obtaining great-power status for India, but also from the fact that China and India have been asymmetrically dependent on certain energy-rich countries for supplies of their much-needed fossil fuels, especially oil. Under these circumstances, China's and India's asymmetrical interdependence has enabled the United States and certain energy suppliers to influence or shape, to various extents, both countries' policy behavior. In other words, faced with asymmetrical interdependence stemming from the distribution of power and energy resources, China's and India's behaviors have been subject to pressures from other state actors.

Conclusion

China's and India's policy behavior has been both proactive and reactive, and has been shaped by pressures arising at both the unit level and the systemic level, identified in this book as two-level pressures. More specifically, both countries' actions in seeking to maximize their domestic wealth through their proactive policy behavior have been moderated in certain instances, either by their ideational interest in enhancing their status and image in the international system, or by their asymmetrical interdependence derived from the asymmetrical distribution of power and/or energy resources in the international system. At the same time, however, both countries' desire to seek status and asymmetrical interdependence has not fundamentally altered their goal to maximize their domestic wealth, so both countries have tended to choose to adjust their proactive measures to a certain extent. In other words, China's and India's policy behavior has been shaped by wealth, status, and asymmetrical interdependence. More specifically, to maximize their wealth and seek status in an asymmetrical, interdependent international system, China and India have adopted some flexible policy measures, both proactive and reactive, based on their cost–benefit calculations. That is, when domestic pressures to seek wealth overwhelm pressures at the systemic level, China's and India's energy and climate policy measures are expected to be proactive. However, when external pressures stemming from their desire to seek status and from their asymmetrical interdependence are powerful enough to overwhelm domestic pressures, China and India are expected to modify their proactive energy and climate policy measures and, accordingly, those measures are reactive. The following chapters will explore this feature in detail by investigating how China and India have tried to address their energy insecurity through an "inside-out" process (Chapters 3 and 4) and how they have addressed climate change through an "outside-in" process (Chapters 5 and 6).

PART II

The Inside-Out

3 | Domestic Energy Policies: Proactive

Proposition 1: The energy policies of China and India are designed to help these countries achieve sustained, fast economic development at the domestic level.

China's and India's energy policies, domestic and foreign, have been driven by their efforts to address energy insecurity at the domestic level, as discussed in Chapter 2. Given this fact, the main point of this chapter, and Chapter 4, is that China's and India's policy measures to ensure energy security have been proactive, as a result of their own domestic pressures to procure sufficient energy to fuel their economic development and poverty reduction, or an "inside-out" process. Specifically, the proactiveness of China's and India's energy security policies, domestic and foreign, stems from the fact that energy security has been prioritized as one of their national security issues that have some strategic imperatives for their economic development. Accordingly, achieving energy security has been a top priority on China's and India's policy agenda. In this chapter, I begin by demonstrating the priority status of energy security in China and India by exploring Chinese and Indian leaders' rhetoric, and both countries' national development plans. The following section examines China's and India's domestic policy measures that address energy insecurity. The differences between those measures, stemming from both countries' different domestic political and economic systems, will then be explored. Finally, I draw a brief conclusion about both countries' domestic energy policy.

Energy Security: A Top Priority

It is widely recognized that energy security and the energy sector as a whole remain vital for a country's national economy and national security.[1] For both China and India, the two largest developing countries in the process of industrialization, energy is of great importance for their economic development, so energy security has become a top priority for both countries. This point is reflected in Chinese and Indian leaders' rhetoric. In other words, both Chinese and Indian leaders have attached great importance to their countries' energy security.[2]

In China's case, for instance, as early as November 2003, the then Chinese President Hu Jintao put forward, for the first time, the concept of finance and oil as two national economic security issues, at the closing session of the Central Economic Work Conference. Specifically, Hu pointed out that given the fact that more than half of China's oil imports came from the Middle East, Africa, and Southeast Asia, and that about four-fifths of its imported crude oil was transported through the Strait of Malacca, where "some big countries" had been trying to control the shipping lane, China was faced with a "Malacca dilemma,"[3] so he required that the Chinese government should figure out a new oil development strategy and take proactive measures to ensure national energy security. In July 2006, Hu Jintao stated clearly that China attached great importance to energy security, and he even put forward three new concepts for global energy security at the summit of the G8 and five leading emerging economies.[4] Energy

---

[1] Daniel Yergin, "Ensuring Energy Security," *Foreign Affairs* 85, no. 2 (March/April 2006), 69–82.

[2] This book focuses only on both countries' energy security challenge in the first 15 years of the 21st century. Historically, China and India have experienced energy insecurity too. For instance, China was faced with energy insecurity in 1950, shortly after the Korean War broke out, when Western countries imposed an oil embargo on China, and once again in 1959, after China's break with the Soviet Union, when energy security became the then Chinese Chairman Mao Zedong's top policy priority. See Michal Meidan, "The Structure of China's Oil Industry: Past Trends and Future Prospects," *OIES Paper*: WPM 66 (The Oxford Institute for Energy Studies, May 2016), 4. In terms of India, the oil price spike stemming from the 1990–91 Persian Gulf crisis led to a balance-of-payments shock, leaving the country with almost no foreign currency reserves, which eventually forced India to carry out economic reforms in 1991. See Yergin, "Ensuring Energy Security," 72.

[3] "China's 'Malacca dilemma'," *China Daily* (June 15, 2004), available at www.chinadaily.com.cn/gb/worldinfo/2004-06/15/content_339435.htm, accessed December 23, 2015.

[4] "胡锦涛阐述全球能源安全 [Hu Jintao Elaborated on Global Energy Security]," 新华网 [*Xinuanet*] (July 18, 2006), available at http://news.xinhuanet.com/politics/2006-07/18/content_4847040.htm, accessed December 23, 2015.

Domestic Energy Policies: Proactive 81

security continues to be a major concern for the current Chinese leadership. For instance, in June 2014, the Central Leading Group on Financial and Economic Affairs held a meeting with an exclusive focus on China's energy security policy. Chinese President Xi Jinping presided over this meeting, and Xi put forward five points to bring about an "energy revolution," focusing on energy conservation, diversifying energy supply, enhancing energy efficiency, and international energy cooperation.[5]

Similarly, India's top leaders have attached great importance to energy security. For instance, in November 2004, Manmohan Singh, India's then prime minister, said in an interview that "Energy security is of critical importance to India. It is second only in our scheme of things to food security."[6] In August 2005, at the first meeting of the Indian Energy Coordination Committee, Singh once again emphasized that "The availability of adequate, reliable and reasonably priced energy supplies is a sine qua non for achieving the growth targets we have set ourselves ... We have to ensure that we build up adequate energy security to insulate the economy from any future shock."[7] In December 2013, when Singh attended the 8th Asia Gas Partnership Summit, he said that India was progressively pursuing all options to achieve energy security.[8] In the same vein, Narendra Modi, India's incumbent prime minister, stated clearly in his media statement, during the official visit of Russian President Vladimir Putin to India in December 2015, that energy security is critical for India's economic development.[9]

---

[5] "习近平：积极推动我国能源生产和消费革命 [Xin Jinping: Actively Pushing Forward National Energy Production and Consumption Revolution]," 新华网 [*Xinhuanet*] (June 13, 2014), available at http://news.xinhuanet.com/politics/2014-06/13/c_1111139161.htm, accessed February 26, 2018.

[6] Edward Luce, "Head to Head in the Quest for National Energy Security," *Financial Times* (November 16, 2004), available at www.ft.com/content/57f0b140-3812-11d9-991f-00000e2511c8, accessed February 26, 2018.

[7] Manmohan Singh, "PM's Opening Remarks at the First Meeting of the Energy Coordination Committee" (August 6, 2005), available at http://archivepmo.nic.in/drmanmohansingh/speech-details.php?nodeid=158, accessed February 26, 2018.

[8] "India Pursuing All Options to Achieve Energy Security: Manmohan Singh," *Daily News & Analysis* (December 3, 2013), available at www.dnaindia.com/india/report-india-pursuing-all-options-to-achieve-energy-security-manmohan-singh-1928968, accessed February 26, 2018.

[9] "Energy Security Is Critical for India's Economic Development: PM Modi," *ANI News* (December 11, 2015), available at http://aninews.in/newsdetail2/

Apart from the importance attached to energy security in their top leaders' rhetoric, the strategic importance of energy security for both China and India has been written down in their official energy policy documents. For instance, in 2010, the Chinese government issued "The State Council's Decision on Accelerating Cultivation and Development Strategic Emerging Industries," listing seven emerging strategic industries, including energy conservation and environmental protection industries, and new energy industries, including the nuclear, solar, wind, and biomass energy technology industries, to which the Chinese government will give strong industrial policy support, such as access to dedicated state industrial funds or preferential loans, or research and development funds.[10] In *China's Energy Policy 2012*, an official White Paper issued by the Chinese government, it states that "Energy is the material basis for the progress of human civilization and an indispensable basic condition for the development of modern society. It remains a major strategic issue for China as the country moves towards its goals of modernization and common prosperity for its people."[11] In other words, to reach its goals of modernization and societal wellbeing, energy is of great importance for China.

Similarly, in 2000, the Indian government issued its *Hydro Carbon Vision 2025*, which set the ambitious goal of building a cleaner and greener India through the development of renewable energy sources, and highlighted oil as central to its energy security.[12] Its main target is "To assure energy security by achieving self-reliance through increased indigenous production and investment in equity oil abroad."[13] In other words, it aimed at not only increasing India's domestic energy production, but also at securing the supplies of external energy sources. In its

story195392/energy-security-is-critical-for-india-039-s-economic-development-pm-modi.html, accessed December 23, 2015.

[10] 国务院 [The State Council], "国务院关于加快培育和发展战略性新兴产业的决定 [The State Council's Decision on Accelerating Cultivation and Development Strategic Emerging Industries]" (October 10, 2010), available at www.gov.cn/zwgk/2010-10/18/content_1724848.htm, accessed February 26, 2018.

[11] Information Office of the State Council of the PRC, *China's Energy Policy 2012* (October 24, 2012), available at www.china.org.cn/government/whitepaper/node_7170375.htm, accessed February 26, 2018.

[12] Bala Bhaskar, *Energy Security and Economic Development in India: A Holistic Approach* (New Delhi: TERI Press, 2013), 15.

[13] Ministry of Petroleum and Natural Gas, Government of India (GOI), *India Hydrocarbon Vision – 2025* (2000), 1, available at http://petroleum.nic.in/sites/default/files/vision.pdf, accessed February 26, 2018.

2006 *Integrated Energy Policy*, the Indian government claimed that "the energy challenge is of fundamental importance" for its economic growth.[14] This document not only defines energy security in the Indian context, but also sets some broad guidelines for achieving energy security for India.[15]

The strategic importance of energy security for China and India has also been inked into both countries' national development plans, known as Five-Year Plans (FYPs), which are a series of economic development initiatives, mapping strategies for economic development, setting growth targets, and launching reforms in the relevant time frames.[16] Since the early 2000s, energy security has become a key priority in China's development plans.[17] For example, the issue of energy security featured prominently in China's 10th FYP (2001–5),[18] which devoted an entire section to illustrating how to deal with energy security, in addition to mentioning the strategic importance of oil. This plan emphasized optimizing the energy mix while ensuring overall energy security.[19] Moreover, this plan put

---

[14] Planning Commission, GOI, *Integrated Energy Policy: Report of the Expert Committee* (August 2006), 1, available at http://planningcommission.nic.in/reports/genrep/rep_intengy.pdf, accessed February 26, 2018.

[15] Ibid., 15–16.

[16] Xueliang Yuan and Jian Zuo, "Transition to Low Carbon Energy Policies in China – from the Five-Year Plan Perspective," *Energy Policy* 39, no. 6 (June 2011), 3855.

[17] For a detailed analysis of the issue of energy security in China's development plans, see Kang Wu, "China's Energy Security: Oil and Gas," *Energy Policy* 73, no. 1 (October 2014), 6–7.

[18] By 2016, the Chinese government had issued 13 Five-Year Plans. In order to show that its economy was no longer centrally planned in nature, the Chinese government has made a subtle change to the Plan's title, that is, it has changed "计划" to "规划," since the former represented the centrally planned economy, while the latter's nature is much more neutral, but both of these Chinese words can be translated as "plan" in English. Given the time period of this study (2000–15), the 13th Five-Year Plan will not be discussed here. The full contents of the previous 12 Five-Year Plans are available on Chinese official websites about the history of the Chinese Communist Party. See 中国共产党新闻网 [Chinese Communist Party News Net], "历次五年规划(计划) [Previous Five-Year Plans (Plans)]," available at http://dangshi.people.com.cn/GB/151935/204121, accessed February 26, 2018.

[19] 国务院 [The State Council], "中华人民共和国国民经济和社会发展第十个五年计划纲要 [The Tenth Five-Year Plan Guidelines for National Economic and Social Development of People's Republic of China]" (March 15, 2001), 12–13, available at http://ghs.ndrc.gov.cn/ghwb/gjwngh/200709/P020070912634253001114.pdf, accessed October 14, 2015.

Table 3.1 *Selected Chinese targets for energy development in Five-Year Plans*

	11th FYP target (2010)	11th FYP result (2010)	12th FYP target (2015)
Share of non-fossil energy in energy consumption (i.e. hydropower, new renewable energy, nuclear power)	10% excluding nuclear	8.6%	11.4%
Energy consumption per unit of GDP (energy intensity)	−20%	−19.1%	−16%

Source: 国务院 [The State Council], 《中华人民共和国国民经济和社会发展第十二个五年规划纲要》 ["The Twelfth Five-Year Plan Guidelines for National Economic and Social Development of People's Republic of China"], March 16, 2011, available at http://news.xinhuanet.com/politics/2011-03/16/c_121193916.htm, accessed December 24, 2015

forward, for the first time, the establishment of strategic petroleum reserves (SPRs) of oil to enhance its national energy security, which will be discussed in more detail in the following section.[20] In its 11th FYP (2006–10), the Chinese government set some specific targets relating to energy development (see Table 3.1). For instance, China set a goal to increase the share of non-fossil energy in its energy consumption by 10 percent, while reducing its energy intensity in its GDP by 20 percent. And the actual results were 8.6 percent and 19.1 percent respectively by the end of this plan. During its 12th FYP (2011–15), China intended to increase by 11.4 percent the share of non-fossil energy in its energy consumption, while reducing its energy intensity by 16 percent. The 12th FYP also reiterated its supportive policy toward the seven emerging strategic industries mentioned above.[21] Furthermore, the 12th FYP set the goal of developing a secure, stable, economic, and clean modern energy industrial system.[22]

[20] Ibid., 12.
[21] 国务院 [The State Council], "'十二五'规划纲要(全文) [The Twelfth Five-Year Plan Guidelines (full text)]" (March 16, 2011), available at www.china.com.cn/policy/txt/2011-03/16/content_22156007.htm, accessed February 28, 2018.
[22] Ibid.

Like the Chinese government, the Indian government has also set some specific targets for its energy development.[23] Specifically, the Indian government intends to reduce its reliance on crude oil, while increasing the share of natural gas, nuclear energy, and renewable energy in its energy mix by 2022, although coal will continue to be the dominant energy source, accounting for half of India's total energy consumption. Moreover, the Indian government has also set some targets for its energy intensity. For instance, in its 11th FYP (2007–12), India set a target to increase energy efficiency by 20 percent by 2016–17.[24] Compared with India's previous FYPs, which usually emphasized energy supply, growth, and socio-economic development, its 11th FYP marks a significant shift because it "highlighted the severe shortages of energy, dominance of coal and the need to expand resources through exploration, energy efficiency, renewables, and research and development."[25] Under its 12th FYP (2012–17), India is targeted to reduce its energy intensity by 10 percent.[26]

In short, energy security has been a top priority for China and India, which has been revealed by the great importance which both countries' top leaders have attached to it through their rhetoric, and the fact that not only have both countries issued some official documents emphasizing the importance of energy security, but they have also included the issue of energy security in their long-term national development plans. Apart from emphasizing the priority status of energy security in rhetoric and in print, both China and India have taken some proactive policy measures at the domestic level to procure their energy security.

Proactive Domestic Measures to Procure Energy Security

To address the challenge of energy insecurity, China and India have taken some proactive measures to diversify their domestic energy

---

[23] Planning Commission, GOI, *Twelfth Five-Year Plan (2012–2017), Economic Sectors*, Volume II (2013), 134, available at http://planningcommission.gov.in/plans/planrel/12thplan/pdf/12fyp_vol2.pdf, accessed February 26, 2018.

[24] Planning Commission, GOI, *Eleventh Five-Year Plan (2007–2012)*, Volume I: *Inclusive Growth* (2008), 207, available at http://planningcommission.nic.in/plans/planrel/fiveyr/11th/11_v1/11th_vol1.pdf, accessed February 26, 2018.

[25] P. Balachandra, Darshini Ravindranath, and N. H. Ravindranath, "Energy Efficiency in India: Assessing the Policy Regimes and their Impacts," *Energy Policy* 38, no. 11 (November 2010), 6433.

[26] Planning Commission, *Twelfth Five-Year Plan (2012–2017)*, Volume II, 131.

supplies, so as to provide sufficient energy to ensure their long-term economic growth. In other words, given the fact that both countries' indigenous fossil fuel resources (i.e. coal, oil, and natural gas) are insufficient to meet their increased domestic demand, they have taken proactive measures to develop non-fossil fuel energy resources, especially renewable energy and nuclear energy, in addition to enhancing their energy efficiency and establishing their own strategic oil reserves.

Developing Renewable Energy

One key element of China's and India's energy security strategy is to promote renewable energy, which is one of the most important measures both countries have adopted to reduce their reliance on fossil fuels.

To do so, both governments have promulgated laws and related policies to support the development of renewable energy. In China's case, since the beginning of the 21st century, the Chinese government has begun to adopt a series of policies to cultivate the development of renewable energy. Specifically, in 2001, China set out its policy to develop new and renewable energy for the very first time in its *Tenth Five-Year Plan for New and Renewable Energy Development*, which clearly specified that the annual utilization of new and renewable energy (excluding small hydro and biomass) would reach 13 million tonnes of coal equivalent (tce) by 2005.[27] In 2005, China launched its first ever National Renewable Energy Law, which came into force on January 1, 2006. This law established a framework for regulating renewable energy, set a target for renewable energy development, and created a cost-sharing mechanism including a special fund for renewable energy development.[28] This law was amended in December

---

[27] 国家经贸委 [National Economic and Trade Commission],《新能源和可再生能源产业发展"十五"规划》 [*Tenth Five-Year Plan for New and Renewable Energy Development*] (2001), available at www.ccchina.gov.cn/Detail.aspx?newsId=27979&TId=60, accessed May 4, 2015.

[28] 中国人民共和国人民代表大会 [National People's Congress of PRC],《中华人民共和国可再生能源法》 [*Renewable Energy Law of People's Republic of China*] (February 28, 2005), available at www.gov.cn/ziliao/flfg/2005-06/21/content_8275.htm, accessed February 26, 2018.

2009 to solve the bottleneck of grid connections and to facilitate the dispatch of electricity generated from renewable energy.[29]

Moreover, in its 11th FYP, China stated that the objective of developing renewable energy was to "encourage the production and consumption of renewable energy and increase its share in total primary energy consumption."[30] In September 2007, the Chinese government issued a new *Medium- and Long-Term National Planning of Renewable Development*, which was refined in its March 2008 *11th Five-Year Renewable Energy Development Plan*.[31] Both plans specified that the objective of developing renewable energy was to raise the share of renewable energy in its total primary energy consumption to 10 percent by 2010.[32] In July 2008, in its *Mid- and Long-Term Renewable Energy Implementation Plan*, China set a target of a 1 percent share of non-hydro renewables in its total power generation by 2010, and 3 percent by 2020 for regions served by centralized power grids.[33] In its 12th FYP, China also set some specific targets for electricity generation from non-fossil fuels by 2015, that is, 40 GW of nuclear, an additional 70 GW of wind, an additional 5 GW of solar, and an additional 120 GW of hydro.[34] Furthermore, in its *12th Five-Year Plan for Renewable Energy Development* issued by China's National Development and Reform Commission (NDRC) in July 2012,

---

[29] 中国人民共和国人民代表大会 [National People's Congress of PRC], 《中华人民共和国可再生能源法（修正案）》 [*Amendments to Renewable Energy Law of People's Republic of China*] (December 26, 2009), available at www.gov.cn/flfg/2009-12/26/content_1497462.htm, accessed February 26, 2018.

[30] 国务院 [The State Council], 《中华人民共和国国民经济和社会发展第十一个五年规划纲要》 [*The Eleventh Five-Year Guideline for National Economic and Social Development of People's Republic of China*] (March 16, 2006), available at www.gov.cn/ztzl/2006-03/16/content_228841.htm, accessed February 26, 2018.

[31] 中华人民共和国国家发展和改革委员 [National Development and Reform Commission of PRC (NDRC)], 《可再生能源中长期发展规划》 [*Medium- and Long-Term Plan for Renewable Energy Development*] (August 2007), available at www.ndrc.gov.cn/zcfb/zcfbghwb/200709/W020140220601800225116.pdf, accessed February 26, 2018; 国家发展和改革委员 [NDRC], 《可再生能源发展"十一五"规划》 [*Eleventh Five-Year Plan for Renewable Energy Development*] (2008), available at www.ndrc.gov.cn/zcfb/zcfbghwb/200803/W020140220604486824065.pdf, accessed February 26, 2018.

[32] Ibid., 18, 13.

[33] Joanna I. Lewis, "Building a National Wind Turbine Industry: Experiences from China, India and South Korea," *International Journal of Technology and Globalisation* 5, no. 3–4 (2011), 284.

[34] The State Council, "The Twelfth Five-Year Plan Guidelines."

the Chinese government specified support measures for renewable energy deployment, such as establishing an evaluation system for renewable energy development, implementing a renewable energy electricity quota system, and improving renewable energy subsidies, and fiscal, taxation, and financial policy.[35]

Among the renewable energies, the Chinese government has given specific support to the development of wind and solar energy by adopting the policy of privileged tax and feed-in tariffs.[36] For instance, in 2008, the Chinese government announced that wind and solar power projects would pay no corporate income tax for the first three years, and the tax rate for the following three years would be 50 percent, in addition to under 50 percent value added tax rebates.[37] In addition, the Chinese central government has encouraged its state-owned commercial banks to provide favorable and low-interest financial loans for those state-owned enterprises (SOEs) engaged in renewable energy industries. Consequently, wind and solar power SOEs can easily get financial loans and credit guarantees from Chinese state-owned banks. According to research, in 2011, bank loans to Longyuan Power, Huaneng New Energy, and Datang, three of China's largest renewable energy SOEs, were 27.795 billion yuan ($4.03 billion), 13.536 billion yuan ($1.96 billion), and 78.77 billion yuan ($11.41 billion) respectively, accounting for 57 percent, 71.5 percent, and 51.2 percent of new energy financing.[38] In Joanna Lewis's words, "Renewable energy technology development is now positioned at the core of China's overarching national economic plan and supported by its industrial policy."[39]

---

[35] 国家发展和改革委员 [NDRC],《可再生能源发展"十二五"规划》[*Twelfth Five-Year Plan for Renewable Energy Development*] (August 7, 2012), available at www.ce.cn/cysc/ny/zcjd/201302/01/t20130201_21331890.shtml, accessed February 26, 2018.

[36] Wenbin Lin, Alun Gu, Xin Wang, and Bin Liu, "Aligning Emissions Trading and Feed-in Tariffs in China," *Climate Policy* 16, no. 4 (2016), 434–455.

[37] Zeng Ming, Liu Ximei, Li Yulong, and Peng Lilin, "Review of Renewable Energy Investment and Financing in China: Status, Mode, Issues and Countermeasures," *Renewable and Sustainable Energy Reviews* 31 (March 2014), 31.

[38] Ibid.

[39] Joanna I. Lewis, *Green Innovation in China: China's Wind Power Industry and the Global Transition to a Low-Carbon Economy* (New York: Columbia University Press, 2013), 2.

Similarly, the Indian government has also promulgated legislation for the development of non-fossil fuel energy sources. For instance, in its 2003 Electricity Act, the Indian government set the target for renewable energy to contribute 10 percent of total power generation capacity by 2012.[40] This legislation "mandates the State Electricity Regulatory Commissions (SERCs) to promote generation of electricity from renewable sources of energy by providing suitable measures for connectivity with the grid and sale of electricity to any person."[41] In addition to this legislation, the Indian government has promulgated policies to facilitate the development of renewable energy. For instance, in 2005, the Indian government issued the *National Electricity Policy*, which allows the SERCs to establish a preferential tariff for electricity generated from renewable sources, to enable them to be cost-competitive.[42] In 2006, the Indian government issued the *National Tariff Policy*, which requires SERCs to fix certain minimum percentages for the purchase of renewable power and to establish the rule of competitive procurement of renewable energy power through a bidding process.[43] The aim of both the *National Electricity Policy* and the *National Tariff Policy* is to increase the share of renewable energy in the total energy supply mix. In its 2008 *Hydro Power Policy*, the Indian government set a goal to harness the full potential of hydroelectricity generation.[44]

Apart from its federal-level policies, at India's state level, a feed-in tariff – a long-term contract of 20–25 years – is employed by the state government to buy the power produced by a renewable project at a pre-determined tariff, which is based on the cost of power production and is higher than the average pooled purchase cost. Accordingly, a

---

[40] Lewis, "Building a National Wind Turbine Industry," 288.
[41] P. R. Krithika and Siddha Mahajan, "Governance of Renewable Energy in India: Issues and Challenges," *TERI-NFA Working Paper Series* No. 14, The Energy and Resources Institute (March 2014), 10, available at www.teriin.org/projects/nfa/pdf/working-paper-14-Governance-of-renewable-energy-in-India-Issues-challenges.pdf, accessed May 4, 2016.
[42] Ibid., 12.   [43] Ibid.
[44] Ministry of Power, GOI, *Hydro Power Policy 2008*, available at https://powermin.nic.in/sites/default/files/uploads/new_hydro_policy.pdf, accessed February 28, 2018.

feed-in tariff includes an implicit subsidy from the state government in the form of a feed-in premium.[45]

In order to promote the development of solar energy, the Indian government launched the Jawaharlal Nehru National Solar Mission in 2009, "which has set ambitious goals on generation capacity additions from solar technology – solar thermal and solar photovoltaic – in terms of both grid-connected and offgrid applications."[46] This mission adopts a three-phase approach, which covers the period from India's 11th FYP to 13th FYP (2017–22); it establishes a national-level policy framework for solar energy utilization, including power generation in India.[47] Moreover, in *Low Carbon Strategies for Inclusive Growth* and *Renewable Energy in India: Progress, Vision and Strategy*, the Indian government has set electricity generation targets from solar energy for 2020 and 2050 in its energy mix, in addition to harnessing wind energy to reach 50 GW before 2025.[48] In its 12th FYP, the Indian government set the targets of 20 GW of solar and 50 GW of wind capacity by 2022. These targets were upgraded to 100 GW of solar and 60 GW of wind capacity in February 2015 in the Indian government's Budget 2015 targets.[49]

Thanks to the aforementioned policy measures, China and India have turned out to be two of the largest clean-energy investors in the world. According to the 9th "Global Trends in Renewable Energy

---

[45] Gireesh Shrimali, Shobhit Goel, Sandhya Srinivasan, and David Nelson, "Solving India's Renewable Energy Financing Challenge: Which Federal Policies Can Be Most Effective?" A CPI-ISB Series (March 2014), 3, available at http://climatepolicyinitiative.org/wp-content/uploads/2014/03/Which-Federal-Policies-can-be-Most-Effective.pdf, accessed February 26, 2018.

[46] Krithika and Mahajan, "Governance of Renewable Energy in India," 12–13.

[47] Ibid., 13.

[48] Planning Commission, GOI, *The Final Report of the Expert Group on Low Carbon Strategies for Inclusive Growth* (April 2014), available at http://planningcommission.nic.in/reports/genrep/rep_carbon2005.pdf, accessed February 26, 2018; Ministry of New and Renewable Energy, GOI, *Renewable Energy in India: Progress, Vision and Strategy* (October 25, 2010), available at www.indiaenvironmentportal.org.in/files/mnre-paper-direc2010-25102010.pdf, accessed February 26, 2018.

[49] Gireesh Shrimali, Sandhya Srinivasan, Shobhit Goel, Saurabh Trivedi, and David Nelson, "Reaching India's Renewable Energy Targets Cost-Effectively," CPI (Climate Policy Initiative) and ISB (The Indian School of Business) Series (April 2015), 9, available at http://climatepolicyinitiative.org/wp-content/uploads/2015/04/Reaching-Indias-Renewable-Energy-Targets-Cost-Effectively.pdf, accessed February 26, 2018.

Investment 2015," a report issued jointly by the Frankfurt School–UNEP Collaborating Centre for Climate and Sustainable Energy Finance and Bloomberg New Energy Finance, China and India are now the largest and seventh-largest investors in renewable energy in the world, particularly in power generation. In 2014, for instance, China witnessed the largest renewable energy investments – a record $83.3 billion, up 39 percent from 2013, much higher than the $38.3 billion invested by the United States, the second-largest investor. In the same year, India invested $7.4 billion, ranking it among the top ten investing countries.[50] In terms of cumulative investment in renewable energy, during the period 2004–14, for instance, China and India invested more than $400 billion and $100 billion respectively in clean energy, and were ranked first and seventh in the world.[51]

Expanding Nuclear Energy

China and India have tried to expand the contribution of nuclear energy to their energy mix in order to increase their energy security. According to the IEA's *World Energy Outlook 2014*, nuclear energy can help energy-importing countries to "reduce their dependence on foreign supplies and limit their exposure to fuel price movements in international markets."[52] In its New Policies scenario, the IEA projects that installed nuclear capacity growth across the world will be 60 percent, from 392 GW in 2013 to 624 GW in 2040, with the increase concentrated heavily in China (46 percent), plus India, Korea, and Russia (30 percent), and the United States (16 percent), countered by a 10 percent drop in the EU.[53]

China and India have set a policy of expanding nuclear energy in their national FYPs. For instance, in its 11th FYP, China stated that it "will proactively promote the nuclear power development with a priority on construction of million kW scale nuclear power stations."[54]

---

[50] Frankfurt School–UNEP Collaborating Centre for Climate and Sustainable Energy Finance, *Global Trends in Renewable Energy Investment 2015* (March 11, 2015), available at http://fs-unep-centre.org/sites/default/files/attachments/unep_fs_globaltrends2015_chartpack.pdf, 10, accessed February 26, 2018.
[51] Ibid.   [52] IEA, *World Energy Outlook 2014*, 27.   [53] Ibid., 383.
[54] 国务院 [The State Council], "中华人民共和国国民经济和社会发展第十一个五年规划纲要 [The Eleventh Five-Year Plan Guidelines for National Economic and Social Development of the People's Republic of China]" (March 16, 2006),

In 2007, the Chinese government promulgated *The Medium- and Long-term Development Plan for Nuclear Energy*, which set the target for nuclear power development of an installation capacity of 40 million kW and annual electricity generation of 260–280 billion kWh by the year 2020.[55] However, the Japanese Fukushima nuclear accident in March 2011 triggered intensive scrutiny by the Chinese public of its nuclear energy development, which led its national policy to refocus on the safety of its nuclear power expansion. For instance, in its 12th FYP, the Chinese government stated that it would develop nuclear power efficiently, with safety as a foundation. In other words, nuclear safety has become a key condition for China's nuclear power expansion.[56] In this plan, the Chinese government still listed nuclear energy as one of the priorities of its energy development strategy, and targeted its nuclear energy expansion in the planned period to reach 40 million kW, through accelerating nuclear power development in its coastal provinces, while stably promoting the construction of nuclear power plants in its provinces located in central areas.[57]

India has set up a three-stage development process to develop its nuclear energy, and its main goal is to make full use of its rich thorium reserves to become truly energy independent beyond 2050.[58] Like China, India has also set some specific targets for the development of nuclear energy in its national FYPs. For instance, in its 10th FYP (2002–7), the Indian government set a target to raise its nuclear power generation capacity from 2,720 MW in 2001 to 7,180 MW by 2009–10.[59] In its 11th FYP, the Indian government planned to increase the growth rate of its domestic supply of nuclear energy from 9.18

available at www.gov.cn/ztzl/2006-03/16/content_228841.htm, accessed February 28, 2018.

[55] 国家发展和改革委员会 [National Development and Reform Commission],《核电中长期发展规划》[*The Medium- and Long-Term Development Plan for Nuclear Energy*] (October 2007), 8, available at www.gov.cn/gzdt/att/att/site1/20071104/00123f3c4787089759a901.pdf, accessed February 26, 2018.

[56] Yuan and Zuo, "Transition to Low Carbon Energy Policies in China," 3857.

[57] The State Council, "The Twelfth Five-Year Plan Guidelines."

[58] Planning Commission, *Integrated Energy Policy*, xxii, 35–36.

[59] Planning Commission, GOI, *Tenth Five-Year Plan (2002–2007)*, Volume II: *Sectoral Policies and Programmes* (2002), 700, available at http://planningcommission.nic.in/plans/planrel/fiveyr/10th/volume2/10th_vol2.pdf, accessed February 26, 2018.

percent in 2002–7, to 19.13 percent in 2007–12.[60] In its 12th FYP, the Indian government regarded nuclear energy as "another important energy source for the country, [which] has the greatest potential over the next 20 years, of providing a substitute for coal-based electricity."[61] However, against the backdrop of the Fukushima nuclear accident, the Indian government stressed the importance of environmental and safety issues in the process of nuclear energy expansion,[62] and acknowledged that "It is unlikely that large nuclear capacity could be added over the Twelfth Plan period."[63] In 2015, the total installed capacity of nuclear power in operation in India was 5.78 GW, in addition to six reactors with an installed capacity of 4.3 GW, being at different stages of commissioning and construction.[64]

Although China and India slowed down the pace of their nuclear energy development after Fukushima, both countries have already resumed their great efforts to push forward nuclear energy development. For instance, in early 2014, China's National Energy Administration issued the 2014 Energy Work Guidance, calling for "steadily pushing forward" approvals for the construction of nuclear power stations in coastal areas, which was halted right after the Fukushima nuclear accident. Moreover, on July 3, 2014, NDRC vice chair, Wu Xinxiong, announced during a research trip to Anhui that the NDRC would accelerate approval procedures for nuclear power stations.[65] According to MIT research, China will soon surpass the traditional leaders in nuclear power, such as Russia and South Korea, and will even surpass Japan by 2020.[66] Similarly, India confirmed its policy

[60] Planning Commission, GOI, *Eleventh Five-Year Plan (2007–2012)*, Volume III: *Agriculture, Rural Development, Industry, Services and Physical Infrastructure* (New Delhi: Oxford University Press, 2008), 347.
[61] Planning Commission, *Eleventh Five-Year Plan (2007–2012)*, Volume I, 8.
[62] Ibid., 22.   [63] Ibid., 122.
[64] GOI, "India's Intended Nationally Determined Contribution: Working Towards Climate Justice" (October 2015), 4, available at www4.unfccc.int/submissions/INDC/Published%20Documents/India/1/INDIA%20INDC%20TO%20UNFCCC.pdf, accessed February 26, 2018.
[65] "能源局：适时启动核电重点项目 [National Energy Administration: Timely Starting Key Nuclear Projects]," 《中国投资指南》 [*Invest in China*] (July 3, 2014), available at www.fdi.gov.cn/1800000121_21_64141_0_7.html, accessed February 26, 2018.
[66] Mike Orcutt, "China Will Soon Leapfrog Traditional Leaders in Nuclear Power," *MIT Technology Review* (July 27, 2015), available at www.technologyreview.com/news/539691/china-will-soon-leapfrog-traditional-leaders-in-nuclear-power, accessed February 26, 2018.

to expand its nuclear energy in its intended nationally determined contribution (INDC), an official document submitted by the Indian government to the Secretariat of the UNFCCC in October 2015, in which the Indian government says that it will achieve 63 GW installed nuclear capacity by 2032, if the supply of fuel is ensured.[67] Moreover, India has set a target for nuclear power to supply 25 percent of its electricity by 2050.[68]

According to the World Nuclear Association (WNA), China and India have, respectively, 29 and 21 civil nuclear reactors under operation, in addition to over 20 units and 6 units respectively under construction. Moreover, in India, over 38 further units are planned.[69] Not surprisingly, WNA claims that "China and India [are] getting up to speed in nuclear energy."[70]

Enhancing Energy Efficiency and Reducing Energy Intensity

Improving energy efficiency and reducing energy intensity are both important measures that China and India have adopted to procure their energy security (and reduce their CO_2 emissions). In this regard, both countries have also set the policy of improving energy efficiency and decreasing energy intensity (referring to the amount of energy required to produce a unit of output and measured by the ratio of total energy consumption to GDP) in their national development plans.

China's energy intensity is much higher than the global. In 2006, for example, Chinese energy consumption per unit of GDP was 2.4 times higher than the global average, 4.9 times higher than in EU countries, and 8.7 times higher than in Japan.[71] In the early 2000s, China's

---

[67] GOI, "India's Intended Nationally Determined Contribution," 4.
[68] Steve Kidd, "India – Could it yet Replicate China in Nuclear Power?," *Nuclear Engineering International* (December 23, 2015), available at www.neimagazine.com/opinion/opinionindia-could-it-yet-replicate-china-in-nuclear-power-4760885, accessed February 26, 2018.
[69] World Nuclear Association, "Plans for New Reactors Worldwide" (October 2015), available at www.world-nuclear.org/info/current-and-future-generation/plans-for-new-reactors-worldwide, accessed December 22, 2015.
[70] Ibid.
[71] Cited in Carmen Richerzhagen and Imme Scholz, "China's Capacities for Mitigating Climate Change," *World Development* 36, no. 2 (2008), 312.

energy consumption surpassed its economic growth.[72] To address this challenge, in its 11th FYP, China set an ambitious target for energy-efficiency improvement – that is, the energy intensity of the country's GDP to be reduced by 20 percent between 2005 and 2010 (see Table 3.1), which was the first time that a quantitative and binding target had been set for energy efficiency in China.[73] In order to reach this target, the Chinese government took some proactive measures to advance and monitor implementation by setting some specific targets for provinces and industrial sectors; and energy efficiency improvement criteria were adopted to evaluate the job performance of its local government officials.[74] Moreover, the Chinese government mobilized a national campaign to promote energy efficiency, focusing particularly on the largest and least efficient energy-consuming enterprises. Specifically, the government initiated the so-called "Ten-Thousand Enterprises Program," targeted at approximately 10,000 companies that consume about one-third of the country's energy.[75] In the process of implementing this energy efficiency policy, thousands of inefficient power plants and factories were closed, which saved the equivalent of 750 million tonnes of coal and 1.5 billion tonnes of CO_2 (5 percent of global CO_2 emissions in 2010).[76] During China's 11th FYP period, a reported 72.1 GW of thermal capacity was closed. The closure of

---

[72] Kevin Lo and Mark Y. Wang, "Energy Conservation in China's Twelfth Five-Year Plan Period: Continuation or Paradigm Shift?," *Renewable and Sustainable Energy Reviews* 18 (2013), 499–507.
[73] Jiang Lin, Nan Zhou, Mark Levine, and David Fridley, "Taking out 1 Billion Tons of CO_2: The Magic of China's 11th Five-Year Plan?," *Energy Policy* 36, no. 3 (March 2008), 954–970.
[74] 国务院 [The State Council], "国务院关于印发节能减排'十二五'规划的通知 [The State Council Notice on Printing and Distributing Twelfth Five-Year Plan on Energy Saving and Emission Reductions]" (August 6, 2012), available at www.gov.cn/zwgk/2012-08/21/content_2207867.htm, accessed February 26, 2018.
[75] Information Office of the State Council of the PRC, *China's Energy Policy 2012*; Kevin Lo, He Li, and Mark Wang, "Energy Conservation in China's Energy-Intensive Enterprises: An Empirical Study of the Ten-Thousand Enterprises Program," *Energy for Sustainable Development* 27 (2015), 105–111.
[76] Zhu Liu, Dabo Guan, Douglas Crawford-Brown, Qiang Zhang, Kebin He, and Jianguo Liu, "A Low-Carbon Road Map for China," *Nature* 500 (August 8, 2013), 143.

these inefficient power and industrial facilities helped contribute to its energy intensity reduction of 19.1 percent mentioned above.[77]

Moreover, China promulgated its revised Energy Conservation Law on October 28, 2007, which set energy conservation as a national policy. Such a policy is very important not only for energy security, but also for emission reduction in the future, according to research.[78] In 2011, China further issued an *Integrated Working Plan on Energy Saving and Emission Reductions*.[79] Through these great efforts, China's energy efficiency has been improved during its 12th FYP. According to research, "Chinese energy efficiency has converged towards that in developed and other developing economies in recent years and its rate of improvement has slowed down."[80] This result was confirmed by the *Global Energy Statistical Yearbook 2015*: based on the criteria of energy intensity of GDP at constant PPPs, China witnessed a strong decline in energy intensity in 2014.[81]

For India, energy efficiency "was first viewed in the context of global concerns on the scarcity of fossil fuels, then as a means to achieving cost effectiveness and finally as critical to mitigating climate change."[82] In order to enhance its energy efficiency, since 2001, the Indian government has enacted the Energy Conservation Act. "The Act provides a regulatory mandate for standards and labelling of equipment and

---

[77] 温家宝 [Wen Jiabao], "政府工作报告——2011年3月5日在第十一届全国人民代表大会第四次会议上 [Report on the Work of the Government – Delivered on March 5, 2011 at the Fourth Session of the Eleventh China National People's Congress]," available at www.gov.cn/2011lh/content_1825233.htm, accessed February 26, 2018.

[78] Bing Jiang, Zhenqing Sun, and Meiqin Liu, "China's Energy Development Strategy under the Low-Carbon Economy," *Energy* 35, no. 11 (November 2010), 4261.

[79] 国务院 [The State Council], "'十二五'节能减排综合性工作方案 [Integrated Working Plan on Energy Saving and Emission Reductions in the Twelfth Five-Year Plan]" (August 31, 2011), available at www.gov.cn/zwgk/2011-09/07/content_1941731.htm, accessed February 26, 2018.

[80] David I. Stern and Frank Jotzo, "How Ambitious Are China and India's Emissions Intensity Targets?," *Energy Policy* 38, no. 11 (November 2010), 6782.

[81] Enerdata, *Global Energy Statistical Yearbook 2015*, available at https://yearbook.enerdata.net/energy-intensity-GDP-by-region.html, accessed December 25, 2015.

[82] P. Balachandra, Darshini Ravindranath, and N. H. Ravindranath, "Energy Efficiency in India: Assessing the Policy Regimes and their Impacts," *Energy Policy* 38, no. 11 (November 2010), 6429.

appliances and also includes provisions for energy conservation building codes for commercial buildings as well as energy consumption norms for energy intensive industries."[83] Based on this Act, India established a new institution to govern its energy conservation – the Bureau of Energy Efficiency (BEE). Since its inception, the BEE has successfully launched a number of programs and initiatives.[84] For instance, the preparation of an Energy Conservation Action Plan was among the first initiatives taken by the BEE. This plan was released in August 2002, being developed for wide dissemination and the implementation of standards set by the Bureau.[85] In addition, this Act set up some appliance standards and labeling, energy consumption norms, and energy-use reporting requirements for energy-intensive industrial units, among other measures, to conserve energy. The standards gave a "thrust on energy efficiency in [the] industrial sector, standard[s] and labeling for appliances, demand-side management, energy efficiency in commercial buildings, capacity building of energy managers and energy auditors, energy performance codes and manual preparation among others."[86] This Act was amended in 2010, the main amendments including a statement that "The Central Government may issue the energy savings certificate to the designated consumer whose energy consumption is less than the prescribed norms and standards in accordance with the procedure as may be prescribed."[87]

Another important policy measure the Indian government has adopted is to issue an Energy Conservation Building Code (ECBC), to guide the design of new commercial buildings where there is the greatest scope for efficiency improvements, which became mandatory from 2017.[88] In addition, India also set specific energy consumption (SEC) targets mandated for 478 units in eight energy-intensive sectors, to be achieved in yearly cycles.[89] These initiatives have resulted in a

---

[83] Sarat Kumar Sahoo, Payal Varma, Krishna Prabhakar Lall, and Chanpreet Kaur Talwar, "Energy Efficiency in India: Achievements, Challenges and Legality," *Energy Policy* 88, no. 1 (January 2016), 495.
[84] Balachandra, Ravindranath, and Ravindranath, "Energy Efficiency in India," 6432.
[85] Sahoo et al., "Energy Efficiency in India," 495.   [86] Ibid.   [87] Ibid.
[88] IEA, *World Energy Outlook 2014*, 283.
[89] Bhaskar Sarma, "Energy Efficiency in India: Challenges and Lessons," In-Session Technical Expert Meeting on Energy Efficiency ADP, Bonn (March 13, 2014), available at https://unfccc.int/files/bodies/awg/application/pdf/2_india_revised.pdf, accessed February 26, 2018.

saving in capacity generation of 10,836 MW during the 11th FYP.[90] Furthermore, India initiated a market-based mechanism known as the "Perform, Achieve, and Trade" (PAT) scheme in 2011, to set benchmark efficiency standards for 563 power plants, steel mills, and cement plants that collectively account for more than half of India's energy consumption. The scheme includes energy savings certificates that can be sold and traded.[91]

Thanks to these efforts, India's energy efficiency has improved significantly. According to its 12th FYP, India's energy intensity has dropped from 1.09 kilograms of oil equivalent (koe) in 1981 to 0.62 koe in 2011.[92]

Establishing Strategic Petroleum Reserves

The establishment of strategic petroleum reserves (SPRs) is the most important policy measure adopted by both the Chinese and the Indian governments, to reduce their vulnerability and sensitivity to supply disruptions and price volatilities stemming from their high dependence on overseas oil supplies. This is because SPRs should be able to provide both countries with protection against potential disruptions and volatilities, at least in the short- and medium-term.

China first set out its policy to establish SPRs in its 10th FYP, as mentioned above. Specifically, the Chinese government declared for the very first time in this plan, the "construction of [a] strategic petroleum reserve system in order to enhance the national energy security."[93] This policy was further developed in its following two FYPs. In its 11th FYP, the Chinese government stated that it would expand the existing and establish new national petroleum reserve bases.[94] In the 12th FYP, the government specified that it would plan and construct energy reserve infrastructure properly; improve the oil reserve system; strengthen the natural gas and coal reserve, and

---

[90] Sahoo et al., "Energy Efficiency in India," 495.
[91] Johnson G. Pryor and Irene Mobley (eds), *India: Domestic Policies, Foreign Relations and Cooperation with the United States* (New York: Nova Science Publishers, Inc., 2012).
[92] Planning Commission, *Twelfth Five-Year Plan (2012–2017)*, Volume II, 130.
[93] The State Council, "The Tenth Five-Year Plan Guidelines."
[94] The State Council, "The Eleventh Five-Year Plan Guidelines."

develop the emergency response system.[95] According to some Chinese researchers, "It has become the critical component of China's energy policy to establish and expand the national petroleum reserves with the aim of enhancing the national energy security."[96]

In practice, China's construction of its SPR program began in 2004 and is expected to be fully completed in three phases over 15 years. In November 2014, the Chinese government made its first official announcement about the progress of its SPRs. According to this announcement, the first phase of the government emergency stockpile had reserved about 91 million barrels of crude oil, equivalent to about nine days of oil use, which is stored in Zhenhai, Zhoushan, Dalian, and Huangdao.[97] It also announced that it planned to complete the construction of the second phase by 2020, which is designed to hold 170 million barrels of oil and has already been partially filled; the third phase will start by 2020.[98] Since the decline of the oil price on the international market in late 2014 and early 2015, China has stepped up efforts to buy oil to increase its storage.[99] Not surprisingly, China has surpassed the United States, becoming the largest oil importer. In order to manage the construction and management of the SPRs, the Chinese government created a new institution – the National Oil Reserve Centre (NORC) in 2007.[100]

India has also started to build its own SPRs to enhance its energy security, which is in accordance with its 2006 *Integrated Energy Policy* suggestion that India should maintain a 90-day reserve of net oil

---

[95] The State Council, "The Twelfth Five-Year Plan Guidelines."
[96] Yuan and Zuo, "Transition to Low Carbon Energy Policies in China," 3856–3857.
[97] "Update 1: China Makes First Announcement on Strategic Oil Reserves," *Reuters* (November 20, 2014), available at www.reuters.com/article/2014/11/20/china-oil-reserves-idUSL3N0TA1QE20141120, accessed February 26, 2018.
[98] Ibid.
[99] 李春莲 [Li Chunlian], "石油战略储备借'低价时段'提速 [Strategic Oil Reserves Accelerated in the Period of Low Oil Prices]," 《证券日报》 [*Securities Times*] (July 16, 2015), available at www.ccstock.cn/finance/hangyedongtai/2015-07-16/A1436983922281.html, accessed October 14, 2015. According to the EIA, China has surpassed the United States as the largest crude oil importer since 2014, when China imported 6.1 million barrels of crude oil per day while the United States imported 5.1 million barrels per day. See EIA, "China" (May 14, 2015), 4.
[100] Y. Bai, C. A. Dahl, D.Q. Zhou, and P. Zhou, "Stockpile Strategy for China's Emergency Oil Reserve: A Dynamic Programming Approach," *Energy Policy* 73 (October 2014), 13.

imports so as to hedge against both supply disruption and/or price volatility in the international energy market.[101] India's progress in building its SPRs was noted in its 12th FYP:

> The Government is in the process of creating strategic crude oil storage capacity for 15 days at Vishakhapatnam (1.33 million tonnes), Mangalore (1.50 million tonnes) and Padur (2.5 million tonnes) through a Special Purpose Vehicle, namely, Indian Strategic Petroleum Reserve Ltd. (ISPRL). The storage would be further upgraded at other suitable locations by an incremental capacity of 12.5 million tonnes during the Twelfth Plan period.[102]

In sum, faced with the challenge of energy insecurity, China and India, the world's largest and third-largest energy consumers, have adopted similar proactive policy measures to assure their energy security. On the supply side, both countries have been promoting greater use of renewables in their energy mix, mainly through solar and wind power generation, in addition to nuclear energy. On the demand side, efforts have been made to make energy consumption more efficient through various policy measures that promote innovative utilization under the ambit of both countries' energy conservation legislation. Moreover, to hedge against any potential supply disruptions, China and India have already begun to establish their own SPRs.

Differences between China's and India's Domestic Energy Policy Approaches

Although China and India have undertaken similar domestic approaches to address their energy insecurity, there are still some subtle differences between their approaches, which have largely stemmed from their different political and economic systems. In terms of their political systems, it is widely known that China is characterized as a one-party, centralized, authoritarian system,[103] whereas India is a

---

[101] Planning Commission, *Integrated Energy Policy*, 127.
[102] Planning Commission, *Twelfth Five-Year Plan (2012–2017)*, Volume II, 173.
[103] Neil Collins and Andrew Cottey, *Understanding Chinese Politics: An Introduction to Government in the People's Republic of China* (Manchester and New York: Manchester University Press, 2012); William A. Joseph (ed.), *Politics in China: An Introduction* (2nd edn) (New York: Oxford University Press, 2014).

democratic, federal system.[104] With regard to their economic systems, the Chinese economy is largely state-owned, while the Indian economy is more market-based. With their different political and economic systems, Chinese and Indian domestic energy policy inevitably exhibit some differences in approach.

In terms of their efforts to improve their energy efficiency, the Chinese government has employed a "top-down" approach, associated with its authoritarian system, while the Indian government has used a more "bottom-up" approach, associated with its more market-oriented system. More specifically, China has used its strong national, provincial, and local management system to implement its strict energy conservation rules, regulations, and laws, while India has relied on market forces, by deregulating energy prices to attract energy efficiency investment and technologies.[105] Moreover, the Chinese government has used its authority to shut down a large number of small, inefficient, and energy-intensive industrial facilities, which has brought about improved energy efficiency in the whole industrial sector.[106] In stark contrast, India's policy practice and policy implementation/monitoring system do not allow such closures to happen, even if the enterprise is inefficient or the operation of the enterprise is in deficit.[107]

China's "top-down" approach can be further illustrated by the above-mentioned Ten-Thousand Enterprises Energy Conservation Low-Carbon Program (or Ten-Thousand Enterprises Program) in its 12th FYP, which regulates the energy consumption and energy conservation behavior of enterprises (those which consume 10,000 tce or more annually) by the government. Essentially, such an approach is to command and control, which "not only allocates energy-saving targets to the regulated enterprises but also demands the enterprises [to] meet a number of energy management requirements."[108]

In order to reach its energy conservation targets, the Chinese government established the so-called "Energy Conservation Target

---

[104] Robert Jenkins, *Democratic Politics and Economic Reform in India* (Cambridge: Cambridge University Press, 1999).
[105] Ming Yang, "Energy Efficiency Policy Impact in India: Case Study of Investment in Industrial Energy Efficiency," *Energy Policy* 34, no. 17 (November 2006), 3108.
[106] Ibid., 3106.   [107] Ibid., 3108.
[108] Kevin Lo, He Li, and Mark Wang, "Energy Conservation in China's Energy-Intensive Enterprises: An Empirical Study of the Ten-Thousand Enterprises Program," *Energy for Sustainable Development* 27 (August 2015), 106.

Responsibility System" in 2007,[109] which imposes mandatory energy intensity reduction requirements on local governments, linking the implementation of central policies to financial bonuses and career advancement of local cadres.[110] This centralized evaluation system has turned out to be "an effective enforcement mechanism when local government scrambled to deliver the targets with whatever means available to them in 2010."[111] During the 12th FYP period, the Chinese government continues to rely on regulatory policies or administrative approaches rather than economic tools to reach its energy conservation target.[112]

In contrast, Indian enterprise managers have fewer worries about energy efficiency than their Chinese counterparts because India does not have this kind of cadre evaluation system. India has adopted more flexible and market-oriented measures to improve its energy efficiency. For instance, as mentioned above, the Government of India developed and promulgated an ECBC for new commercial buildings. Although Indian central government has the authority to implement the ECBC, state governments have "the flexibility to modify the code to suit local or regional needs and apply them as seen fit."[113] Moreover, in order to make full use of market forces to enhance the implementation of the ECBC, the BEE developed a "voluntary Star Rating Programme," which is "based on evaluation of individual performances of buildings under the programme in terms of the energy used over the area as found in kWh/sq.m/year."[114]

There are also some differences between these two countries as far as the process of the development of renewable energy is concerned. One of the major differences is in the role of state-owned enterprises (SOEs) and the extent to which the state, rather than the private sector, organizes and manages the mobilization of resources.[115] More

[109] 国务院 [The State Council], "国务院批转节能减排统计监测及考核实施方案和办法的通知 [Notice on Implementation Plan for Energy Conservation Accounting, Monitoring and Evaluation]" (November 23, 2007), available at www.gov.cn/zwgk/2007-11/23/content_813617.htm, accessed February 27, 2018.
[110] Lo and Wang, "Energy Conservation in China's Twelfth Five-Year Plan Period," 501.
[111] Ibid.   [112] Ibid.   [113] Sahoo et al., "Energy Efficiency in India," 495.
[114] Ibid.
[115] Kavita Suranaa and Laura Diaz Anadon, "Public Policy and Financial Resource Mobilization for Wind Energy in Developing Countries: A Comparison of

specifically, China has developed wind energy largely under direct governmental control, within the existing set of large state-owned electric utilities, and encouraged a number of SOEs to engage in wind turbine manufacturing. Private capital has played only a minor role in the form of foreign minority investment in joint ventures for wind turbine manufacturing. Thus, "In China, ownership of wind farms is highly concentrated in a small set of homogeneous investors, with over 90% of the market share associated with the public sector. The top five investors are large power generating companies who had over 90% of the market share in 1997 and continued to hold a substantial market share of over 50% in 2012."[116]

In contrast, the growth of wind power in India was driven primarily by state-level incentives, in conjunction with an accelerated depreciation benefit extended by the national Government of India.[117] India has developed wind energy largely outside of the existing partially state-owned electric utility sector, by opening up wind energy development and wind turbine manufacturing to private capital.[118] According to research, "In India, the private sector mobilized almost all financial resources for wind farm investments, with over 90% of the market under highly fragmented ownership throughout the four diffusion stages."[119] Moreover, in comparison with Chinese renewables, SOEs' easy access to Chinese state-owned bank loans, supported by the Chinese government as well as government subsidies, India's renewable energy companies have great difficulty getting public financial support. For instance, research shows that less than one-third of public sector banks in India lend to renewable energy projects, and the situation is worse for private sector banks, where less than one-fifth lend to renewable projects.[120]

Approaches and Outcomes in China and India," *Global Environmental Change* 35 (November 2015), 343.

[116] Ibid., 348.

[117] David Nelson, Gireesh Shrimali, Shobhit Goel, Charith Konda, and Raj Kumar, "Meeting India's Renewable Energy Targets: The Financing Challenge," *CPI-ISB Report* (December 2012), 2, available at http://climatepolicyinitiative.org/wp-content/uploads/2012/12/Meeting-Indias-Renewable-Targets-The-Financing-Challenge.pdf, accessed February 27, 2018.

[118] Suranaa and Anadon, "Public Policy and Financial Resource Mobilization for Wind Energy in Developing Countries," 343.

[119] Ibid.   [120] Nelson et al., "Meeting India's Renewable Energy Targets," 12.

In sum, while China and India face similar challenges in terms of energy security, their governments' energy policies, while similar, bear some subtle differences. The Chinese government adopts centralized policies and uses the development of alternative energy to fossil fuels, including renewable energy and nuclear energy, as a strategic tool for promoting its economic development, while India's approach to renewables and energy efficiency is less coordinated and more in the hands of market forces.

Conclusion

Energy security has been both China's and India's top policy priority, illustrated not only by their leaders' rhetoric on the importance of energy security, but also by both countries' long-term development plans in which energy security has an independent agenda with special emphasis. Accordingly, China and India have taken some proactive measures at the domestic level in order to achieve energy security. These include measures to diversify their energy supplies in non-fossil fuels, especially renewable energy and nuclear energy, while promoting energy conservation through improving energy efficiency and reducing energy intensity. In other words, the core of China's and India's energy security strategy is to develop low-carbon energy and utilize fossil fuel cleanly and efficiently.[121] In parallel, both countries have also taken measures to prepare for supply emergencies by establishing their own SPRs. All these measures will play a significant role in enhancing China's and India's efforts to address their energy insecurity challenge, and the first three measures concurrently represent their efforts to address climate change. Both countries' motivation for taking domestic measures to address energy insecurity will be more thoroughly explored, together with the motivations for their domestic efforts to address climate change, in Chapter 6, given the fact that China's and India's climate policies have been imbedded in their energy policies.

---

[121] Bing Jiang, Zhenqing Sun, and Meiqin Liu, "China's Energy Development Strategy under the Low-Carbon Economy," *Energy* 35, 11 (November 2010), 4261.

Despite the domestic measures they have taken to reduce their reliance on fossil fuels, China and India will continue to rely predominantly on those fuels – oil, natural gas, and coal – to meet their energy demand. Both countries' fossil fuel reserves are so limited, however, that they cannot rely on their own resources to meet increased demand. As a result, both countries have encouraged their NOCs to seek overseas energy resources to meet their increased domestic energy demand.

4 Energy Diplomacy: Proactively Preempting and Reactively Restraining

Proposition 2: The energy policies that China and India have pursued globally are constrained by the patterns of asymmetrical interdependence and international norms in which China and India are enmeshed. In other words, their energy diplomacy is increasingly constrained by forces emanating from the systemic level.

This chapter uses the analytical framework developed in Chapter 2 to explore China's and India's foreign policy or diplomacy targeted at procuring overseas energy supplies. It first illuminates the proactive and reactive features of such diplomacy. It then uses the examples of China's and India's energy diplomacy toward Iran, Sudan, and Myanmar to prove how China and India have been proactively preempting overseas energy resources, while at the same time reactively restraining their behavior as a result of external pressures stemming from both state and non-state actors. A brief discussion of the differences between the three cases is included. The chapter then considers how two-level pressures have shaped China's and India's proactive and reactive energy diplomacy. A brief conclusion is drawn in the final section.

Proactive and Reactive

As discussed in Chapter 2, China's and India's indigenous energy resources are insufficient to meet their increased energy demand, so both countries have had to encourage their NOCs to embark on "going-out" to seek overseas equity energy resources, especially oil and natural gas.[1] As far as China is concerned, in December 1993,

---

[1] In China, going-out to seek overseas energy resources was first initiated by the CNPC and was then adopted by the Chinese government. See Bo Kong, *China's International Petroleum Policy* (Santa Barbara, CA: Praeger Security International, 2010), 37–60. In contrast, it was the Indian government that first encouraged its NOCs to invest in overseas energy resources. See Luke Patey,

when the country became a net oil importer, the Chinese government put forward the guideline for its oil industry to "fully utilize domestic and international resources and markets."[2] China's three major NOCs, namely China National Petroleum Corporation (CNPC), China Petrochemical Corporation (Sinopec), and China National Offshore Oil Corporation (CNOOC), began investing in overseas upstream oil and natural gas in the 1990s and have intensified their efforts since the latter part of that decade, after the Chinese government's formulation of its "going-out strategy." More specifically, in January 1997, the then Chinese President Jiang Zemin encouraged Chinese oil companies to begin "going-out" to seek overseas oil, and he specifically named two regions – Africa and Central Asia – that Chinese oil companies should target, because of China's friendly relationships with the majority of the countries in these two regions. Moreover, Jiang emphasized that it was impossible for Chinese oil companies not to engage in "going-out" or to utilize international markets. Chinese oil companies, he said, should "base firmly on domestic market while actively participating in and utilizing international oil resources, walking on 'two legs'."[3]

"Going-out" as a national strategy was first officially written into China's 10th FYP (2001–5), and then re-emphasized in its 11th FYP (2006–10) and 12th FYP (2011–15). Specifically, in its 10th FYP, China stated that its priorities were, among others, to "explore overseas energy resources, establish [an] overseas petroleum and natural gas supply base and diversify oil supplies,"[4] all of which is regarded as a critical component of China's energy security strategy. Its 11th FYP stated clearly that the "going-out strategy" should be carried out and that domestic firms should expand their exploration of overseas oil and natural gas, based on adherence to the principle of equal cooperation and mutual benefit.[5] In its 12th FYP, the Chinese government encouraged its various companies to speed up their "going-out strategy," based on the principle of market orientation and companies'

The New Kings of Crude: China and India's Global Struggle for Oil in Sudan (London: Hurst & Company, 2014), 124.

[2] Cited in 童晓光 [Tong Xiaoguang], "实施'走出去'战略,充分利用国外油气资源 [To Carry out 'Going-Out' for Fully Using Oil and Natural Gases Abroad]," 《国土资源》 [*Land & Resources*] 31, no. 2 (February 2004), 6.
[3] Ibid.   [4] Ibid.
[5] The State Council, "The Eleventh Five-Year Plan Guidelines."

autonomous decision-making, because this strategy would deepen mutually beneficial cooperation in international energy resource exploration and refinery.[6]

The history of the Indian NOC's – Oil and Natural Gas Corporation (ONGC) Videsh – "going-out" experience dates back to 1965, when ONGC Videsh was incorporated as Hydrocarbons India Pvt. Ltd. to carry out exploration and development of the Rostam and Raksh oil fields in Iran, and to undertake a service contract in Iraq.[7] However, this company's experience turned out to be a failure.[8] It was only from mid-1990s that Videsh began to re-orient its focus toward acquiring overseas oil and gas assets.[9] Since the late 1990s, apart from ONGC Videsh Limited (OVL), some other NOCs, such as Oil India Limited (OIL), Indian Oil Corporation (IOC), and Gas Authority of India Limited (GAIL), have begun to engage in overseas exploration and production of oil and gas assets.[10] Thus, in the summary of the performance of its 9th FYP (1997–2002), India's 10th FYP (2002–7) states that its oil companies have "secured equity oil abroad by participating in the oil and gas project in Vietnam and in Sakhalin (Russia) and signing an agreement with Iraq for oil exploration."[11]

Since the early 2000s, when the Indian government set out its going-out strategy in its FYPs, India's NOCs have begun to systemically go abroad to seek overseas oil and gas assets. The Indian government does not explicitly use the term "going-out strategy" in its FYPs, as the Chinese government has done, but it has set out its energy security strategy, especially oil security strategy, in its FYPs, that is, "diversification of sources for crude supplies, strategic storage and globalization measures to bring equity oil and gas/LNG from abroad."[12] The Indian government also encourages its NOCs to seek overseas energy assets. For instance, in its 10th FYP, increasing the production of equity oil

---

[6] The State Council, "The Twelfth Five-Year Plan Guidelines."
[7] Oil and Natural Gas Corporation Limited, "ONGC Videsh Limited," available at www.ongcindia.com/wps/wcm/connect/ongcindia/Home/SubsidiariesJVs/Subsidiaries/ONGC+Videsh+Limited, accessed February 27, 2018.
[8] Patey, *The New Kings of Crude*, 124.   [9] Ibid.
[10] Jonas Meckling, Bo Kong, and Tanvi Madan, "Oil and State Capitalism: Government–Firm Coopetition in China and India," *Review of International Political Economy* 22, no. 6 (2015), 1172.
[11] Ibid., 770.
[12] Planning Commission, *Tenth Five-Year Plan (2002–2007)*, Volume II, 773.

and natural gas abroad is listed as one component of its energy security strategy,[13] namely, "In view of the stagnating domestic production of crude and the widening gap between demand and supply of oil and gas, there is a need to diversify oil supply sources, and acquire equity oil and gas abroad. This would be an important component of the strategy to achieve oil security."[14] To do so, the Indian government encourages its state-owned and private oil companies "to tap opportunities available abroad for acquiring exploration acreages, either on their own or through strategic alliances."[15] With the Indian government's policy encouragement, the Indian NOCs mentioned above have been actively seeking overseas equity oil and gas. As a result, during the 10th FYP, OVL acquired 22.24 million tonnes of oil and oil equivalent of gas (O&OEG) from its overseas activities, which is more than 12 million tonnes more than the original acquisition target of 10.14 million tonnes set out in the plan.[16]

In its 11th FYP (2007–12), the Indian government continued to emphasize the significance of equity oil in meeting its energy security, that is, "Maximizing supply of crude oil and gas at the least possible cost as part of energy security by … getting equity oil abroad."[17] Therefore, Indian "NOCs and private companies are being encouraged to venture abroad."[18] Furthermore, to ensure its energy security, one of the strategies the Indian government stated in its 12th FYP (2012–17) is that "investments in energy assets in foreign countries, especially for coal, oil and gas and uranium should be stepped up."[19]

Not only have the Chinese and Indian governments written a going-out strategy into their national development plans, but they have also employed their state power to facilitate their NOCs going-out to seek overseas energy resources through energy diplomacy,[20] which bears proactive and reactive features. For China and India, the proactive feature of energy diplomacy refers to the fact that both governments employ all measures at their disposal to foster and/or build up their

---

[13] Ibid., 773. [14] Ibid., 774. [15] Ibid.
[16] Planning Commission, *Eleventh Five-Year Plan (2007–2012)*, Volume III, 364.
[17] Ibid., 363. [18] Ibid., 366.
[19] Planning Commission, *Twelfth Five-Year Plan (2012–2017)*, Volume II, 135.
[20] So far, there is no unified definition for "energy diplomacy" in the existing literature. In this book, "energy diplomacy" refers to energy-consuming countries' employment of political, economic, military and diplomatic measures to facilitate their NOCs seeking overseas energy assets.

bilateral ties with energy-rich countries across the world, in order either to procure long-term contracts for a privileged supply of energy, or to facilitate their NOCs in gaining oil and gas equities in those energy-rich countries' energy sector. These proactive measures include the following activities.

To begin with, both countries employed some political tools to forge closer ties with energy-rich countries. The first tool most frequently adopted by both countries is to enhance their bilateral relations with energy-rich countries across the world during high-level official visit exchanges with the targeted countries. In China's case, central government officials, from the president to ministers from a variety of ministries, carry out official visits to energy-rich countries. In fact, some of the Chinese president's state visits are directly aimed at signing energy contracts to secure overseas supplies.[21] The same is true for India. For instance, according to a report by Tanvi Madan, Indian officials have increased their visits to oil-rich countries, whether from the Prime Minister's Office, the Ministry of Petroleum and Natural Gas (MPNG), or the Ministry of External Affairs (MEA). At the same time, the Indian government has increased the number of invitations issued to leaders and officials from those countries to visit India.[22] Through these frequent, high-level official exchange visits, China and India have sought to establish strategic relationships with some of the energy-rich countries.[23]

Apart from their use of proactive political tools to facilitate their ties with energy-rich countries, China and India have also utilized some economic measures, including enhancing trade interdependence and increasing direct investment, as well as providing economic aid and loans to those countries. Specifically, China and India have reached free trade agreements (FTAs) with some energy-rich countries and regions. For instance, China has signed a bilateral FTA with energy-rich

---

[21] Amy Myers Jaffe and Steven W. Lewis, "Beijing's Oil Diplomacy," *Survival* 44, no. 1 (2002), 115–134; Charles E. Ziegler, "The Energy Factor in China's Foreign Policy," *Journal of Chinese Political Science* 11, no. 1 (March 2006), 1–23; Ian Taylor, "China's Oil Diplomacy in Africa," *International Affairs* 82, no. 5 (September 2006), 937–959.

[22] Madan, *India*, 47.

[23] Domingos Jardo Muekalia, "Africa and China's Strategic Partnership," *African Security Review* 13, no. 1 (2004), 5–11; Feng Zhongping and Huang Jing, "China's Strategic Partnership Diplomacy: Engaging with a Changing World," *ESPO Working Paper*, no. 8 (June 2014).

countries such as Peru and Australia, while India is still in the process of negotiating an FTA with Australia. Both China and India have signed FTAs with the Association of Southeast Asian Nations (ASEAN), and both have been in the process of negotiating FTAs with the Gulf Cooperation Council (GCC) – consisting of Bahrain, Kuwait, Oman, Qatar, Saudi Arabia, and the United Arab Emirates – since 2004.[24] In terms of other economic stimulants, loan/aid-for-oil is one of the most efficient tools used by the two countries to acquire equity supplies. For instance, in the competition for Angola's oil deal in October 2004, China offered $2 billion in aid for various projects in Angola, which put India's offer of $200 million for constructing railways into the shade.[25] Moreover, according to IEA research, between 2009 and 2010 alone, China signed 12 long-term loans-for-oil and loans-for-gas supply agreements with Angola, Bolivia, Brazil, Ecuador, Ghana, Kazakhstan, Russia (two agreements), Turkmenistan, and Venezuela (two agreements).[26]

Furthermore, China and India have employed some military-related measures to boost their energy ties with the energy-rich countries. Specifically, both China and India usually offer military cooperation to those countries by either providing them with weapons or providing training for their military personnel, or by participating in UN Peacekeeping Missions in those countries. In addition to their proactive political, economic, and military means, both China and India have also been willing to provide some energy-rich countries with diplomatic support at both international and regional levels so as to bolster their energy relations with those countries.

At the same time, however, China and India have had to adjust these aforementioned proactive measures toward certain energy-rich

---

[24] China FTA Network, "China–GCC FTA," available at http://fta.mofcom.gov.cn/topic/engcc.shtml, accessed February 27, 2018; Ministry of Commerce and Industry, GOI, "International Trade," available at http://commerce.nic.in/trade/international_ta_current_details.asp, accessed February 27, 2018.
[25] "中印激战西非石油 安哥拉油田竞争印度败北 [China and India Fiercely Fighting for West African Oil, India Being Defeated in the Competition for Angolan Oil Fields]," 《大纪元》 [Epoch Times], October 19, 2004, available at www.epochtimes.com/gb/4/10/19/n694691.htm, accessed February 27, 2018.
[26] Julie Jiang and Jonathan Sinton, Overseas Investments by Chinese National Oil Companies: Assessing the Drivers and Impacts (Paris: OECD/IEA, 2011), 41.

countries, particularly the so-called "pariah states,"[27] as a result of external pressures either from state actors or non-state actors, which leads us to the reactive feature of both countries' energy diplomacy. To explore such proactive and reactive features in more detail, the following three sections use case studies of China's and India's energy diplomacy toward Iran, Sudan, and Myanmar.

Iran[28]

Iran has long been known for its rich oil and natural gas reserves. According to BP's statistics, by the end of 2014, Iran accounted for 9.4 percent and 18.0 percent respectively of global oil and natural gas proved reserves, being the world's fifth-largest oil producer and having the largest gas reserves.[29] Moreover, its reserve-to-production ratio of both oil and gas is more than 100 years, the highest in the world. It is, therefore, no surprise that Iran has been an attractive target for China's and India's energy diplomacy since as early as the 1990s.[30]

Proactive Measures

During the 2000s, although Western countries, especially the United States, stepped up efforts to impose increasingly comprehensive and

---

[27] David Zweig and Bi Jianhai, "China's Global Hunt for Energy," *Foreign Affairs* 84, no. 5 (September/October 2005), 25–38; Stephanie Kleine-Ahlbrandt and Andrew Small, "China's New Dictatorship Diplomacy: Is Beijing Parting with Pariahs," *Foreign Affairs* 87, no. 1 (January/February 2008), 23–37.

[28] The content of China's energy diplomacy toward Iran in this section draws heavily on my article. See Fuzuo Wu, "China's Puzzling Energy Diplomacy toward Iran," *Asian Perspective* 39, no. 1 (2015), 47–69.

[29] BP, *BP Statistical Review of World Energy* (June 2015), 6, 20, available at www.bp.com/content/dam/bp/pdf/Energy-economics/statistical-review-2015/bp-statistical-review-of-world-energy-2015-full-report.pdf, accessed August 19, 2015.

[30] On China's and India's relations with Iran, see Dingli Shen, "Iran's Nuclear Ambitions Test China's Wisdom," *The Washington Quarterly* 29, no. 2 (2006), 55–66; John W. Garver, *China and Iran: Ancient Partners in a Post-Imperial World* (Seattle: University of Washington Press, 2006); Sujata Ashwarya Cheema, "India–Iran Relations: Progress, Challenges and Prospects," *India Quarterly* 66, no. 4 (2010), 383–396.

tough sanctions against Iran for its controversial nuclear program,[31] China and India employed political, economic, military, and diplomatic measures to foster their closer energy ties with Iran.

Politically, China and India have frequently exchanged high-level official visits with Iran, regardless of US efforts to isolate Tehran.[32] For instance, in April 2002, Chinese President Jiang Zemin paid a five-day state visit to Iran, even though the US President George W. Bush had listed Iran as a member of the "Axis of Evil" in January of that year.[33] During his visit, Jiang emphasized the importance of the Sino-Iranian relationship and signed six agreements with Iranian President Mohammad Khatami, including one on oil and gas cooperation.[34] Ten years later, in June 2012, when the United States and its Western allies tried to impose tough sanctions against Iran, Chinese President Hu Jintao not only invited Iranian President Ahmadinejad to attend the 12th summit of the Shanghai Cooperation Organization (SCO), but also held an official meeting with him.[35] Moreover, in September 2012, China's top legislator, Wu Bangguo, Chairman of China's National People's Congress Standing Committee, paid a four-day official visit to Iran, the first time in 16 years that China's top legislator had visited Iran.[36]

---

[31] Kenneth Katzman, *Iran Sanctions* (Congressional Research Service report, June 13, 2013), available at www.fas.org/sgp/crs/mideast/RS20871.pdf, accessed July 29, 2013.
[32] Ministry of Foreign Affairs of PRC, "Bilateral Relations" (August 22, 2011), available at www.fmprc.gov.cn/chn/pds/gjhdq/gj/yz/1206_40/sbgx, accessed September 19, 2012.
[33] George W. Bush, "State of the Union Address" (January 29, 2002), available at https://georgewbush-whitehouse.archives.gov/news/releases/2002/01/20020129-11.html, accessed March 6, 2018.
[34] Ministry of Foreign Affairs of PRC, "President Jiang Zemin Held Talks with Iranian President Mohammad Khatami" (April 22, 2002), available at www.fmprc.gov.cn/eng/topics/3740/3745/t19162.htm, accessed September 20, 2012.
[35] "Iranian President Arrives in Beijing for China Visit, SCO Summit," *Xinhuanet* (June 5, 2012), available at http://news.xinhuanet.com/english/china/2012-06/05/c_123240001.htm, accessed November 10, 2012.
[36] "China's Top Legislator Starts Visit to Iran for Closer Ties," *Xinhuanet* (September 9, 2012), available at http://news.xinhuanet.com/english/china/2012-09/09/c_131838742.htm, accessed November 10, 2012; Zhu Zhe, "Top Legislator Meets with Iranian President," *China Daily* (September 12, 2012), available at www.chinadaily.com.cn/china/2012-09/12/content_15751362.htm, accessed February 27, 2018.

Regarding India's high-level official visit exchanges with Iran, in April 2001, for instance, Indian Prime Minister Vajpayee visited Iran and signed the Tehran Declaration with his Iranian counterpart, Muhammad Khatami. This Declaration stated:

The geographical situation of Iran and its abundant energy resources along with the rapidly expanding Indian economy and energy market on the other, create a unique complementarity which the sides agree to harness for mutual benefit. In this context they agreed to accelerate the process of working out an appropriate scheme for the pipeline options and finalising the agreement reached on LNG.[37]

Two years after this visit, in January 2003, Iranian President Khatami visited India as a Chief Guest at India's Republic Day parade, and signed the New Delhi Declaration with Prime Minister Vajpayee, entitled "Vision of a Strategic Partnership for a More Stable, Secure and Prosperous Region and for Enhanced Regional and Global Cooperation."[38] Among other areas of economic cooperation, such as transit, transport, industries, science and technology, and agricultural fields, the New Delhi Declaration emphasized energy cooperation between the two countries, in addition to enhancing their cooperation on defense, including training and exchange of visits.[39] Moreover, this Declaration announced that a strategic partnership was established between India and Iran. As an India scholar commented, both declarations recognize the complementary interests of both countries in the energy sector: "The former envisaged energy as a strategic area in the bilateral relationship, whereas the latter emphasised mutual benefits that would accrue from enhanced cooperation."[40] Five years after Khatami's state visit to India, in April 2008, Iranian President Ahmadinejad briefly visited India. In August 2012, Indian Prime Minister Manmohan Singh visited Iran to attend the 16th Non-Aligned Movement (NAM) Summit held in Tehran,

---

[37] GOI, "Text of Tehran Declaration," available at http://pib.nic.in/archieve/pmvisit/pm_visit_iran/pm_iran_rel4.html, accessed August 29, 2015.
[38] MEA, "The New Delhi Declaration" (January 25, 2003), available at http://mea.gov.in/bilateraldocuments.htm?dtl/7544/The+Republic+of+India+and+the+Islamic+Republic+of+Iran+quotThe+New+Delhi+Declarationquot, accessed April 20, 2015.
[39] Ibid.   [40] Cheema, "India–Iran Relations," 384.

followed by bilateral talks with Iranian leaders, which was "the first [visit] by an Indian PM in over a decade."[41]

China and India have also adopted some economic measures to boost their energy ties with Iran. The first economic means China and India have employed to enhance their relations with Iran have been to increase their bilateral trade with Iran. For instance, in 2009, China was Iran's largest trade partner, with nearly $27 billion in total trade;[42] India was Iran's fourth-largest trade partner, after China, Japan, and the United Arab Emirates, with about $14 billion in total trade.[43]

In the energy sector, China's and India's NOCs have been actively seeking and developing oil and gas assets in Iran, despite Western economic sanctions against Tehran.[44] For example, in October 2004, China and Iran signed a memorandum of understanding (MOU) in which Iran agreed to sell to China 250 million tonnes of LNG over the next 30 years, and export 150,000 barrels of oil per day for 25 years at market prices. Sinopec was also given the privilege of developing Yadavaran oil field, Iran's largest undeveloped oil field.[45] In addition, Chinese NOCs have tended to fill the vacuum left by Western oil companies in the Iranian energy sector, especially upstream exploring and mining. For instance, after Inpex, a Japanese oil company, gave up its stake in Iran's Azadegan oil field in October 2006 and October 2010, due to persistent pressures from the United States, CNPC stepped in, signed a contract with Iran, and obtained a 70 percent stake in this oil field in September 2009.[46] Also in 2009, CNPC signed a $4.7 billion contract with Iran's National Iranian Oil Company (NIOC), to help develop phase 11 of South Pars gas field, filling the

[41] BBC, "India PM Manmohan Singh Arrives in Iran" (August 29, 2012), available at www.bbc.com/news/world-asia-india-19406936, accessed February 27, 2018.
[42] Shayerah Ilias, "Iran's Economic Conditions: U.S. Policy Issues," (Congressional Research Service report, April 22, 2010), 23, available at www.fas.org/sgp/crs/mideast/RL34525.pdf, accessed February 27, 2018.
[43] Ibid.   [44] Katzman, *Iran Sanctions* (June 13, 2013).
[45] "China, Iran Sign Biggest Oil & Gas Deal," *China Daily* (October 31, 2004), available at www.chinadaily.com.cn/english/doc/2004-10/31/content_387140.htm, accessed February 27, 2018.
[46] "中石油与伊朗签油田开发协议 持其70%股权 [CNPC Signed an Oil Agreement with Iran Holding 70% of its Shares]," *China Daily* (September 30, 2009), available at www.chinadaily.com.cn/hqcj/2009-09/30/content_8754043.htm, accessed February 27, 2018.

void left by French Total SA.[47] Furthermore, it was reported that since September 2009, Chinese state-run companies have begun to sell 30,000–40,000 barrels a day of petrol from the Asian spot market to Iran through third parties, accounting for one-third of Iranian total imports.[48] Thus, Chinese NOCs have become the major players in Iran's energy sector.[49]

As for India, in January 2005, it signed an energy cooperation agreement with Iran. In this agreement, Iran's state National Iranian Gas Export Co. agreed to sell up to 7.5 million tonnes per year of LNG for a period of 25 years from the South Pars gas fields, starting in 2008–9, to the IOC and GAIL. Moreover, Iran offered to give Indian ONGC a 20 percent stake in the 300,000 barrels-per-day (bpd) Yadavaran oil development, and a 100 percent stake in the 30,000 bpd Juffair field.[50] India had once intended to build a gas pipeline to import Iran's natural gas. Specifically, since the late 1980s, India and Iran had been discussing the construction of a natural gas pipeline, starting from Iran's natural gas field, transiting through Pakistan, and terminating in India, later known as the Iran–Pakistan–India (IPI) natural gas pipeline. In the early 2000s, India and Iran conducted some further negotiations on the construction of this gas pipeline.[51] Aside from these two initiatives, several Indian oil companies have been engaged in exploring Iran's oil and gas reserves. For example, in 2008, a joint venture comprised of India's OVL – the overseas arm of ONGC, IOC, and OIL – discovered a massive gas field, with recoverable reserves of 12.8 trillion cubic feet, which was subsequently named

---

[47] "China Pulls out of South Pars Project: Report," *Reuters* (July 29, 2012), available at www.reuters.com/article/2012/07/29/iranchina-southpars-idUSL6E8IT0T020120729, accessed February 27, 2018; Zhang Junmian, "Delays Could End CNPC's South Pars Involvement" (August 1, 2012), available at www.china.org.cn/business/2012-08/01/content_26090271.htm, accessed February 27, 2018.

[48] Javier Blas, Carola Hoyos, and Daniel Dombey, "Chinese Begin Petrol Supplies to Iran," *Financial Times* (September 23, 2009), available at www.ft.com/intl/cms/s/0/b858ace8-a7a4-11de-b0ee-00144feabdc0.html#axzz49Cycbkor, accessed July 3, 2015.

[49] For details of Chinese NOCs' investment in oil and gas projects in Iran, see Zhao, *China and India*, 91–93.

[50] "Iran, India Sign Broad Energy Pact," *Oil Daily* (January 7, 2005).

[51] S. Pandian, "The Political Economy of Trans-Pakistan Gas Pipeline Project: Assessing the Political and Economic Risks for India," *Energy Policy* 33, no. 5 (March 2005), 659–670; Shiv Kumar Verma, "Energy Geopolitics and Iran–Pakistan–India Gas Pipeline," *Energy Policy* 35, no. 6 (June 2007), 3280–3301.

Farzad-B gas field, and the joint venture intended to invest $5 billion over a period of seven to eight years.[52]

In addition to their active involvement in Iran's energy industry, China and India have been involved in developing Iran's infrastructure. For instance, China has built metros, dams, and ports in Iran;[53] India is constructing Chahbahar, a port in Iran, and is laying railway tracks to connect this port to Zaranj in Afghanistan. India is also an important participant in the building of the North–South Transport Corridor, which links Indian ports to Iran's port at Bandar Abbas and, in the future, Chahbahar.[54] This corridor will facilitate goods transport from India to Iran, then via rail to the Caspian Sea, and onward to northern Europe and Russia.[55]

Militarily, it has been well-documented that China has been a major military goods supplier to Iran since the 1980s, and played a role in Iran's nuclear program until the late 1990s,[56] and in its missile program in the 2000s.[57] Although China stopped its nuclear-related exports to Iran in the 2000s,[58] its traditional military goods sales to Iran continue. According to the Stockholm International Peace Research Institute's (SIPRI) Arms Transfer Database, during 2002–11, the total value of conventional weapons sold to Iran by China reached $727 million, making China Iran's second-biggest arms supplier, after Russia. During

[52] "OVL, Partners Drop Plans to Develop Oil Field in Iran," *Live Mint* (September 15, 2009), available at www.livemint.com/Companies/n4PcsOMEsf7yBLlYERb5OK/OVL-partners-drop-plans-to-develop-oil-field-in-Iran.html, accessed March 1, 2018.
[53] Scott Harold and Alireza Nader, "China and Iran: Economic, Political, and Military Relations," RAND *Occasional Paper* (2012), 13, available at www.rand.org/content/dam/rand/pubs/occasional_papers/2012/RAND_OP351.pdf, accessed March 1, 2018.
[54] Regine A. Spector, "The North–South Trade Corridor" (Brookings Institution, July 2002), available at www.brookings.edu/research/articles/2002/07/03russia-spector, accessed March 1, 2018.
[55] C. Christine Fair, "India and Iran: New Delhi's Balancing Act," *The Washington Quarterly* 30, no. 3 (Summer 2007), 149.
[56] Evan S. Medeiros, *Reluctant Restraint: The Evolution of Chinese Nonproliferation Policies and Practices 1980–2004* (Stanford, CA: Stanford University Press, 2007), 131–174; Bates Gill, *Rising Star: China's New Security Diplomacy* (rev. edn), (Washington, D.C.: Brookings Institution Press, 2010), 79–80.
[57] Shirley A. Kan, *China and Proliferation of Weapons of Mass Destruction and Missiles: Policy Issues* (Congressional Research Service report, March 11, 2013), 18–19, available at www.fas.org/sgp/crs/nuke/RL31555.pdf, accessed August 1, 2013.
[58] Medeiros, *Reluctant Restraint*, 58–65.

the same period, Iran was China's second-largest arms importer, after Pakistan. Furthermore, between 2008 and 2011, when Russia's arms exports to Iran declined dramatically, China replaced Russia as the biggest annual arms supplier to Iran.[59]

India has also tried to enhance its military and defense ties with Iran. For instance, in 2001, India and Iran signed an MOU on defense cooperation, focusing on training and an exchange of visits.[60] Following on from this, the aforementioned New Delhi Declaration of January 2003 committed India and Iran to "explore opportunities for cooperation in defense and agreed areas, including training and exchange of visits."[61] In addition, both countries' navies conducted joint exercises. The first one was held in March 2003 in the Arabian Sea, which was "coincident with the expanding U.S. military presence in the Persian Gulf and Arabian Sea in advance of the Iraq invasion."[62] In March 2006, India conducted a second naval exercise with Iran, which took place concurrently with US President Bush's visit to Afghanistan, India, and Pakistan.[63]

In addition to the political, economic, and military measures taken to facilitate their ties with Iran, China and India have tended to employ their diplomatic power to support Iran at both international and regional levels. At the international level, for instance, China and India utilized their membership of the Board of Governors at the IAEA, the UN atomic watchdog, to postpone Western countries' efforts to refer the Iranian nuclear issue to the UNSC. More specifically, from 2003, when Iran's secret nuclear program was revealed, the United States tried to seek China's and India's cooperation at the IAEA in order to refer the Iranian nuclear issue to the UNSC for sanctions.[64] However, China and India strongly opposed the US approach and steadfastly insisted that this issue should be resolved within the IAEA framework through negotiations, rather than being referred to the UNSC, a stance declared by China to be "a fundamental position of ours on this issue,"

---

[59] Stockholm International Peace Research Institute (SIPRI), *SIPRI Arms Transfer Database*, available at http://arms trade.sipri.org/armstrade/page/values.php, accessed November 17, 2012.

[60] Harsh V. Pant, "A Fine Balance: India Walks a Tightrope between Iran and the United States," *Orbis* 51, no. 3 (2007), 498.

[61] MEA, "The New Delhi Declaration."   [62] Fair, "India and Iran," 151.

[63] Ibid.

[64] Kan, *China and Proliferation of Weapons of Mass Destruction and Missiles*, 9.

Energy Diplomacy: Proactive and Reactive   119

Table 4.1 *China's and India's votes on the resolutions on the Iranian nuclear issue at the IAEA, 2003–2012*

Date of resolution	China's votes	India's votes
September 12, 2003	Adopted without voting	
November 26, 2003	Against	Not-for
March 13, 2004	Against	Not-for
June 18, 2004	Against	Not-for
September 18, 2004	Against	Not-for
November 29, 2004	Against	Not-for
August 11, 2005	Against	Not-for
September 24, 2005	Abstention	For
February 4, 2006	For	For
November 27, 2009	For	For
November 18, 2011	For	For
September 13, 2012	For	For

Note: Voting records are not available for some resolutions; it is not clear whether or not India voted against the resolutions or abstained its votes, so I use "Not-for" to represent these two possibilities, namely "against" or "abstention."
Source: IAEA, "IAEA and Iran: IAEA Resolutions," 2003–12, available at www.iaea.org, accessed November 23, 2012

in a statement made by Chinese ambassador Zhang Yan at the Board of Governors meeting at the IAEA on September 18, 2004.[65] In practice, China had consistently voted against any resolutions adopted by the IAEA on the Iranian nuclear issue before its abstention on September 24, 2005, while India had either voted against the resolution or abstained from voting (see Table 4.1). Moreover, China and India had previously opposed sanctions against Iran. For instance, Jiang Yu, a spokesperson of the Ministry of Foreign Affairs (MOFA), clearly stated, "We believe that imposition of sanctions and pressure is neither a solution nor conducive to the diplomatic efforts to solve the Iranian

[65] "Statement by Ambassador Zhang Yan on Iranian Nuclear Issue," Board of Governors Meeting, IAEA (September 18, 2004), available at www.chinesemission-vienna.at/eng/hyyfy/t181914.htm, accessed March 1, 2018.

nuclear issue."[66] In July 2010, against the backdrop of the UNSC imposing the fourth round of sanctions against Iran on June 9, in a speech at a seminar in New Delhi, jointly organized by India's Institute for Defence Studies and Analyses and Iran's Institute of Political and International Studies, Nirupama Rao, the then Indian Foreign Secretary, stated straightforwardly that "Iran is a country extremely important to India from the perspective of energy security," expressing India's opposition to further sanctions against Iran because those sanctions "can have a direct and adverse impact on Indian companies and more importantly, on our energy security and our attempts to meet the development needs of our people."[67]

At the regional level, in July 2005, together with India and Pakistan, Iran was granted observer status at the SCO,[68] which was regarded as "another dimension of China's effort to counter US efforts to isolate Tehran."[69] In the same vein, India supported Iran's entry into the South Asian Association for Regional Cooperation (SAARC) as an observer in 2007, so as to deepen its bilateral link with Iran.[70]

Through these efforts, China's and India's energy ties with Iran, despite the escalation of the Iranian nuclear crisis between 2003 and 2015, were boosted dramatically, which is best illustrated by their bilateral crude oil trade. According to the IEA's *Oil Market Report* (*OMR*) issued in February 2012, China's imports from Iran had been growing steadily up until 2011, and China was then Iran's largest crude oil importer, accounting for one-fifth of Iranian oil exports.[71] On China's side, Iran was China's third-largest crude oil supplier after

---

[66] MOFA, "Foreign Ministry Spokesperson Jiang Yu's Regular Press Conference on September 24, 2009" (September 25, 2009), available at www.china-un.org/eng/fyrth/t606680.htm, accessed March 1, 2018.

[67] Nirupama Rao, "Speech at IDSA–IPIS Strategic Dialogue on India and Iran: An Enduring Relationship" (Institute for Defence Studies and Analyses, July 5, 2010), available at www.idsa.in/KeynoteAddressIndiaandIrananenduringrelationship_nirupamaroy, accessed March 1, 2018.

[68] "上海合作组织 [Shanghai Cooperation Organization]," 新华网 [*Xinhuanet*] (June 1, 2006), available at http://news.xinhuanet.com/ziliao/2002-06/01/content_418824.htm, accessed March 1, 2018.

[69] John Garver, "China–Iran Relations: Cautious Friendship with America's Nemesis," *China Report* 49, no. 1 (February 2013), 84.

[70] Satyajit Mohanty, "SAARC Observer Status for Iran: Regional Implications," available at www.ipcs.org/article/india-the-world/saarc-observer-status-for-iran-regional-implications-2261.html, accessed March 1, 2018.

[71] IEA, *Oil Market Report* (Paris: OECD/IEA, February 10, 2012), 18.

Saudi Arabia and Angola, accounting for nearly 11 percent of China's total oil imports in 2011.[72] When, in 2011, the United States' economic sanctions against Iran made it difficult to trade oil, China paid for Iranian oil with its goods, a kind of barter trade between the two countries.[73] Similarly, India had become Iran's second-largest oil importer, after China. In 2008, Iranian oil imports accounted for 16 percent of India's total imports. Even in 2011, India still imported 350,000 bpd from Iran.[74] When faced with the negative impact of US and EU sanctions on the Iranian oil sector, India "reached an agreement with Iran to pay for 45% of its crude purchases in rupees."[75] Like China, India had also conducted a barter trade with Iran. For instance, India paid for its oil imports from Iran through the sale of wheat, pharmaceuticals, rice, sugar, soybeans, and other products to Iran.[76]

In sum, despite the efforts of Western countries, especially the United States, to isolate and impose sanctions against Iran, China and India stepped up their efforts to strengthen their energy ties with Iran through political, economic, military, as well as diplomatic measures. As a result, together with India, China once became "the only major player still active in the Iranian oil patch."[77]

Reactive Measures

As noted above, China and India have tried all means at their disposal to forge closer energy ties with Iran. In recent years, however, the Chinese and Indian governments and their NOCs have taken some reactive measures that obviously run counter to their efforts to strengthen energy relations with Iran.

---

[72] EIA, "China" (September 4, 2012), 9, available at www.eia.gov/countries/analysisbriefs/China/china.pdf, accessed November 20, 2012.
[73] Najmeh Bozorgmehr, Anna Fifield, and Leslie Hook, "China and Iran Plan Oil Barter," *Financial Times* (July 24, 2011), available at www.ft.com/cms/s/0/2082e954-b604-11e0-8bed-00144feabdc0.html#axzz276JYqcID, accessed September 15, 2012.
[74] IEA, *Oil Market Report*, 18.   [75] Ibid.
[76] Kenneth Katzman, "Iran Sanctions," *Current Politics and Economics of the Middle East* 4, no. 2 (July 2013), 238.
[77] Erica Downs and Suzanne Maloney, "Getting China to Sanction Iran: The Chinese-Iranian Oil Connection," *Foreign Affairs* 90, no. 2 (March/April 2011), 15.

First and most importantly, China not only gave up its "fundamental position" on how to resolve the Iranian nuclear issue, but also voted against Iran at both the IAEA and the UNSC on this issue. More specifically, although China at first voted against any IAEA resolutions on the Iranian nuclear issue and insisted that it should be resolved under the framework of the IAEA, since its abstention on September 24, 2005, China has consistently voted in favor of all resolutions on this issue, including one on February 4, 2006, which eventually referred it to the UNSC.[78] Since then, moreover, China has consistently sided with the other permanent members of the UNSC in voting for the resolutions on this issue (see Table 4.2), which has enabled the UNSC to impose multiple rounds of escalating tough sanctions against Iran.[79]

Like China, India has more recently sided with the United States and other big powers at the IAEA to vote against Iran on the Iranian nuclear issue. As Table 4.1 shows, India has voted in favor of the resolutions on the Iranian nuclear issue at the IAEA since September 24, 2005, which led to criticism and a warning from Iran. Specifically, an Iranian Foreign Ministry spokesman said that "India's vote came as a great surprise to us," and warned India that "We will reconsider our economic cooperation with those countries that voted against us."[80] This warning came at a time when India and Iran were in the middle of the process of negotiating the IPI gas pipeline. India's vote also provoked domestic criticism from both the opposition parties and the Singh government's Communist allies.[81] Despite Iran's warnings and domestic criticisms, India voted against Iran five times in total at the IAEA Board of Governors, from September 2005 to September 2012 (see Table 4.1).

In addition, Chinese and Indian diplomatic rhetoric ran against Iran's nuclear ambitions. Specifically, Chinese leaders made it clear that China opposes nuclear proliferation and even sought to persuade Iran to cooperate with the international community to solve this issue.

---

[78] Permanent Mission of the PRC in Vienna, "Statement by Ambassador Wu Hailong at the IAEA Board Meeting Regarding the Iranian Nuclear Issue" (September 24, 2005), available at www.chinesemission-vienna.at/eng/hyyfy/t223174.htm, accessed March 1, 2018.
[79] Downs and Maloney, "Getting China to Sanction Iran," 15–21.
[80] BBC, "Iran Warns India on Nuclear Vote" (September 27, 2005), available at http://news.bbc.co.uk/2/hi/south_asia/4285868.stm, accessed May 20, 2013.
[81] Ibid.

Table 4.2 UN Security Council resolutions on the Iranian nuclear issue, 2006–2012

Resolution	1696	1737	1747	1803	1835	1887	1929	1984	2049
Date	July 31, 2006	Dec. 23, 2006	Mar. 24, 2007	Mar. 3, 2008	Sept. 27, 2008	Sept. 24, 2009	June 9, 2010	June 9, 2011	June 7, 2012
For	China France Russia Britain USA	China France Russia Britain USA	China France Russia Britain USA	China France Russia Britain USA	China France Russia Britain USA	China France Russia Britain USA	China France Russia Britain USA	China France Russia Britain USA	China France Russia Britain USA
	+ 9	+ 10	+ 10	+ 9	+ 10	+ 10	+ 7	+ 9	+ 10
Against	0	0	0	0	0	0	Brazil Turkey	0	0
Abstention	Qatar	0	0	Indonesia	0	0	Lebanon	Lebanon	0

Source: The United Nations Security Council, "Security Council Resolutions," available at www.un.org/documents/scres.htm, accessed November 23, 2012

For instance, Chinese Premier Wen Jiabao reiterated, during his six-day Middle East tour in January 2012, that China "adamantly opposes Iran developing and possessing nuclear weapons."[82] In the same vein, when Chinese President Hu Jintao held a meeting with Iranian President Ahmadinejad, in June 2012 at the SCO summit in Beijing, Hu told his counterpart, "China hopes the Iranian side can weigh up the situation, take a flexible and pragmatic approach, have serious talks with all six related nations, and enhance dialogues and cooperation with the International Atomic Energy Agency so as to ensure the tensions can be eased through negotiations."[83] Similarly, Indian leaders employed diplomatic rhetoric to express India's non-favor policy toward Iran's nuclear ambitions. For instance, the then Indian Prime Minister Manmohan Singh stated clearly at the annual India–EU summit in September 2008: "We don't support a nuclear weapon state emerging in our region. So there is no question of supporting nuclear weapon ambitions of Iran."[84] Singh repeated this rhetoric during his attendance at the Nuclear Summit in April 2010 in Washington, D.C.: "we do not favour Iran's nuclear weapons ambitions."[85] By the same token, in January 2015, Indian Prime Minister Modi, jointly with the visiting US President Obama, emphasized "the criticality of Iran taking steps to verifiably assure the international community of the exclusively peaceful nature of its nuclear programme."[86]

Moreover, China and India have also taken concrete measures to reduce their close energy ties with Iran, which is evidenced not only by

[82] Michael Wines, "China Leader Warns Iran Not to Make Nuclear Arms," *New York Times* (January 20, 2012), available at www.nytimes.com/2012/01/21/world/asia/chinese-leader-wen-criticizes-iran-on-nuclear-program.html?_r=0, accessed June 25, 2012.

[83] "Chinese President Advocates Iranian Nuclear Dialogues," *People's Daily Online* (June 8, 2012), available at http://english.people.com.cn/90883/7840518.html, accessed March 1, 2018.

[84] "PM's Clear Message: India Won't Allow a Nuclear Iran" (September 29, 2008), available at www.ibnlive.com/news/india/pm-on-france-nuke-pm-read-pm-on-iran-298314.html, accessed June 25, 2012.

[85] "Sanctions Not Answer to Iran's Nuclear Ambitions: Manmohan," *The Hindu* (April 14, 2010), available at www.thehindu.com/news/national/sanctions-not-answer-to-irans-nuclear-ambitions-manmohan/article396742.ece, accessed March 6, 2018.

[86] The White House, "U.S.–India Joint Statement: 'Shared Efforts, Progress for All'" (January 25, 2015), available at www.whitehouse.gov/the-press-office/2015/01/25/us-india-joint-statement-shared-effort-progress-all, accessed March 1, 2018.

the reduction of both countries' oil imports from Iran, but also by Chinese and Indian NOCs' more cautious activities in the Iranian energy sector. According to the EIA, China's total oil imports in 2013 reached record highs, but its imports from Iran actually dropped to 429,000 bpd, from 439,000 bpd in 2012 and 555,000 bpd in 2011. Accordingly, "Iran fell to the sixth-largest crude oil exporter to China behind Saudi Arabia, Angola, Oman, Russia, and Iraq, and constituted 8% of China's crude oil imports in 2012 and 2013 compared to 11% in 2011."[87] Moreover, according to the US Congressional Research Service (CRS) report on Iranian sanctions issued in August 2015, "during 2012–13 China cut its buys of oil from Iran to about 435,000 barrels per day from its 2011 average of about 550,000 barrels per day."[88] In India's case, according to the same CRS report, "India has reduced its imports of Iranian oil substantially since 2011."[89] Specifically, India imported 320,000 bpd in 2011 on average, but by the time of the start of the Joint Plan of Action (JPA) in November 2013, India's oil imports had dropped to 190,000 bpd, which implies that "by the time of the JPA, Iran was only supplying about 6% of India's oil imports (down from over 16% in 2008)."[90] By the end of 2015, Iran had dropped from India's second-largest oil supplier to its seventh.[91]

Another concrete move taken by India to reduce its energy ties with Iran occurred in December 2010, when the Reserve Bank of India declared that a regional clearinghouse – the Asian Clearing Union (ACU) – could no longer be used to settle energy trade transactions. The ACU, established by the UN in 1947, had once been the primary channel through which Iran and India completed oil transactions.[92] This move played a significant role in tightening the web of sanctions

---

[87] EIA, "China" (May 14, 2015), 8.
[88] Kenneth Katzman, *Iran Sanctions* (Congressional Research Service report, August 4, 2015), 36, available at www.fas.org/sgp/crs/mideast/RS20871.pdf, accessed August 26, 2015.
[89] Ibid., 37.   [90] Ibid., 22, 37.
[91] Nidhi Verma, "India's 2015 Iran Oil Imports Fall by a Quarter – Trade," *Reuters* (January 7, 2016), available at http://in.reuters.com/article/india-iran-oil-import-idINKBN0UL0NV20160107, accessed March 1, 2018.
[92] "US Welcomes India Bid to Restrict Trade with Iran," *The Indian Express* (December 30, 2010), available at http://indianexpress.com/article/news-archive/web/us-welcomes-india-bid-to-restrict-trade-with-iran, accessed March 1, 2018.

against Iran by barring Indian companies from a range of deals, including oil transacted through the ACU.[93]

In an example of Chinese and Indian NOCs' more cautious activities in Iran's energy sector, although CNPC signed the contract to develop Iran's phase 11 of the South Pars natural gas field in 2009, it did not carry it out, as a result of delays and in July 2012, CNPC finally decided to withdraw from this project,[94] terminating its contract with NIOC in April 2013.[95] Although CNPC played a major role in the South and North Azadegan fields in January 2009, to fill the void left by the withdrawal of the Japanese firm Inpex, on April 29, 2014, Iran cancelled the South Azadegan contract with CNPC on the grounds that CNPC had carried out "no effective work" since it took the stake in 2009.[96] Similarly, although CNOOC signed a $16 billion contract with Iran to develop its North Pars gas field, the company did not carry out this project, using the excuse of a lack of funding, which led the Iranian government to demand CNOOC cease involvement in this project in 2011.[97]

Like Chinese NOCs, Indian oil companies became very cautious about their activities in the Iranian energy sector. As mentioned above, India reached an energy agreement with Iran in 2005, but this agreement remained largely on paper. Although OVL and another two

---

[93] Jay Solomon and Subhadip Sircar, "India Joins U.S. Effort to Stifle Iran Trade," *Wall Street Journal* (December 29, 2010), www.wsj.com/articles/SB10001424052970203513204576046893652486616, accessed March 1, 2018.

[94] Judy Hua and Chen Aizhu, "China's July Oil Imports from Iran Fall 28 Percent on Month," *Reuters* (August 21, 2012), available at http://in.reuters.com/article/2012/08/21/us-china-oil-iran-idINBRE87K0LK20120821, accessed March 1, 2018; Yeganeh Salehi, "CNPC Withdraws from South Pars Gas Field Project, Shargh Reports," *Bloomberg* (July 31, 2012), available at www.bloomberg.com/news/2012-07-30/cnpc-withdraws-from-south-pars-gas-field-project-shargh-reports.html, accessed March 1, 2018.

[95] Mehr News Agency, "Iran, China's CNPC Terminate Contract in South Pars" (April 23, 2013), available at https://en.mehrnews.com/news/54942/Iran-China-s-CNPC-terminate-contract-in-South-Pars, accessed March 1, 2018.

[96] "Iran–CNPC Breakup: Tehran Eyes the West," *Christian Science Monitor* (May 5, 2014), cited in Katzman, *Iran Sanctions* (August 4, 2015), 53.

[97] "中海油伊朗项目被叫停 三油企海外并购喜忧并存 [CNOOC's Iranian Project Was Demanded to Stop: Joys Accompanied with Concerns for Three NOCs' Overseas Mergers and Acquisitions]," 《证券日报》 [*Securities Daily*] (October 17, 2011), available at http://finance.jrj.com.cn/industry/2011/10/17024911302678.shtml, accessed March 1, 2018.

Indian oil companies discovered the Farzad-B natural gas field mentioned above, these Indian companies delayed and ultimately relinquished the development of this gas field. In addition, India withdrew from the IPI gas pipeline.[98] Reliance Industries of India, an Indian refinery company, had stopped not only selling gasoline to Iran since January 2009, but also buying crude oil from Iran since February 2010, which lasted until April 2016, after the international sanctions against Iran were partially lifted.[99]

Driving Forces Behind China's and India's Reactive Measures

Given China's and India's high dependence on overseas energy supplies, it is reasonable and rational for both countries to take various measures to strengthen their ties with an energy-rich country such as Iran. In other words, China's and India's domestic energy demand requires them to forge closer relations with Iran. Under such circumstances, the driving forces behind China's and India's seemingly "irrational" reactive policy toward Iran stems from some external factors. I argue that it is pressures from the United States and its allies that have driven China and India to change their policy toward Iran. To test the causal mechanism between the pressures exerted by the United States and its allies on China and India on the Iranian nuclear issue, and the reactive measures taken by China and India under those pressures, I use the method of "process tracing"[100] to empirically manifest the process of how the pressures exerted on China and India led them to shift their policy toward Iran. As we will see, process tracing proves that China's and India's reactive measures were taken as a result of their willingness to accommodate the United States and its allies' policy preference toward Iran – namely, to impose sanctions on Iran – rather than their own preference to strengthen their energy ties with Iran.

---

[98] "India/Iran: India Seeks 'Auspicious Re-Birth' in Iran's Energy Sector," *Asia News Monitor* (August 25, 2015).

[99] "Reliance Buys Iranian Oil after Six-Year Hiatus," *The Economic Times* (April 24, 2016), available at http://economictimes.indiatimes.com/industry/energy/oil-gas/reliance-buys-iranian-oil-after-six-year-hiatus/articleshow/51963025.cms, accessed March 1, 2018.

[100] Gary King, Robert O. Keohane, and Sidney Verba, *Design Social Inquiry: Scientific Influence in Qualitative Research* (Princeton, NJ: Princeton University Press, 1994), 226.

The United States and its Allies' Pressures on China's Iran Policy at the IAEA and UNSC

It is clear that persuasion and criticism from the United States led China to change its voting behavior at the IAEA on the Iranian nuclear issue. More specifically, on September 13, 2005, President George W. Bush, at a bilateral meeting with Chinese President Hu Jingtao on the sidelines of a UN summit in New York, tried to persuade Hu not to block the IAEA from referring the Iranian nuclear issue to the UNSC, and asked for Hu's cooperation to prevent Iran from developing nuclear weapons.[101] On September 21, 2005, Robert Zoellick, the US Deputy Secretary of State, in his remarks to the National Committee on US–China Relations, not only directly criticized China's mercantilist ways in seeking energy around the world, but also linked its stance on Iran's nuclear program with the seriousness of its commitment to non-proliferation and its reputation as "a responsible stakeholder."[102] Three days later, on September 24, 2005, China abstained from voting on the resolution adopted at the IAEA (see Table 4.1),[103] which was regarded by the United States as progress in China's policy attitudes toward Iran.[104]

In addition, increased pressures from the United States led China to side with the United States and other big powers in voting for the resolutions on the Iranian nuclear issue adopted at the IAEA and the UNSC from 2006 onwards (see Tables 4.1 and 4.2). Specifically, on January 24–26, 2006, Zoellick visited Beijing to stress the importance of the Iran problem, and convinced China to support the USA and its

---

[101] William Roberts, "Bush Seeks Hu's Help on Iran, Accepts Invitation to Visit China," *Bloomberg* (September 13, 2005), available at www.bloomberg.com/apps/news?pid=newsarchive&sid=advaqdBGetHU&refer=us, accessed November 21, 2011.

[102] Robert B. Zoellick, "Whither China: From Membership to Responsibility?" (September 21, 2005), available at www.ncuscr.org/files/2005Gala_RobertZoellick_Whither_China1.pdf, accessed September 21, 2009.

[103] IAEA, "Implementation of the NPT Safeguards Agreement in the Islamic Republic of Iran: Resolution Adopted on 24 September 2005" (September 24, 2005), available at www.iaea.org/Publications/Documents/ Board/2005/gov2005-77.pdf, accessed March 10, 2012.

[104] Anna Langenbach, Lars Olberg, and Jean DuPreez, "The New IAEA Resolution: A Milestone in the Iran–IAEA Saga," Nuclear Threat Initiative (November 2005), available at www.nti.org/e_research/e3_69a.html, accessed March 2, 2018.

Western allies in their policy toward Iran.[105] Barely ten days after Zoellick's visit, on February 4, 2006, China voted for the resolution adopted at the IAEA Board of Governors, which eventually referred the Iranian nuclear issue to the UNSC. On April 18–21, 2006, Chinese President Hu Jintao paid a state visit to Washington. During the summit, Bush urged Hu to use China's relationship with Iran to persuade it to give up its nuclear ambitions and comply with its international obligations.[106] After the summit, on June 1, 2006, at Bush's invitation, Hu and Bush held a telephone conversation to further discuss the Iranian nuclear issue, in which Hu told Bush that China would "play a constructive role" in addressing this issue.[107] One month later, on July 31, 2006, China voted for the UNSC Resolution 1696, the first resolution adopted at the UNSC on the Iranian nuclear issue, which demanded that Iran suspend uranium enrichment by August 31, 2006; otherwise, it would face potential economic and diplomatic sanctions.[108]

In September 2009, US President Barack Obama, in his discussion of the Iranian nuclear issue with Hu Jintao on the sidelines at the UN General Assembly summit, told Hu that the Iranian nuclear issue was "a core national interest" for the United States.[109] Before Obama paid a state visit to Beijing on November 15–19, 2009, two senior US National Security Council officials visited Beijing in person to press home the United States' concerns about Iran's nuclear program. During the summit in Beijing, Obama "promised 'consequences'," if

---

[105] "Top US Official Begins China Visit," *China Daily* (January 24, 2006), available at www.chinadaily.com.cn/english/doc/2006-01/24/content_515046.htm, accessed March 2, 2018.

[106] Bonnie S. Glaser, "U.S.–China Relations: Pomp, Blunders, and Substance: Hu's Visit to the U.S.," *Comparative Connections* 8, no. 2 (2006), 4.

[107] Ministry of Foreign Affairs of PRC, "President Hu Jintao Holds a Telephone Conversation with President Bush" (June 1, 2006), available at www.fmprc.gov.cn/eng/zxxx/t256401.htm, accessed November 14, 2012.

[108] UNSC, "SC/8792: Security Council Demands Iran Suspend Uranium Enrichment by 31 August, or Face Possible Economic, Diplomatic Sanctions" (July 31, 2006), available at www.un.org/press/en/2006/sc8792.doc.htm, accessed March 2, 2018.

[109] John Pomfret and Joby Warrick, "China's Backing on Iran Followed Dire Predictions," *Washington Post* (November 26, 2009), available at www.washingtonpost.com/wp-dyn/content/article/2009/11/25/AR2009112504112.html, accessed March 2, 2018.

Iran failed to comply with international non-proliferation efforts.[110] Before the end of the summit, a joint statement was issued, in which both sides stated that they had reached a consensus on how to address the Iranian nuclear issue.[111] Consequently, on November 27, 2009, China voted for the IAEA resolution on "censuring Iran's nuclear enrichment facility at Qom," an action that was regarded by US officials as "a direct result" of Obama's visit.[112]

Not only the United States, but also its allies, following requests made by the former, have applied pressure on China. Saudi Arabia, one of the United States' allies in the Middle East and China's largest oil supplier, is a case in point. More specifically, in mid-December 2009, the United States initiated at the UNSC a further resolution to impose tougher and more comprehensive sanctions against Iran, which was strongly opposed by China. Against this backdrop, the United States tried to enlist Saudi Arabia's influence on Beijing. To do so, the US Secretary of State, Hillary Clinton, in mid-February 2010, and Secretary of Defense, Robert Gates, in early March 2010, visited Saudi Arabia to request the Saudis to persuade China to support US efforts at the UNSC.[113] In response to their request, Prince Saud al-Faisal, the Saudi foreign minister, visited Beijing in early March 2010, to discuss the issue, applying implicit pressure on China through his rhetoric in an interview in mid-March 2010: "China is perfectly aware of the scope of its responsibilities and its obligations, including in the position it holds on the international stage and as a permanent member of the [UN] Security Council."[114] Under the indirect pressure applied

---

[110] Edward Luce and Geoff Dyer, "Obama Visit Yields Few Concrete Results," *Financial Times* (November 18, 2009), available at www.ft.com/cms/s/0/cb35fe02-d372-11de-9607-00144feabdc0.html#axzz1C0KoXGrT, accessed June 22, 2012.

[111] Ministry of Foreign Affairs of PRC, "China–US Joint Statement" www.fmprc.gov.cn/eng/wjdt/2649/t629497.htm, accessed June 22, 2012.

[112] Helene Cooper and William J. Broad, "Russia and China Endorse Agency's Rebuke of Iran," *New York Times* (November 27, 2009), available at www.nytimes.com/2009/11/28/world/28nuke.html, accessed March 2, 2018.

[113] Mark Landler, "Clinton Raises U.S. Concerns of Military Power in Iran," *New York Times* (February 15, 2010), available at www.nytimes.com/2010/02/16/world/middleeast/16diplo.html?_r=0, accessed March 2, 2018; Jim Garamone, "U.S. Seeks Saudis' Help for U.N. Sanctions on Iran" (March 10, 2010), available at http://archive.defense.gov/news/newsarticle.aspx?id=58271, accessed March 2, 2018.

[114] AFP, "China 'Knows its Duties' in Iran Nuclear Tussle: Saudi" (March 15, 2010), available at www.iranfocus.com/en/index.php?option=com_content&

by the United States through Saudi Arabia, in March 2010, China shifted its opposition and agreed to the US initiative to adopt another resolution at the UNSC to impose further sanctions against Iran.[115]

Despite this, China still tried to postpone the adoption of such a resolution at the UNSC.[116] As a result, the United States continued to apply pressure on China. For instance, during the Nuclear Security Summit in April 2010, Obama continued to urge Beijing to support US efforts to impose further sanctions against Iran at the UNSC. Hu Jintao replied that "China and the United States shared the same overall goal,"[117] which implies that China would support US policy. Consequently, two months later, in June 2010, China voted for the UNSC Resolution 1929, after it had intentionally delayed the adoption of such a resolution by as long as six months, and successfully watered down some harsher terms that the United States and its allies intended to include.[118] This resolution authorized the UNSC to impose the fourth round of sanctions against Iran over its suspected nuclear weapons program. Later, Alaeddin Boroujerdi, chief of the Iranian Parliament's National Security and Foreign Policy Commission, publicly criticized China (and Russia) for being influenced by the US measures.[119]

The United States and its Allies' Pressures on India's Iran Policy at the IAEA

India's five votes against Iran at the IAEA were also a corollary of pressures applied by the United States and its allies. More specifically, India's first-ever vote against Iran took place against the backdrop of

view=article&id=19909:china-knows-its-duties-in-iran-nuclear-tussle-saudi& catid=8&Itemid=113, accessed March 2, 2018.
[115] Garver, "China–Iran Relations," 84; Parris H. Chang, "China's Policy toward Iran and the Middle East," *The Journal of East Asian Affairs* 25, no. 1 (Spring/Summer 2011), 8–9.
[116] Harvey Morris, "China Bolsters US Push for Iran Sanctions," *Financial Times* (April 9, 2010), available at www.ft.com/cms/s/0/53163bd0-43f7-11df-9235-00144feab49a.html#axzz27TLtlBLU, accessed June 22, 2012.
[117] Steve Holland and Ross Colvin, "Obama, China Discuss Iran at Nuclear Summit," *Reuters* (April 12, 2010), available at www.reuters.com/article/idUSTRE63B0PZ20100412, accessed March 2, 2018.
[118] Garver, "China–Iran Relations," 84.
[119] "Iran Dismisses UN Resolution, Threatening Counter Measures," *Xinhuanet* (June 10, 2010), available at http://news.xinhuanet.com/english2010/world/2010-06/10/c_13342230.htm, accessed June 25, 2012.

intense efforts by George W. Bush's administration to secure the US Congressional passage of the Indo-US civil nuclear deal that was announced in the joint statement issued by Bush and the visiting Indian Prime Minister Manmohan Singh on July 18, 2005.[120] Since then, the United States has, on several occasions, applied pressure on India by linking its voting behavior at the IAEA with the fate of the US–India civil nuclear deal in the US Congress. For instance, on September 8, 2005, the US Congress held a hearing entitled "The U.S.–India: An Emerging Entente?," with the Indo-US nuclear deal as a main focus. At the hearing, several congressmen reminded India that its policy toward Iran would directly impact the fate of the Indo-US civil nuclear agreement in the US Congress. Tom Lantos, the late democratic Congressman, warned India that:

There is a degree of reciprocity ... we expect of India, which has not been forthcoming ... we agree to undertake a tremendous range of path-breaking measures to accommodate India, while India blithely pursues what it sees should be its goal and policy vis-à-vis Iran. There is a quid pro quo in international relations. And if our Indian friends are interested in receiving all of the benefits of U.S. support ... we have every right to expect that India will reciprocate in taking into account our concerns ... and without reciprocity, India will get very little help from the Congress.[121]

Faced with the pressures of this explicit issue-linkage, on September 24, 2005, India voted for the resolution at the IAEA – the first time that India had voted against Iran. Right after this vote, the United States expressed its appreciation to India via the White House spokesman, Scott McClellan: "We appreciate the support. The world is saying to Iran that it is time to come clean."[122]

---

[120] White House Press Release, Office of the Press Secretary, "Joint Statement by President George W. Bush and Prime Minister Manmohan Singh" (July 18, 2005), available at http://2001-2009.state.gov/p/sca/rls/pr/2005/49763.htm, accessed March 2, 2018; MEA, "Civil Nuclear Cooperation," available at www.mea.gov.in/in-focus-topic.htm?24/Civil+Nuclear+Cooperation, accessed June 20, 2012. For research on this topic, see Asheley J. Tellis, "What Should We Expect from India as a Strategic Partner?" in Henry Sokolski (ed.), *United States and India Strategic Cooperation* (New York: Nova Science Publishers, Inc., 2010), 125–137.

[121] Cited in R. Ramachandran, "Iran Policy Was Key to Nuclear Deal with U.S.," *The Hindu* (September 30, 2005), available at www.thehindu.com/2005/09/30/stories/2005093008341200.html, accessed June 21, 2012.

[122] BBC, "Iran Warns India on Nuclear Vote."

India's second vote against Iran on February 4, 2006 at the IAEA Board of Governors, which eventually referred the Iranian nuclear issue to the UNSC, due to concerns over Iran's uranium enrichment program, was also a reactive measure, resulting from pressures applied on India by both the United States and its ally, Saudi Arabia. More specifically, this vote took place against the backdrop of the US Congress debate over the Henry J. Hyde United States–India Peaceful Atomic Energy Cooperation Act (or the Hyde Act), in January 2006. During this debate, India was warned once again, by several congressmen, that the deal hinged on India's support for America's Iran policy.[123] Moreover, the Act's text unambiguously made the US export of nuclear fuel or technology to India conditional on the latter's cooperation with the United States on the Iranian nuclear issue, which was written into Section 103 (b)(4) of this Act as, "Secure India's full and active participation in United States efforts to dissuade, isolate, and, if necessary, sanction and contain Iran for its efforts to acquire weapons of mass destruction, including a nuclear weapons capability and the capability to enrich uranium or reprocess nuclear fuel, and the means to deliver weapons of mass destruction."[124] Also in January 2006, Saudi Arabia's King Abdullah visited India and attended India's Republic Day parade as the chief guest. During this visit, Singh and Abdullah signed the "Delhi Declaration," in which both sides announced their intention to "develop a strategic energy partnership." Through this partnership, Saudi promised to provide India with "reliable, stable and increased volume of crude oil supplies, through 'evergreen' long-term contracts," as well as strengthening their bilateral energy cooperation and investments in both oil and gas sectors.[125] Moreover, just a week before the IAEA's meeting to discuss Iran's nuclear program, David C. Mulford, then US ambassador to India, warned India that if it did not vote along with the United States at the IAEA, the effect on the Indo-US nuclear deal would be "devastating,"

---

[123] Uma Purushothaman, "American Shadow over India–Iran Relations," *Strategic Analysis* 36, no. 6 (November 2012), 904.
[124] Authenticated U.S. Government Information, GPO, *An Act: United States and India Nuclear Cooperation*, 4, available at www.gpo.gov/fdsys/pkg/BILLS-109hr5682enr/pdf/BILLS-109hr5682enr.pdf, accessed March 2, 2018.
[125] "Delhi Declaration, Signed by King Abdullah bin Abdulaziz Al Saud of the Kingdom of Saudi Arabia and Prime Minister Dr. Manmohan Singh of India" (January 27, 2006), 2, available at www.indianembassy.org.sa/WebFiles/Delhi%20Declaration.pdf, accessed February 10, 2016.

as the US Congress would stop considering the matter, so the nuclear deal would die.[126] With the fate of the Indo-US civil nuclear deal at risk in the US Congress, India had to vote with the United States against Iran for the second time at the IAEA.

India's third vote against Iran at the IAEA, on November 27, 2009, was also subject to US pressures, especially those stemming from the bilateral summit. More specifically, on November 22–24, 2009, Indian Prime Minister Singh paid a state visit to the United States, with US President Obama hosting the first state dinner of his presidency for Singh. In their bilateral joint statement, it was stated that:

Prime Minister Singh and President Obama reaffirmed their shared vision of a world free of nuclear weapons and pledged to work together, as leaders of responsible states with advanced nuclear technology, for global non-proliferation, and universal, non-discriminatory and complete nuclear disarmament. Part of that vision is working together to ensure that all nations live up to their international obligations.[127]

This implicitly stated that India would support US policy toward the Iranian nuclear issue at the IAEA. Unsurprisingly, three days after Singh's visit, India voted along with the United States against Iran for the third time.

India's fourth vote against Iran at the IAEA, on November 18, 2011 was also subject to pressures from a summit meeting with the United States. More specifically, the timing of this vote coincided with the meeting of Indian Prime Minister Singh and US President Obama on the sidelines of the East Asia summit in Bali, on November 18, 2011. During the meeting, Singh advocated a diplomatic approach by continuing to engage with Iran, rather than resorting to confrontation.[128] However, as before, India's diplomats voted

---

[126] "World in Brief: U.S.–India Nuclear Deal Could Die, Envoy Warns," *Washington Post* (January 26, 2006), available at www.washingtonpost.com/archive/politics/2006/01/26/world-in-brief/c04faf51-3e20-4a40-b2b8-04f210a20413, accessed March 2, 2018.

[127] The White House, "India and the United States: Partnership for a Better World," Joint Statement between Prime Minister Dr. Singh and President Obama (November 24, 2009), available at www.whitehouse.gov/the-press-office/joint-statement-between-prime-minister-dr-singh-and-president-obama, accessed September 1, 2015.

[128] Devirupa Mitra, "India Votes against Iran in IAEA," *The New Indian Express* (November 20, 2011), available at www.newindianexpress.com/nation/article246038.ece?service=print, accessed March 2, 2018.

Energy Diplomacy: Proactive and Reactive 135

against Iran at the IAEA on the same day when Singh and Obama held this informal meeting.

Apart from pressures from the United States, India was also faced with pressures from US allies in the Middle East, especially Saudi Arabia and Israel. More specifically, in March 2010, during Singh's state visit to Saudi Arabia, in a declaration signed by Singh and Saudi King Abdullah, both countries jointly "encouraged Iran to respond to those efforts in order to remove regional and international doubts about its nuclear programme, especially as these ensure the right of Iran and other countries to peaceful uses of nuclear energy according to the yardsticks and procedures of International Atomic Energy Agency and under its supervision."[129] In June 2012, an Israeli delegation visited India to persuade the country to accept Israel's assessment of Iran's nuclear capabilities.[130] Under these various pressures, India voted against Iran for the fifth time at the IAEA, on September 13, 2012.

The United States and its Allies' Pressures on China and India to Reduce their Energy Ties with Iran

As mentioned above, China and India have been Iran's two largest oil importers and important investors in the Iranian energy sector, which has indirectly provided Iran with the funds to support its nuclear program, so the United States and its allies have tried to impose various sanctions, targeted at waning Iranian energy revenues. These measures have placed huge pressures on China and India to reduce their close energy ties with Iran.

First, US legal sanctions against the Iranian energy sector have forced China and India to cut their oil imports from Iran. Specifically, the US government sought to reduce Iranian oil income by imposing sanctions against its oil importers. On January 19, 2012, several US legislators wrote to its Treasury Secretary to suggest that the

---

[129] "India, S Arabia for Peaceful Resolution of Iran's N-issue," *Hindustantimes* (March 1, 2010), available at www.hindustantimes.com/world-news/india-s-arabia-for-peaceful-resolution-of-iran-s-n-issue/article1-514206.aspx, accessed March 2, 2018.

[130] Sujan Dutta, "Israel Notice on Iran: Nudge to Delhi to Toe Line on Tehran," *The Telegraph* (May 24, 2012), available at www.telegraphindia.com/1120524/jsp/frontpage/story_15526098.jsp#.VeUZyXlzPIU, accessed March 2, 2018.

Department of the Treasury should define a "significant reduction" as an 18 percent purchase reduction based on total price paid (not just volumes), when determining whether or not to impose sanctions against Iranian oil importers.[131] Under the threat of sanctions, China and India, as well as US allies such as Japan and South Korea, had to cut their oil imports from Iran.[132] In return, the United States granted China and India a sanction waiver on June 28, 2012, renewing this waiver in late 2012 and mid-2013, given the fact that China's and India's imports from Iran dropped significantly, as discussed above. The reason for such a significant drop can hardly be attributed to market forces since, according to the *OPEC Monthly Oil Market Report* issued in January 2014, the Iranian crude oil price fell from $109.06 per barrel in 2012 to $105.73 per barrel in 2013.[133] Under such circumstances, based on the supply–demand rule, China's and India's oil imports from Iran should have risen, or at least stabilized, at the 2011 level, given their high reliance on overseas oil supplies and their record highs of oil imports in 2013.

The Reserve Bank of India forbidding Indian companies to use the ACU to process their transactions with Iran was also a move that was directly reactive to US pressures. According to the *Wall Street Journal* in December 2010, the US Treasury had regularly raised the issue with India for more than a year. US officials had warned India that Indian firms conducting transactions through the ACU ran the risk of violating a law signed by Obama in July 2010 that banned international firms from doing business with 17 Iranian banks and much of Tehran's oil and gas sector, as well as the Revolutionary Guard. If Indian companies were found to be in violation, they could be banned from doing business in the United States.[134] Moreover, in order to win India's support on the Iranian nuclear issue, US President Obama, during his visit to India in November 2010, announced the United States' endorsement of India's bid to become a permanent member on

---

[131] Katzman, *Iran Sanctions* (August 4, 2015), 21.   [132] Ibid., 22.
[133] Organization of the Petroleum Exporting Countries, *OPEC Monthly Oil Market Report* (January 16, 2014), 6, available at www.opec.org, accessed September 10, 2015.
[134] Jay Solomon and Subhadip Sircar, "India Joins U.S. Effort to Stifle Iran Trade," *Wall Street Journal* (December 29, 2010), available at www.wsj.com/articles/SB10001424052970203513204576046893652486616, accessed March 2, 2018.

Energy Diplomacy: Proactive and Reactive 137

the UN Security Council and to join the Nuclear Suppliers Group, the informal body that controls the trade in nuclear technologies. Faced with this pressure from the United States, India had to ban its companies from using the ACU to conduct a range of transactions with Iran.[135]

Pressures stemming from the bilateral summit with the United States also stimulated India to reduce its energy ties with Iran. For instance, according to the CRS report on sanctions on Iran, the Indian government requested its refiners, in advance of President Obama's visit to India in January 2015, to further cut purchases from Iran, so "some Indian firms have ended or slowed work on investments in Iranian oil and gas fields."[136]

Although the United States granted sanction waivers to China and India, it has imposed many sanctions against Chinese entities since the mid-1990s.[137] In January 2012, for instance, the United States imposed sanctions against Zhuhai Zhenrong, a Chinese state-owned company, for being Iran's largest supplier of refined petroleum products. This was regarded as a specific signal to CNPC, Sinopec, and CNOOC, as, compared with Zhuhai Zhenrong, these three Chinese NOCs are much more exposed to the impact of potential sanctions, since they have invested billions of dollars in the US energy sector.[138] In August 2012, the United States imposed sanctions against the Bank of Kunlun Co., a unit of CNPC, for its "significant transactions for Iran's banks."[139] The United States has also imposed sanctions against Indian individuals for their relation with the Iranian nuclear program. Specifically, in September 2004, two of India's nuclear scientists were sanctioned by the United States under the Iran, North Korea, and Syria Nonproliferation Act, or Executive Order 12938.[140] Against this backdrop, Chinese and Indian NOCs have become more cautious

---

[135] Ibid.   [136] Katzman, *Iran Sanctions* (August 4, 2015), 37.
[137] Kan, *China and Proliferation of Weapons of Mass Destruction and Missiles*, 65–73.
[138] Andrew Quinn, "U.S. Slaps Sanctions on China State Energy Trader over Iran," *Reuters* (January 12, 2012), available at www.reuters.com/article/us-iran-usa-sanctions-idUSTRE80B1DW20120112, accessed March 2, 2018.
[139] Kan, *China and Proliferation of Weapons of Mass Destruction and Missiles*, 73.
[140] Katzman, *Iran Sanctions* (August 4, 2015), 72.

about their activities in the Iranian energy sector, to avoid incurring US sanctions. According to a Western scholar, "Acquiring Western oil and gas companies was so important to Chinese NOCs that they stalled their work in Iran to avoid penalties from US sanctions that might upset their activities in the US."[141]

Sudan

Sudan provides China and India with "an outstanding chance" to bolster "their presence in the international oil industry."[142] So although Western companies, notably the US oil company Chevron, withdrew from Sudan as a result of the US government's sanctions on this country, China's CNPC, in the mid-1990s,[143] and then India's NOCs, in the early 2000s, have invested in Sudan's energy sector.[144] Since then, China and India have employed political, economic, military, and diplomatic means to build up their energy-centered relationships with Sudan.

Proactive measures

Politically, China and India have used high-level official visits and meetings to enhance their bilateral ties with Sudan. For instance, Sudanese President Omar al-Bashir officially visited China in 1990, 1995, and 2011, and attended the Forum on China–Africa Cooperation (FCAC) summit in Beijing in 2006. Chinese President Hu Jintao paid a state visit to Sudan in February 2007.[145] In terms of meetings, in

---

[141] Patey, *The New Kings of Crude*, 251.

[142] Daniel Large and Luke A. Patey, "Sudan 'Looks East'," in Daniel Large and Luke A. Patey (eds), *Sudan Looks East: China, India and the Politics of Asian Alternatives* (Rochester, NY: James Currey, 2011), 2.

[143] 王璐 [Wang Lu] and 孙学峰 [Sun Xuefeng], "从'走出去'到'融进去'— 中石油开发建设融入苏丹的条件 [From 'Going-Out' to 'Fitting In': How CNPC's Projects Can Win the Sudanese Support]," 《世界经济与政治》 [*World Economy and Politics*] no. 3 (2013), 100–114; 伍 福 佐 [Fuzuo Wu], "中国能源外交与国际责任—以达尔富尔问题为例 [China's Energy Diplomacy and International Responsibility: A Case Study of Darfur]," 《阿拉伯世界研究》 [*Arab World Studies*] no. 3 (May 2010), 59–66.

[144] Øystein Tunsjø, *Security and Profit in China's Energy Policy* (New York: Columbia University Press, 2013), 92; Patey, *The New Kings of Crude*, 31–51.

[145] 中华人民共和国外交部 [MOFA], "中国同苏丹的关系 [China's Relations with Sudan]" (July 2015), available at www.fmprc.gov.cn/mfa_chn/gjhdq_603914/gj_603916/fz_605026/1206_606236/sbgx_606240, accessed September 3, 2015.

April 2005, during the Golden Jubilee celebrations in Jakarta of the fiftieth anniversary of the Bandung conference, Hu held a meeting with Bashir, with a focus "predictably placed on the achievements of mutually beneficial cooperation, together with a reaffirmation of common support for sovereignty and non-interference."[146]

In October 2003, then Indian President A. P. J. Abdul Kalam paid a state visit to Sudan – the first visit by an Indian president in 28 years – accompanied by a number of high-level Indian officials.[147] During his visit, Kalam announced a concessional line of credit of $50 million to Sudan, and a grant of medicines for $50,000 for the flood victims in Kassala.[148] Moreover, Kalam and his Sudanese counterpart signed the Bilateral Investment Promotion and Protection Agreement, the Double Taxation Avoidance Agreement, and the MOU on Cooperation in the field of information technology.[149] Soon after this historic state visit, several Sudanese ministers in different fields paid separate visits to India during December 2003. For instance, on December 5–9, 2003, Sudan's Energy and Mining Minister Ahmed Al-Jaz visited India, while Sudan's Minister of National Defence Lieutenant General Bakri Hassan Salih visited India on December 12–15, 2003.[150] Moreover, Sudan's Transportation Minister Mohd Elsamani Elwasila, accompanied by State Minister of Finance Hassan Ahmed Taha, visited India on December 22–29, 2003.[151] In 2008, after the International Criminal Court (ICC) indicted Sudanese President Bashir in July of that year, Ahmed Al-Jaz led a delegation to New Delhi to meet the Indian ministers of External Affairs, Finance, Railways, and Petroleum and Natural Gas.[152]

Both countries have employed economic tools to facilitate their relations with Sudan. The most important economic measure China and India have used to enhance their ties with Sudan is bilateral trade. According to Sudan Data Portal, during the period of 1995 and 2014, China and India were Sudan's largest trading partners, accounting for

---

[146] Daniel Large and Luke A. Patey, "Conclusion: China, India and the Politics of Sudan's Asian Alternatives," in Large and Patey (eds), *Sudan Looks East*, 183.
[147] MEA, *Annual Report (1 January 2003–31 March 2004)* (2004), 60, available at https://mea.gov.in/Uploads/PublicationDocs/165_Annual-Report-2003-2004.pdf, accessed March 2, 2018.
[148] This news article is available at www.indembsdn.com/eng/ipresident.html, accessed January 20, 2011.
[149] MEA, *Annual Report*, 60.   [150] Ibid.   [151] Ibid.
[152] Large and Patey, "Sudan 'Looks East'," 32.

nearly 22 percent and 8 percent respectively of Sudan's annual total trade.[153] Apart from trade, China and India have employed some other economic tools to court Sudan. For instance, according to the Chinese Ministry of Foreign Affairs' records, since the 1970s, China has provided Sudan with a certain amount of economic assistance. The economic and technological cooperation between the two countries covers sectors such as oil, mineral exploration, construction, road and bridge, agriculture, textile, health care, and education.[154] Since 2005, India's economic assistance to Sudan has been broadened to include infrastructure, agriculture, human resource development, information and communication technologies, as well as small and medium-sized industries.[155]

Moreover, both China and India have employed the tool of loan-for-oil in their relationship with Sudan – that is, China and India have provided loans to Sudan to extract its energy reserves and build energy infrastructure, in exchange for oil imports at established prices.[156] For instance, in February 2007, during Chinese President Hu Jintao's visit to Sudan, China granted Sudan an interest-free loan of 100 million yuan ($14.5 million) to build a new presidential palace, in addition to writing off up to $70 million in Sudanese debts to China.[157] In January 2013, the state-run China Development Bank granted a $1.5 billion loan to Sudan which will not have to be repaid until after a grace period of five years.[158] In July 2013, China's Export Import Bank, also a state-run bank, granted Sudan a $700 million loan to build a new airport for the capital Khartoum, in exchange for Sudan's

---

[153] Sudan Data Portal, "Sudan Imports, Major Trade Partners" (February 17, 2016), available at http://sudan.opendataforafrica.org/xftuxsc/sudan-imports-major-trade-partners, accessed July 20, 2016.
[154] MOFA, "China's Relations with Sudan."
[155] Devika Sharma and Pragya Jaswal, "The India–Africa Energy Partnership: Prospects and Challenges," *Energy Security Insight* 2, no. 3 (October 2007), 14.
[156] EIA, "China" (May 14, 2015).
[157] Opheera McDoom, "China's Hu Tells Sudan it Must Solve Darfur Issue," *Washington Post* (February 2, 2007), available at www.washingtonpost.com/wpdyn/content/article/2007/02/02/AR2007020200462.html, accessed January 26, 2011.
[158] "Sudan Receives $1.5 Billion Loan from China Development Bank," *Reuters* (January 17, 2013), available at www.reuters.com/article/sudan-china-loan-idUSL6N0AM9UY20130117, accessed March 2, 2018.

agreement to allow South Sudan's oil exports to transit through Port Sudan.[159] India, in January 2004, signed an agreement with Sudan on the Exim Bank Line of Credit of $50 million pledged by the Indian president during his state visit to Sudan, as mentioned above.[160] In January 2006, India's Exim Bank extended lines of credit to Sudan worth $392 million, far beyond the lines of credit that India had ever extended to any other individual African country.[161]

Militarily, both countries sell weapons to Sudan. According to a report issued by Human Rights First, a human rights NGO, China not only sold small arms, aircraft, and heavy weapons to Sudan, but also engaged in military cooperation with Sudan, in addition to supporting Sudan's arms manufacturing industry.[162] More specifically, from 2003 to 2006, China sold over $55 million worth of small arms to Sudan. China has been the largest arms supplier to Sudan. Since 2004, China has been the near-exclusive provider of small arms to Sudan, supplying approximately 90 percent of Sudan's small arms purchases each year.[163] Moreover, the UN Panel of Experts, established by the UN Security Council resolution 1591 in 2005, documented the "prominence of Chinese manufactured arms and ammunition" in Darfur, citing the role of companies such as China North Industries Cooperation.[164] Human Rights First also reported that, in 2005, an Indian defense firm entered into contracts with the Sudanese government worth more than $17 million, including provision of battlefield surveillance radar, communication equipment, and night-vision equipment. In addition, Sudan claimed to have received armored fighting vehicles from India since 2004, worth over

---

[159] "China Grants Sudan $700 mln Loan to Build New Khartoum Airport," *Reuters* (July 8, 2013), available at www.reuters.com/article/sudan-china-airport-idUSL6N0FE3RC20130708, accessed March 2, 2018.

[160] MEA, *Annual Report*, 60.

[161] "India Gives US$392 mln Loan to Sudan for Power Plant," *Sudan Tribune* (January 24, 2006), available at www.sudantribune.com/spip.php?article13708, accessed March 2, 2018.

[162] Human Rights First, "China's Arms Sales to Sudan," available at www.humanrightsfirst.org/wp-content/uploads/pdf/080311-cah-arms-sales-fact-sheet.pdf, accessed March 2, 2018.

[163] Ibid.   [164] Large and Patey, "Sudan 'Looks East'," 26.

$1.5 million, although India claimed its total arms sales to Sudan amounted to just over $200,000.[165]

Diplomatically, both countries, but more especially China, have tried to provide diplomatic support to the Sudanese government at the UN Security Council. Since the Darfur humanitarian crisis broke out in 2003, Western countries, especially the United States, have been repeatedly pressing for UN resolutions to authorize a peacekeeping effort and impose stiff sanctions on the Sudanese government, apart from the United States' own sanctions on the Sudan government which have been in place since the mid-1990s.[166] Since 2004, the UN Security Council has passed a number of resolutions concerning the Darfur issue, and China has almost always abstained from its vote on all the resolutions. According to a research study carried out by SIPRI, China abstained from voting on UN Security Council Resolutions 1556 (2004), 1591 (2005), and 1945 (2010), relating to an arms embargo on Sudan (Darfur).[167] On August 31, 2006, China abstained at the UNSC when voting on a resolution that called for a transition from "the African Union Mission in the Sudan" to "the United Nations force in Darfur," endorsed by the United States and Britain.[168] In explaining the reason for its abstention, Wang Guangya, China's ambassador to the UN, said it was "because Sudan's government was not yet ready to accept U.N. peacekeepers on its soil."[169] Moreover, China argued that its Sudan policy was to separate business from politics. For instance, in an interview in the *New York Times*, Zhou Wenzhong, Chinese Deputy Foreign Minister, said "Business is

---

[165] Human Rights First, "Arms Transfers to Sudan, 2004–2006," 2, available at www.humanrightsfirst.org/wp-content/uploads/pdf/CAH-081001-arms-table.pdf, accessed March 2, 2018.

[166] "Darfur Update" (June 6, 2008), available at http://darfur.3cdn.net/46c257b8e3959746d5_ttm6bnau2.pdf, accessed January 23, 2011.

[167] SIPRI, "UN Security Council Voting Record on Resolutions 1556, 1591 and 1945 Relating to an Arms Embargo on Sudan (Darfur)," available at www.sipri.org/databases/embargoes/un_arms_embargoes/sudan/un-security-council-voting-record-sudan, accessed September 3, 2015.

[168] UN, "Security Council Expands Mandate of UN Mission in Sudan to Include Darfur, Adopting Resolution 1706 by Vote of 12 in Favour, with 3 Abstaining" (August 31, 2006), available at www.un.org/press/en/2006/sc8821.doc.htm, accessed March 6, 2018.

[169] "Responsible China? Darfur Exposes Chinese Hypocrisy," *Washington Post* (September 6, 2006), available at www.washingtonpost.com/wp-dyn/content/article/2006/09/05/AR2006090501187.html?sub=AR, accessed March 2, 2018.

business," and "We try to separate politics from business." He also emphasized China's non-interference policy: "I think the internal situation in the Sudan is an internal affair, and we are not in a position to impose upon them."[170]

China had also supported the Sudanese government against the International Criminal Court (ICC). Specifically, on March 4, 2009, the ICC issued arrest warrants for Sudanese President Hassan Ahmad al Bashir for genocide, crimes against humanity and war crimes in Darfur. On March 22, 2009, in his visit to Sudan, Li Jinjun, Vice Minister of the International Department of the Chinese Communist Party Central Committee, expressed China's rejection of the ICC's arrest warrants and pledged its support to the Sudanese government.[171]

India is not a permanent member on the UNSC, so it has only limited influence on this issue. Even so, based on its long-established stance of non-interference, India has persistently opposed sanctions against Sudan over the Darfur conflict and maintained a political line that underlined "Sudan's territorial integrity and sovereignty."[172]

With both countries' political, economic, military, and diplomatic efforts to foster closer energy ties with Sudan, it has turned out that Sudan represents the most successful case in terms of China's and India's efforts to seek overseas energy equities. Before southern Sudan's independence in July 2011, CNPC had eight investment projects in Sudan: block 1/2/4 project (now mostly located in southern Sudan), block 3/7 project (now located in South Sudan), block 6 project, the Red Sea block 15 exploration projects, block 13 projects, the Khartoum Refinery Project, the Khartoum chemical plant, and refinery oil sales. CNPC, therefore, had already built an integrated oil industrial chain, including the exploration of oil and gas, development, production, pipeline, oil refining, and the chemical industry, as well as sales.[173] In 2009, Sudan was China's fifth-largest oil supplier, accounting

---

[170] Howard W. French, "*China in Africa: All Trade, with no Political Baggage*," *New York Times* (August 8, 2004), available at www.nytimes.com/2004/08/08/international/asia/08china.html, accessed March 2, 2018.

[171] "Chinese Delegation in Sudan for Golden Gala," *Sudanese Media Center*, Khartoum (March 22, 2009), cited in Large and Patey, "Sudan 'Looks East'," 25.

[172] Large and Patey, "Sudan 'Looks East'," 32.

[173] MOFA, "China's Relations with Sudan."

for 6 percent of its total imports in that year.[174] However, China's oil imports from Sudan dropped significantly in early 2012, when southern Sudan's oil production was shut down due to the political conflicts between Sudan and South Sudan over their oil resources; it was not resumed until 2013.[175]

India's ONGC OVL has also engaged in developing Sudan's oil. In 2003, OVL acquired a 25 percent share of the Greater Nile Petroleum Operating Company, a joint venture with CNPC (40 percent), Petronas (30 percent), and the Sudan National Oil Company (Sudapet, 5 percent), having been operating in Sudan's oil sector since 1997. This consortium holds proved crude oil reserves of more than one billion barrels, with current production levels at roughly 300,000 bpd from 10 fields, and operates a 935-mile crude oil pipeline.[176] In May 2003, when the first oil from OVL's concessions in Sudan arrived by tanker at a Mangalore port, then Indian Deputy Prime Minister Shri L. K. Advani claimed that "this is not imported oil, this is India's oil."[177] Since then, and before its separation in July 2011, Sudan became an important oil supplier to India.

China's reactive measures toward the Sudan Darfur crisis[178]

As previously mentioned, China's fundamental position on the Sudan Darfur crisis was that this was a domestic issue that should be solved

---

[174] 沈汝发,李晓慧,刘雪 [Shen Rufa, Li Xiaohui, and Liu Xue], "中国原油进口依存度首超警戒线 [China's Oil Imports Surpassed the Cordon]," *Xinhuanet* (March 29, 2010), available at http://news.xinhuanet.com/fortune/2010-03/29/content_13265670.htm, accessed January 13, 2011.

[175] Ibid.

[176] IEA, "Country Analysis Brief: India," 4, available at www.eia.doe.gov/emeu/cabs/India/pdf.pdf, accessed January 19, 2011.

[177] "Talisman Marks End of Era with Completion of Sudan Sale," *Oil and Gas Journal* 101, no. 14 (April 7, 2003), cited in Luke A. Patey, "India in Sudan: Troubles in an African Oil 'Paradise'," in Large and Patey (eds), *Sudan Looks East*, 88.

[178] In this section, I only discuss China's reactive response to international pressures, since until now, most of the pressures have been targeted at China rather than India. This is a view I have drawn from the answer to my question on India's role in Sudan by a Sudanese diplomat, who was on a panel on the rise of China and India organized by the Yale World Fellow Program on November 12, 2010, Yale University. Moreover, my view echoes observations in the existing literature. For instance, Large and Patey observe that "Attention to China, furthermore, has overshadowed the role of India ... in Sudan"

by the Sudan government itself. In other words, China maintained its non-interference policy on this issue. However, from late 2006, China's policy toward this issue began to shift away from passively leaving it alone to actively "maneuvering," when China began to work with the UN Secretary General Kofi Annan on the deployment of a hybrid United Nations–African Union (UN–AU) peacekeeping mission.[179] In other words, since 2006, China has been tacitly interfering in Sudan's Darfur issue.

More specifically, China has played a significant role in persuading Sudan's Bashir government to accept the deployment of the UN peacekeeping force in Darfur. In August 2006, the UNSC adopted UN resolution 1706, which mandated the deployment of a UN force of 20,000 troops to Darfur, under the leadership of UN Secretary General Kofi Annan. On September 9, 2006, a joint statement following the EU–China Summit, on the Darfur issue, stated: "Leaders emphasized that transition from an AU- to a UN-led operation would be conducive to the peace in Darfur,"[180] which implies that China officially accepted the idea of UN peacekeepers being deployed in Darfur. In other words, China accepted the idea of using international forces to help Sudan resolve the domestic Darfur crisis. Since then, Chinese diplomats have been engaged in convincing or persuading the Sudanese government to accept the deployment of UN peacekeeping troops in Darfur. For instance, during the November 2006 UNSC meeting on Darfur, then Chinese ambassador Wang Guangya worked behind the scenes to obtain the Sudanese government's support for a plan that would involve a hybrid UN–AU force.[181]

(Large and Patey, "Sudan 'Looks East'," 2). Also, Patey observes that "India has been able to avoid much of the negative publicity directed at China by activists and the media in the US" (Patey, "India in Sudan: Troubles in an African Oil 'Paradise'," 96).

[179] Jonathan Holslag, "China's Diplomatic Manoeuvring on the Question of Darfur," *Journal of Contemporary China* 17, no. 54 (February 2008), 71–84; International Crisis Group, "China's Thirst for Oil" (June 6, 2008), 26, available at www.crisisgroup.org/~/media/Files/asia/north-east-asia/153_china_s_thirst_for_oil.ashx, accessed January 23, 2011.

[180] "China, EU Issue Joint Statement after Summit (full text)," *Xinhuanet* (September 9, 2006), available at http://news.xinhuanet.com/english/2006-09/10/content_5071239.htm, accessed January 25, 2011.

[181] Alexandra Cosima Budabin, "Genocide Olympics: How Activists Linked China, Darfur and Beijing 2008," in Large and Patey (eds), *Sudan Looks East*, 142.

Chinese leaders made use of different diplomatic channels to persuade the Sudanese government on this issue. For instance, both Chinese President Hu Jintao and Premier Wen Jiabao held talks with Sudanese President Omar al-Bashir during the Forum on China–Africa Cooperation in Beijing in November 2006, with a special focus on the Darfur issue.[182] In order to further persuade Bashir, Hu Jintao visited Khartoum in February 2007. During his visit, Hu not only put forward a four-point principle for solving the Darfur issue during his meeting with Bashir, but he also told Bashir that "the African Union and the United Nations should play constructive roles in a peacekeeping mission in Darfur," and that "what is important right now is to achieve a comprehensive ceasefire in Darfur and speed up the process of political negotiations and let those who have not signed the Darfur Peace Accord join the peace process as soon as possible."[183]

Obviously, Hu's tacit interference in the Sudan Darfur issue had already departed from the rigid state-based sovereignty framework underpinning China's traditional approach. As to Hu's trip to Sudan, Ambassador Wang Guangya acknowledged that President Hu Jintao broke with his government's traditional policy of non-interference in a nation's internal affairs, by telling Sudan to accept a joint UN–AU peacekeeping mission to Darfur. Wang said, "Usually China doesn't send messages, but this time they did. It was a clear strong message that the proposal from Kofi Annan is a good one and Sudan has to accept it."[184] In terms of the way in which Hu sent this message to Bashir, according to Wang, Hu did not use China's trade relations to pressure Sudan, as China "never twists arms." Chinese envoy Li Junhua said that although Bashir did not unambiguously accept the AU–UN force, he was "very forthcoming" in his private talks with Hu.[185]

[182] "Hu Urges to Maintain Stability in Darfur in Talk with Sudanese Counterpart," *Xinhuanet* (November 2, 2006), available at http://news.xinhuanet.com/english/2006-11/02/content_5282826.htm, accessed January 25, 2011; "Chinese Premier Meets Sudanese President," *Xinhuanet* (November 3, 2006), available at http://news.xinhuanet.com/english/2006-11/03/content_5287545_1.htm, accessed January 25, 2011.
[183] "Hu Puts Forward Four-Point Principle on Solving Darfur Issue," *Xinhua* (February 3, 2007), available at www.china.org.cn/english/infernational/198792.htm, accessed March 2, 2018.
[184] "China Told Sudan to Adopt UN's Darfur Plan – Envoy," *Bloomberg* (February 6, 2007), available at www.sudantribune.com/spip.php?article20137, accessed March 2, 2018.
[185] Ibid.

On April 6–9, 2007, the Chinese government sent its assistant foreign minister for African affairs, Zhai Jun, to Khartoum, to urge Sudan to accept the UN troops, while visiting three Darfur refugee camps.[186] After his trip, on April 11, 2007, the Chinese Ministry of Foreign Affairs held a briefing for Chinese and foreign journalists on the Darfur issue, hosted by Qin Gang, then deputy director general of the Information Department. Zhai detailed how he had tacitly interfered in this issue: first of all, during his meetings with Sudanese government leaders, including President Bashir, Zhai reiterated the four proposals on this issue made by President Hu Jintao during his visit to Sudan in February of that same year, and introduced China's opinions on the current status of the Darfur issue; second, he told the Sudanese side of the international community's optimism about the "dual track" strategy proposed by the UN Secretary General's Special Representative; third, he hoped the Sudanese government would show further flexibility toward the plan. In the summary of his mission, Zhai confirmed China's willingness to interfere in resolving this crisis: "China is willing to continue to play a constructive role on the issue of Darfur. It is the basic starting point of the Chinese government to realize the peace, stability and economic reconstruction in Darfur through negotiations."[187] In addition, he drew attention to the ways in which China had engaged in resolving this crisis before his trip: "Since the Darfur issue appeared China has contacted various parties in such forms as exchanging visits of heads of state, dispatching special envoys, holding telephone conversations, exchanging letters and conducting coordination in the UN to narrow down differences and push forward dialogue on an equal footing."[188]

In May 2007, China appointed Liu Guijin, a former ambassador to Zimbabwe and South Africa, as a special envoy to Darfur, to increase

---

[186] Mohamed Osman, "China Urges Sudan to Accept U.N. Troops," *Washington Post* (April 9, 2007), available at www.washingtonpost.com/wp-dyn/content/article/2007/04/09/AR2007040901060.html, accessed March 2, 2018.
[187] Consulate General of the People's Republic of China in San Francisco, "Assistant Foreign Minister Zhai Jun Holds a Briefing for Chinese and Foreign Journalists on the Darfur Issue of Sudan" (April 12, 2007), available at www.chinaconsulatesf.org/eng/xw/t311008.htm, accessed March 2, 2018.
[188] Ibid.

its serious efforts to solve the Darfur crisis.[189] In mid-June 2007, the Sudanese government finally agreed to "unconditionally" accept the deployment of a 23,000-strong hybrid UN–AU peacekeeping force in the western region of Darfur.[190] On this result, then US Deputy Secretary of State, John Negroponte, commented that China played "a pivotal role in brokering the agreement."[191]

Apart from its tacit interference in the Sudanese Darfur issue, China has also tacitly interfered in Sudan's separation by playing an important role in facilitating a peaceful referendum and then a peaceful secession of the south.[192] China took some concrete actions to support the peaceful referendum on South Sudan, accompanied by some diplomatic rhetoric. For instance, in July 2010, Chinese Special Envoy to Darfur, Liu Guijin, visited Sudan, with the purpose of "acquaintance with plans of the new Sudanese government concerning the Darfur peace together with the progress in the implementation of the CPA, inked between north and south Sudan."[193] During this trip, Liu had a meeting with Sudanese Minister of Cabinet Affairs, Luka Biong, about the referendum issue, and told the latter that "we are keen on stability and peace in Sudan. We desire to see south Sudan referendum conducted in a smooth and credible manner because it is in the interest of Sudan and serves the stability in the region."[194] In response to Liu's

[189] Qiang Pen, "China's Special Envoy Visits Darfur," *China Daily* (May 24, 2007), available at www.chinadaily.com.cn/china/2007-05/24/content_879188.htm, accessed March 2, 2018.
[190] "Sudan Accepts UN–AU Force 'Unconditionally'," *Africa News* (June 20, 2007), available at www.africanews.com/site/Sudan_accepts_UNAU_force_unconditionally/list_messages/4496, accessed January 26, 2011.
[191] Cited in Budabin, "Genocide Olympics," 155.
[192] The Sudanese central government and the Sudan People's Liberation Army in the south had engaged in civil war for more than two decades until 2005, when the Comprehensive Peace Agreement (CPA) was signed between the two sides. According to the CPA, two referenda would be conducted in January 2011: one on self-determination for South Sudan and one to decide whether the Abyei area should be part of Bahral-Ghazal State in southern Sudan or the North Kordofan State in northern Sudan; see "China Hopes for Transparent, Fair Referendum in Southern Sudan," *Xinhua* (September 14, 2010), available at http://china.org.cn/world/2010-09/14/content_20931077.htm, accessed March 2, 2018.
[193] "China Keen on Sudan's Stability and Peace: Envoy," *Xinhua* (July 4, 2010), available at www.chinadaily.com.cn/china/2010-07/04/content_10055838.htm, accessed January 12, 2016.
[194] Ibid.

wishes, Luka Biong said the Sudanese government counted on China's role in supporting the two government partners in the implementation of the remaining items of the Comprehensive Peace Agreement (CPA) and the conducting of the referendum according to its fixed timetable.[195] This implied that the Sudanese government welcomed Chinese interference in its referendum.

Moreover, Chinese diplomats and leaders repeatedly stressed the importance of a peaceful referendum through various diplomatic channels, to send clear signals to Khartoum about Chinese policy on this issue. For instance, in September 2010, during his meeting with the visiting Sudanese Foreign Minister Ahmed Ali Karti, then Chinese Vice President Xi Jinping said that no matter what the results of the referendum were, China believed the top priority lay in maintaining peace and stability in Sudan and the region, and Xi reiterated China's stance on supporting the north and south sides of Sudan in steadily pushing forward the north–south peace process through dialogue and friendly consultation. In response to Xi's view, Karti said the Sudanese government was ready to live up to its duty, to continue to advance the north–south peace process and properly resolve the Darfur issue.[196] This rhetoric was iterated one month later, in October 2010, at a regular press conference by the then Chinese Foreign Ministry spokesman Ma Zhaoxu who said that China hoped the referendums planned for January 2011 in Sudan would be conducted in a transparent and fair manner and reflect the Sudanese people's will, and "Whatever the result of the referendums, it is of paramount importance to safeguard the peace and stability of Sudan."[197] In the same vein, at the November 2010 UN Security Council meeting on Sudan, then Chinese ambassador Li Baodong stated that "The south Sudan referendum is a key step in the implementation of the Comprehensive Peace Agreement (CPA)," so "we hope that the referendum will be held in a peaceful, free, transparent and fair manner in accordance with the CPA and reflect the will of the Sudanese people, and that the outcome of the

---

[195] Ibid.
[196] "China Hopes for Transparent, Fair Referendum in Southern Sudan."
[197] "China Hopes Sudan's Referendums to Reflect People's Will," *Xinhua* (October 19, 2010), available at www.chinadaily.com.cn/china/2010-10/19/content_11431372.htm, accessed March 2, 2018.

referendum will be respected by all parties."[198] Along with this diplomatic rhetoric, China not only provided financial assistance for the vote, but also sent an observation mission to monitor the referendum.[199] It turned out that the referendum in South Sudan went smoothly and the result of the referendum was that the people voted overwhelmingly for independence. Later, Liu Guijin acknowledged that China had "done a lot of work to persuade" the north to implement the peace agreement of 2005 and referendum in January 2011, in which the south voted by an overwhelming majority to split from the north.[200]

After the successful referendum, China went on to tacitly interfere in the process of the secession of South Sudan in order to ensure that it would be carried out peacefully. Specifically, in June 2011, one week before the secession of South Sudan, Sudanese President Bashir paid a state visit to China. In interviews with official Chinese media, Bashir reassured the Chinese about his commitment to a peaceful secession of the south, and assured Chinese leaders that their investments and energy stake in Sudan would not be threatened by the north–south split.[201] With this promise, on July 9, 2011, South Sudan declared its independence from the north and became an independent country.

In summary, China tacitly interfered in Sudan's domestic affairs. Specifically, China played a significant role in persuading the Sudanese government to accept the deployment of UN–AU peacekeeping troops in the Darfur region. China also played an important role in ensuring a peaceful referendum on South Sudan, and then, its successful and peaceful secession from the north.

[198] "China Hopes for Peaceful, Transparent Referendum in South Sudan," *Xinhuanet* (November 17, 2011), available at http://news.xinhuanet.com/english2010/china/2010-11/17/c_13609765.htm, accessed January 12, 2016.
[199] "China Says it Will Send Observers to Sudan Referendum," *Reuters* (January 4, 2011), available at http://af.reuters.com/article/topNews/idAFJOE70309C20110104, accessed March 2, 2018.
[200] Cited in Chris Buckley, "Update 1: Sudan's Bashir Warns, Reassures China on South Split", *Reuters* (June 26, 2011), available at www.reuters.com/article/2011/06/27/china-sudan-idAFL3E7HR03S20110627, accessed March 2, 2018.
[201] Ibid.

External Pressures on China's Sudan Policy

Given China's longstanding position on non-interference in other countries' domestic affairs, its tacit interference in Sudan's domestic affairs, especially its efforts to engage in solving the Sudan Darfur issue, had largely been stimulated by external pressures. I would argue that international opprobrium of China's Sudan policy applied by human rights-related NGOs, certain UN officials, some of the US news media, and even certain US celebrities, played an important role in pressuring China to modify its non-interference policy toward the Sudan Darfur issue.

To begin with, human rights-related NGOs applied some noticeable pressures on China on its policy toward Sudan Darfur. Specifically, some of these NGOs were highly critical of China for not having done enough to stop the "genocide" in Darfur, the Save Darfur Coalition being the most influential. This NGO made the strongest criticism of China's policy on Darfur.[202] The Coalition even lobbied the US government to increase pressure on China, to make it do whatever it could to end the crisis in Darfur. For instance, in April 2006, just a few days before Chinese President Hu Jintao paid a first state visit to the United States, the Coalition, together with over 160 faith-based, advocacy, and humanitarian aid organizations, called upon US President Bush to make the genocide in Darfur a main topic on his meeting agenda with President Hu.[203] According to CNN news, during the summit, Bush did remark that the United States and China intended to deepen their cooperation in addressing threats to global security, including "the genocide in Darfur, Sudan." In response, Hu said, "We're ready to enhance dialogue and exchanges with the U.S. side on the basis of

---

[202] This NGO was founded in July 2004, and has grown into an alliance of more than 180 religious, political, and human rights organizations committed to ending the alleged genocide in Darfur. See Wikipedia, "Save Darfur Coalition," available at http://en.wikipedia.org/wiki/Save_Darfur_Coalition, accessed March 2, 2018.

[203] Save Darfur Coalition, "Faith-Based and Human Rights Coalition Calls on President Bush to Raise Darfur Issues with Chinese President Hu Jintao" (April 20, 2006), available at www.savedarfur.org/pages/press/faith_based_and_human_rights_coalition_calls_on_president_bush_to_raise_dar, accessed January 24, 2011.

mutual respect and equality to promote the world's cause of human rights."[204] Hu's response implied that his government would do something to solve the Darfur crisis.

Human Rights Watch, another influential human rights NGO based in the United States, wrote a letter called, "Letter to China on the Crisis in Darfur," to Chinese President Hu Jintao in January 2007, in which Beijing was asked to do "a great deal more" on Sudan if it wanted to be seen as "a responsible international power."[205] A few days after this letter was written, President Hu visited several African countries, including Sudan. On February 2, 2007, during his meeting with his Sudanese counterpart, Hu was reported to have told Bashir: "Darfur is a part of Sudan and you have to resolve this problem."[206]

Second, certain UN officials played an important role in pressuring China to play an active role in resolving the Darfur crisis. For instance, in mid-September 2006, after an eight-day visit to Congo, Uganda, and southern Sudan, UN humanitarian chief, Jan Egeland, said that it was vital that a UN peacekeeping force be allowed into the western Sudan region of Darfur to improve the security situation. Therefore, he urged China and some other countries to help persuade the Sudanese government into accepting the deployment of a UN peacekeeping force,[207] which was a direct pressure on China by a UN official.

In addition, some of the US news media played a role in pressuring China to change its stance. The *Washington Post* is a good case in point. After China abstained from voting on the UNSC Resolution 1706, the *Washington Post* published some editorials in September 2006, directly casting doubt on China's role as a "responsible stakeholder" in the international system. For instance, an editorial titled, "Responsible China? Darfur Exposes Chinese Hypocrisy" (September 6,

---

[204] CNN, "Bush Welcomes Chinese President for Talks" (April 20, 2006), available at http://edition.cnn.com/2006/WORLD/asiapcf/04/20/bush.china/index.html, accessed March 2, 2018.
[205] Human Rights Watch, "Letter to China on the Crisis in Darfur" (January 29, 2007), available at www.hrw.org/en/news/2007/01/29/letter-china-crisis-darfur, accessed March 2, 2018.
[206] Opheera McDoom, "China's Hu Tells Sudan It Must Solve Darfur Issue," *Washington Post* (February 2, 2007), available at www.washingtonpost.com/wpdyn/content/article/2007/02/02/AR2007020200462.html, accessed January 26, 2011.
[207] Save Darfur Coalition, "U.N. Humanitarian Chief: Darfur Near Collapse" (September 13, 2006), available at www.savedarfur.org/pages/clips/un_humanitarian_chief_darfur_near_collapse, accessed January 24, 2011.

2006), doubted that China would be "treated as a responsible international player," because China refused to support the UNSC in passing a resolution on the deployment of a UN peacekeeping force in Darfur, regardless of the fact that failure to deploy UN peacekeepers might result in "tens of thousands of civilian deaths." This editorial also called on China to change its position, by not only supporting the UNSC resolution, but also asking the Sudanese government to "accept U.N. peacekeepers immediately."[208] Two weeks later, on September 19, 2006, another editorial, titled, "The Genocide Test: Surely China Does not Believe Sudan's Brazen Lies," pointed out that "If it wants to be regarded as a responsible power, China should use its leverage" to persuade Sudan's president to accept the deployment of a UN peacekeeping force in Darfur. And the editorial concluded by asking, "Is China's a responsible government?"[209]

Another very noticeable pressure on China that persuaded it to alter its non-interference policy toward Sudan stemmed from the so-called "Genocide Olympics" campaign, initiated by certain US celebrities, which linked the Darfur humanitarian crisis with the Beijing 2008 Olympics.[210] This campaign began in February 2007, shortly after Chinese President Hu's visit to Sudan.[211] The campaign had a great impact on the international community after the *Wall Street Journal* published an article entitled "The 'Genocide Olympics'," co-authored by Hollywood actress Mia Farrow, who threatened to launch a worldwide campaign against the Beijing "Genocide Olympics" because of China's stance on Sudan.[212] This campaign greatly shocked Beijing. In a heavier blow to Beijing's Olympic efforts, the US film director Steven Spielberg withdrew from his participation in the planning of the Beijing

---

[208] "Responsible China? Darfur Exposes Chinese Hypocrisy," *Washington Post* (September 6, 2006), available at www.washingtonpost.com/wp-dyn/content/article/2006/09/05/AR2006090501187.html?sub=AR, accessed March 2, 2018.

[209] "The Genocide Test: Surely China Does not Believe Sudan's Brazen Lies," *Washington Post* (September 19, 2006), available at www.washingtonpost.com/wp-dyn/content/article/2006/09/18/AR2006091801030_pf.html, accessed March 2, 2018.

[210] Budabin, "Genocide Olympics," 139–156.

[211] Eric Reeves, "China and the 2008 Olympic Games" (February 10, 2007), available at www.sudanreeves.org/Page-10.html, accessed January 26, 2011.

[212] Ronan Farrow and Mia Farrow, "The 'Genocide Olympics'," *Wall Street Journal* (March 28, 2007), available at www.wsj.com/articles/SB117505109799351409, accessed March 6, 2018.

Olympics opening ceremonies and wrote a personal letter to President Hu Jintao, asking him to use his influence to bring an end to the human suffering in Darfur. A week later, the Chinese government sent its Deputy Foreign Minister Zhai to visit refugee camps in Darfur, the first high-ranking Chinese official to do so, as discussed above.[213]

In sum, faced with these external pressures from mainly non-state actors, the Chinese government had to reactively adjust its non-interference policy toward Sudan on the Darfur crisis, and to actively engage with the Sudanese leaders and persuade them to solve the crisis.

Myanmar

Both China and India have been conducting energy diplomacy in Southeast Asia, which has some energy resources. Myanmar, for example, is a country that is an energy source for both China and India, and an important energy transit country for China; it also has geopolitical significance for both China and India.[214]

Proactive Measures

Both China and India have adopted proactive policy measures to forge closer energy ties with Myanmar. Politically, both countries have exchanged high-level official visits with Myanmar. On China's side, in 2001, Chinese President Jiang Zemin visited Myanmar, which was the first time that a Chinese president had visited this country. During this visit, both sides identified agriculture, human and natural resource development, and infrastructure as priority areas for cooperation. Since the late 2000s, a number of Chinese high-level officials, including vice presidents, premiers and vice premiers, ministers of foreign affairs, chiefs of general staff, as well as vice chairmen of the National People's Congress (NPC), have visited Myanmar. In return, Myanmar's high-level generals and officials have paid visits to China, including Than

---

[213] Linda Jakobson, "The Burden of 'Non-Interference'," *China Economic Quarterly* 11, no. 2 (2007), 16.

[214] See Chenyang Li and Liang Fook Lye, "China's Policies towards Myanmar: A Successful Model for Dealing with the Myanmar Issue?" *China: An International Journal* 7, no. 2 (September 2009), 255–287; Marie Lall, "Indo-Myanmar Relations in the Era of Pipeline Diplomacy," *Contemporary Southeast Asia* 28, no. 3 (2006), 424–446.

Shwe (once), Maung Aye (three times), Khin Nyunt (once), Soe Win (three times), and Thein Sein (three times). In 2009, Member of the Standing Committee of the Political Bureau of the Chinese Communist Party Central Committee, Li Changchun, and Chinese Vice Chairman Xi Jinping visited Myanmar.[215] In May 2011, Thein Sein, the first President of Myanmar, paid a state visit to China. Not only did this visit upgrade the countries' relationship to a strategic partnership, but both countries also signed further economic agreements, regarded as "the latest sign of stronger ties between the two neighbors."[216]

India, with its "Looking East" policy launched in the early 1990s, has been developing and sustaining good relationships with Southeast Asian countries in general, and Myanmar in particular. There have been many high-level exchange visits between India and Myanmar since the late 1990s, when India changed its policy toward Myanmar.[217] In 2003, Indian Vice President Shri Bhairon Singh Shekhawat visited Myanmar. In return, in October 2004, Senior General Than Shwe, head of Myanmar's military-run government, visited India. After that, the two countries began a series of high-level bilateral political and military exchanges. In March 2006, Indian President Abdul Kalam paid a state visit to Myanmar, followed by the visit of Myanmar's third-highest official, the State Peace and Development Council General Thura Shwe Mann, and the Home Minister Maung Oo. In January 2007, Indian External Affairs Minister Mukherjee visited Myanmar. In the same year, Indian Prime Minister Singh and Myanmar Premier General Thein Sein had a meeting in Singapore during the ASEAN summit. In addition, there have been a number of meetings between the two countries' foreign and defense ministers and military chiefs.[218] Moreover, in order to boost its energy cooperation

---

[215] Ministry of Foreign Affairs of PRC, "(China–Myanmar) Bilateral Relations," available at www.fmprc.gov.cn/eng/wjb/zzjg/yzs/gjlb/2747, accessed January 19, 2011.

[216] "China, Myanmar Forge Partnership, Ink Deals on Myanmar President's Maiden Visit," *Xinhuanet* (May 27, 2011), available at http://news.xinhuanet.com/english2010/china/2011-05/27/c_13897797.htm, accessed September 1, 2015.

[217] Lall, "Indo-Myanmar Relations in the Era of Pipeline Diplomacy," 424–446; Daniel Twining, "India's Relations with Iran and Myanmar: 'Rogue State' or Responsible Democratic Stakeholder?" *India Review* 7, no. 1 (January 2008), 1–37.

[218] Twining, "India's Relations with Iran and Myanmar," 16.

with Myanmar, in September 2007, when the crackdown by Myanmar's military regime on peaceful demonstrations by monks and pro-democracy protesters was at its peak, the then Indian petroleum minister Murli Deora visited Yangon and signed a deal worth $150 million, for the exploration of natural gas off the Rakhine coast.[219]

Economically, both China and India have provided Myanmar with different kinds of aid, with the aim of bolstering their energy cooperation with Myanmar. Since the late 1980s, China has been providing Myanmar with assistance such as foreign exchange, equipment, and technology, as well as low-interest or interest-free loans, which have been increasing as China's economy grows. China's economic and trade ties with Myanmar have also been expanding. For instance, within the China–ASEAN FTAs, more than 100 kinds of goods from Myanmar have been entitled to duty-free access to China.[220] China has been Myanmar's largest export market, and their bilateral trade volume has increased over the past several decades. For instance, in 2011, two-way trade between China and Myanmar totaled $4.44 billion.[221] In terms of investment, "China is Myanmar's largest investor with an investment totaling 12.32 billion U.S. dollars in 2010."[222]

According to the Indian MEA, the Indian government has provided economic assistance to Myanmar, including $250,000 provided as a cash grant to the Myanmar government. In addition, India has engaged in more than a dozen major projects in Myanmar, most of which related to improving that country's transportation and communication links.[223] In May 2006, the Indian government approved an extension of its credit line of $20 million to Myanmar for the

---

[219] "Deora Visit Sparks Anger," *The Telegraph* (September 25, 2007), available at www.telegraphindia.com/1070925/asp/foreign/story_8358013.asp, accessed March 2, 2018; Praful Bidwai, "India Shamed into Revising Stand on Junta," *Inter Press Service News Agency* (October 5, 2007), available at http://ipsnews.net/news.asp?idnews=39536, accessed January 15, 2011.
[220] Li and Lye, "China's Policies towards Myanmar," 265.
[221] "China, Myanmar Forge Partnership, Ink Deals on Myanmar President's Maiden Visit."
[222] Ibid.
[223] MEA, "India–Myanmar Relations" (July 2012), available at https://mea.gov.in/Portal/ForeignRelation/myanmar-july-2012.pdf, accessed March 2, 2018.

renovation of the Thanlyin Refinery.[224] Moreover, India's trade with Myanmar has increased significantly. For instance, in 2009–10, India was Myanmar's fourth-largest trading partner, after Thailand, China, and Singapore, with total trade of a record $1.5 billion, up 27 percent over the previous fiscal year.[225]

Militarily, both China and India have strengthened their military ties with this country. For instance, since the late 1980s, when the military came into power in Myanmar, China has been actively seeking military and security cooperation with the country, which has been mainly focused on two areas: first, sale of weapons and military equipment, such as missiles, fighter planes, warships, tanks, armored vehicles, artillery and radar; second, staff training, including training Myanmar air and naval officers how to use the weapons and military equipment imported from China.[226] Between 1990 and 2008, Myanmar accounted for 14 percent of China's arms exports.[227] With regard to India, according to the 2006 report published by Human Rights Watch, India increased its military cooperation with Myanmar's military government by providing the latter with arms and military training. In 2006, for instance, regardless of objections from the British government, India sold Myanmar two BN-2 Islander maritime surveillance aircraft it bought from Britain in the 1980s, in addition to T-55 tanks and 105 mm artillery pieces. Moreover, India offered to train Myanmar's Special Forces in counterinsurgency tactics.[228] According to research, "Myanmar's army chief of staff visited New Delhi in late 2006 in one of many senior officer exchanges, and India has provided valuable military training to Myanmar's armed forces."[229]

Diplomatically, China, with permanent membership of the UNSC, has occasionally made use of its privilege to protect the Myanmar government from being condemned by the UNSC Resolutions for its

---

[224] Suvrokamal Dutta, "India–Myanmar Economic and Strategic Ties: Towards New Direction," *Indo-Burma News* (January 28, 2007), available at www.indoburmanews.net/archives-1/2007/January/economic, accessed January 19, 2011.
[225] MEA, "India–Myanmar Relations."
[226] Li and Lye, "China's Policies towards Myanmar," 266.
[227] Cited in Tunsjø, *Security and Profit in China's Energy Policy*, 170.
[228] Human Rights Watch, "India: Military Aid to Burma Fuels Abuses" (December 7, 2006), available at https://reliefweb.int/report/myanmar/india-military-aid-burma-fuels-abuses, accessed March 6, 2018.
[229] Twining, "India's Relations with Iran and Myanmar," 22.

human rights abuses. For instance, in January 2007, China vetoed a Draft Resolution on Myanmar prepared at the UNSC, and later that year China blocked Western countries' efforts to have the UNSC condemn Myanmar's crackdown on protestors in September 2007.[230]

Although India cannot provide Myanmar with the same diplomatic protection as China, it has tried to provide Myanmar with rhetorical support. For instance, in January 2007, when the Indian External Affairs Minister Mukherjee visited Myanmar, he said, in his comments on the UNSC resolution accusing the Myanmar military regime of suppressing democracy, that Indians were not interested in exporting their own ideology, even though as a democratic country, India "would like democracy to flourish everywhere," democracy "is for every country to decide for itself."[231] Mukherjee's comments implied that India would not interfere in Myanmar's domestic affairs. The motivation of India's pragmatic policy toward Myanmar was vividly revealed by a Western scholar's observation that "in pursuit of energy resources, New Delhi has felt compelled over the past 15 years to greatly expand its political, economic, and military ties with Myanmar, despite its distaste for the military regime there and its historic sympathies for the democratic opposition, led by the imprisoned Daw Aung San Suu Kyi."[232] Furthermore, in August 2008, India granted Myanmar the status of observer in SAARC.[233]

Thanks to these proactive measures, China's and India's oil companies have gained access to Myanmar's rich oil and natural gas resources. Take China's CNPC, for example: this company purchased the Bagan project in central Myanmar from TG World in November 2001; in December 2001, a contract was signed with Myanmar's Ministry of Energy for exploration of an oil field in southern Myanmar, with CNPC obtaining 100 percent holdings in both projects; in 2005, CNPC obtained a high-yield gas exploration project; and in January 2007, CNPC signed another contract with Myanmar's Ministry of Energy, acquiring oil and gas exploration licenses for three deepwater blocks located in offshore Rakhine, covering an area of

---

[230] Li and Lye, "China's Policies towards Myanmar," 267.
[231] Dutta, "India–Myanmar Economic and Strategic Ties."
[232] Twining, "India's Relations with Iran and Myanmar," 14.
[233] MEA, "India–Myanmar Relations."

ten thousand square kilometers.[234] Moreover, CNPC has jointly invested in and built, with Myanmar Oil and Gas Enterprise, the Myanmar–China oil and natural gas pipeline.[235] In terms of India's NOCs in Myanmar, for instance, ONGC's OVL and GAIL hold a 30 percent stake in a consortium, headed by South Korea's Daewoo International, in A-1 and A-3 in the Shwe & Shwe Phyu and Mya fields, which, according to some estimates, hold a reserve of 20 trillion cubic feet of natural gas and can produce 2 billion cubic feet of gas for 25 years.[236]

Reactive Measures

The Chinese government began quietly adjusting its policy toward Myanmar. In February 2007, during Chinese State Councilor Tang Jiaxuan's visit to Myanmar, Tang expressed the Chinese government's concerns to Myanmar's Senior General Than Shwe, that is, China sincerely hoped that Myanmar would be "politically stable, economically developed and nationally harmonious with its people living and working in peace and contentment."[237] Tang's message could be read as an implicit request to the Myanmar government to improve its record on stability, development, and the way it ruled over its people. The result of Tang Jiaxuan's visit was that the Myanmar government accepted a new agreement with the International Labor Organization (ILO), which had almost been thrown out of the country by the junta.[238] After Myanmar's Prime Minister Thein Sein's visit to Beijing in May 2007, the military government announced the resumption of the constitutional national convention, which had been postponed for a long time. In addition, China convened a meeting attended by some of Myanmar's ethnic groups to discuss the issue of restoring peace; and this meeting urged the military government to hold talks with Aung

---

[234] For the details of CNPC's projects in Myanmar, see Tunsjø, *Security and Profit in China's Energy Policy*, 170.
[235] Ibid.   [236] Dutta, "India–Myanmar Economic and Strategic Ties."
[237] Ministry of Foreign Affairs of PRC, "Chairman of the Myanmar State Peace and Development Council (SPDC) Meets with Tang Jiaxuan" (February 27, 2007), available at www.mfa.gov.cn/eng/wjb/zzjg/yzs/gjlb/2747/2749/t300384.htm, accessed January 28, 2011.
[238] International Crisis Group, *China's Myanmar Dilemma*, Asia Report No. 177 (September 14, 2009), 5, available at www.crisisgroup.org/~/media/Files/asia/north-east-asia/177_chinas_myanmar_dilemma.pdf, accessed June 19, 2012.

San Suu Kyi and others in opposition. Moreover, thanks to China's brokering, a ministerial-level meeting between a US official and several Myanmar senior officials was held in Beijing in July 2007.[239] In September 2007, when the military junta cracked down on a peaceful demonstration by Buddhist monks, Wen Jiabao, then Chinese prime minister, urged Myanmar to resolve the crisis peacefully: "China hopes that all parties concerned in Myanmar show restraint, resume stability through peaceful means as soon as possible, promote domestic reconciliation and achieve democracy and development."[240] This was clearly an implicit attempt by China to influence Myanmar's domestic policy. Not surprisingly, a Western scholar commented: "China have moved beyond the principle of non-interference in their neighbor's affairs and urged the Myanmar regime to pursue internal reform for the sake of regional stability."[241]

Similarly, the Indian government made certain changes in its policy toward Myanmar. In November 2007, at an ASEAN summit meeting in Singapore, Indian Prime Minister Manmohan Singh appealed to his Myanmar counterpart to pursue political negotiations with Aung San Suu Kyi and the democratic opposition.[242] In December 2007, the Indian government halted all of its arms sales and transfers to Myanmar, which was "privately confirmed by New Delhi to top U.S. officials."[243]

External Pressures on China's and India's Policy toward Myanmar

Myanmar, labeled a "pariah state" by the West, has been subject to sanctions imposed by Western countries, especially the United States and the EU, due to its military regime's violation of human rights since

---

[239] Ibid.
[240] Cited in Seth Mydans, "U.N. Envoy Brings Appeal for Restraint to Myanmar," *New York Times* (September 30, 2007), available at www.nytimes.com/2007/09/30/world/asia/30myanmar.html?_r=0, accessed March 2, 2018.
[241] Twining, "India's Relations with Iran and Myanmar," 15.   [242] Ibid., 25.
[243] Glenn Kessler, "India's Halt to Burma Arms Sales May Pressure Junta," *Washington Post* (December 30, 2007), available at www.washingtonpost.com/wp-dyn/content/article/2007/12/29/AR2007122901345.html, accessed March 2, 2018.

Energy Diplomacy: Proactive and Reactive 161

it came into power in the late 1980s.[244] The most recent proof of Myanmar's violation of human rights was its behavior in what became known as the "Saffron Revolution" in September 2007, when the military regime brutally cracked down on monks and dissidents. As noted above, both China and India have invested in Myanmar's energy sector and have boosted their ties with the country through different measures. Against this backdrop, the driving force that has stimulated both countries to adjust their foreign policy toward Myanmar stems from external pressures rather than domestic ones. More specifically, international criticism and pressures from the Western countries have played an important role in influencing China's and India's policy shift toward Myanmar.

As mentioned above, in January 2007, China used its power to veto a UNSC resolution on Myanmar, which incurred strong international criticism of its use of its UNSC permanent membership privilege to protect the junta. For instance, the president of the International Federation for Human Rights, a human rights NGO, publicly criticized China's (and Russia's) veto behavior concerning Myanmar as "signing a blank check to the Burmese military regime, enabling the continuation of blatant human rights violations, affecting all Burmese people and the Southeast Asian region, in almost complete impunity."[245]

When Myanmar's military government cracked down on the peaceful demonstrations by monks and protesters in the "Saffron Revolution" mentioned above, both the United States and the EU imposed some tacit pressure on the Chinese government. Even before the crackdown took place, on September 6, 2007, Myanmar was one of the topics discussed in a meeting between the US President George W. Bush and Chinese President Hu Jintao at the Asia-Pacific Economic Cooperation meeting in Sydney.[246] After the crackdown, on September 27, 2007, Bush held a brief meeting with Chinese Foreign Minister Yang

---

[244] The United States has imposed a series of trade and financial sanctions against the military government, while the EU has maintained a Common Position which includes sanctions, travel bans on key military leaders and their families, and a ban on exports of military equipment. See Morten B. Pedersen, *Promoting Human Rights in Burma: A Critique of Western Sanctions Policy* (Lanham: Rowman & Littlefield, 2007).

[245] Cited in International Crisis Group, *China's Myanmar Dilemma*, 5.

[246] Drew Thompson, "US Turns to China to Influence Myanmar," *Asian Times Online* (September 21, 2007), available at www.atimes.com/atimes/China/II21Ad01.html, accessed March 2, 2018.

Jiechi in the White House, to urge "China to put pressure on Burma's military rulers to end their bloody crackdown on protesters," which was an explicit pressure applied by the United States on China to "use its influence in the region to help bring a peaceful transition to democracy in Burma."[247] On September 28, 2007, British Prime Minister Gordon Brown held telephone talks with Chinese Premier Wen Jiabao, to discuss the Myanmar situation. Wen told Brown, "China is very much concerned with the situation in Myanmar," so China would "continue to work with the international community to actively facilitate the proper solution to the problem in Myanmar."[248] EU foreign policy head, Javier Solana, also called on China to exert more pressure on Myanmar in an interview with Germany's *Bild am Sonntag*, by saying, "All countries which have influence on the decisions of those in power in Myanmar should act now. This applies particularly to its direct neighbours such as China."[249] Furthermore, a leading European Parliament lawmaker suggested that European countries should boycott the 2008 Beijing Olympics unless China did more to resolve the Myanmar crisis. Under these pressures from the United States and the EU, China had to respond by taking some policy measures. Thanks to China's persuasion discussed above, the Myanmar government agreed to allow the UN Secretary General's Special Envoy, Ibrahim Gambari, to visit the country in October 2007. China's role in brokering this visit was appreciated by US President Bush when he met with Chinese Foreign Minister Yang Jiechi.[250]

India was also faced with external pressures on its Myanmar policy. One source of pressure came from Amnesty International, an influential

---

[247] Alex Spillius, "George Bush Urges China to Pressure Burma," *The Telegraph* (September 27, 2007), available at www.telegraph.co.uk/news/worldnews/1564491/George-Bush-urges-China-to-pressure-Burma.html, accessed March 2, 2018.

[248] Consulate-General of the People's Republic of China in Manchester, "Premier Wen Jiabao Holds Telephone Talks with His British Counterpart Brown" (September 28, 2007), available at http://manchester.china-consulate.org/eng/sbgx/zzjw/t368380.htm, accessed March 2, 2018.

[249] "China Joins Calls for End to Burmese Violence," *ABC News* (September 30, 2007), available at www.abc.net.au/news/stories/2007/09/30/2047112.htm?section=justin, accessed March 2, 2018.

[250] Matt Spetalnick, "Bush Appeals to China to Pressure Myanmar," *Reuters* (September 27, 2007), available at www.reuters.com/article/2007/09/27/us-myanmar-usa-idUSN2733340120070927?pageNumber=1, accessed March 2, 2018.

NGO, on India's military sales to Myanmar, since this weakened the United States' and the EU's arms embargo against the military regime in this country. More specifically, in March 2007, when there were reports about India's promise to the Myanmar government of a further transfer of military equipment, including the prospective transfer of advanced light helicopters (ALHs) from Hindustan Aeronautics Limited (HAL), the manufacturer of ALHs, to Myanmar, Amnesty International wrote a letter to both HAL and the Indian Government. HAL did not respond, although the letter was sent to the company twice by Amnesty International. While the Indian government did reply, it simply denied the existence of any kind of transfer proposal.[251] So, in July 2007, Amnesty International, together with 13 EU-based NGOs, publicly issued a report entitled, "Indian Helicopters for Myanmar: Making a Mockery of the EU Arms Embargo?," which detailed the agreement reached between Myanmar and India on India's transfer of ALHs to Myanmar, and explained that the reason they were so concerned about this "reported" transfer agreement lay in the fact that ALHs might contain "components, technology and munitions originating from EU Member States and the US," on the one hand, and its "wide range of military capabilities" on the other.[252] This report obviously played a role in leading the Indian government to end its military sales to Myanmar.

Like China, India was also subject to pressure from Western governments. For instance, in July 2007, Australian Foreign Minister Alexander Downer said Western sanctions and threats, as well as ASEAN's approach of "constructive engagement," had failed to convince Myanmar's military regime to end its human rights violations and make progress on democracy, so China and India, with their important economic ties with Myanmar, should pressure the latter to "end rights abuses and democratize faster."[253] This diplomatic rhetoric placed India in the same position as China, in terms of responsibility for

---

[251] Amnesty International, "Indian Helicopters for Myanmar: Making a Mockery of the EU Arms Embargo?," 4, available at www.amnesty.org/en/library/asset/ASA20/014/2007/en/5ada7141-d381-11dd-a329-2f46302a8cc6/asa200142007en.pdf, accessed January 28, 2011.
[252] Ibid.
[253] "Australia Urges China, India to Pressure Burma to End Abuses," *The Irrawaddy* (July 31, 2007), available at www.irrawaddymedia.com/cartoon.php?art_id=8056, accessed January 28, 2011.

the Myanmar issue. In the same vein, in late September 2007, when the military crackdown took place, the United States and the EU issued statements at the United Nations, calling on both China and India to use their influence to force the junta to open talks with opponents.[254] In early October 2007, the EU issued a démarche to India, urging it to use its relationship with the junta to pressure it to reconcile with the democrats.[255] In December 2007, a senior European diplomat singled out India as more blameworthy than China, since India had reached a deal with the junta, giving Myanmar further control over the porous border between the two countries.[256]

In sum, both Chinese and Indian governments were forced to adopt a reactive foreign policy toward Myanmar, following external pressures from some NGOs and Western governments.

Comparison of the Three Cases

These three case studies prove that China's and India's energy diplomacy bears proactive and reactive features. On the one hand, both countries have taken proactive policy measures – political, economic, military, and diplomatic – to boost their energy ties with Iran, Sudan, and Myanmar, the so-called "pariah states" in the international system. On the other hand, however, both countries have had to alter the trajectory of their proactive policy measures as a result of external pressures applied by state and non-state actors.

Aside from these similarities, there are, nonetheless, some subtle differences in terms of the intensity or degree of their reactiveness (see Table 4.3), which has stemmed from the fact that there are differences in the level of pressures faced by China and India in each case. In Iran's case, the degree or level of both China's and India's reactiveness to external pressures, especially from the United States and

[254] David Gordon Smith, "A Double Game: India Suffering Fallout from Burma Crisis," *Spiegel Online International* (September 28, 2007), available at www.spiegel.de/international/world/0,1518,508491,00.html, accessed March 2, 2018.
[255] Manjeet Kripalani, "India's Role in Burma's Crisis," *Business Week* (October 19, 2007), available at www.businessweek.com/globalbiz/content/oct2007/gb20071019_332887.htm, accessed January 28, 2011.
[256] Benny Avni, "Activists Urge More Pressure on Burma," *New York Sun* (December 27, 2007), available at www.nysun.com/foreign/activists-urge-more-pressure-on-burma/68615, accessed March 2, 2018.

Table 4.3 *Comparison of Chinese and Indian reactiveness toward Iran, Sudan, and Myanmar*

	China		India	
Targeted countries	Reactive level	External pressure level	Reactive level	External pressure level
Iran	High	High	High	High
Sudan	High	High	Low	Low
Myanmar	High	High	Modest	Modest

Source: Author

Note: High = top leaders engaged in the process and intense social opprobrium; Low = no top leaders engaged in the process and low social opprobrium; Modest = between high and low

its allies, is high. That is, both China's and India's top leaders were directly engaged in carrying out some reactive policy measures. In the case of Sudan and Myanmar, the intensity or level of China's reactiveness is high, given the fact that the Chinese top leaders were pressured into taking actions to adjust their policies toward these two countries. However, India's reactiveness in the case of Sudan is low, because the international community's main target was China, allowing the Indian government to shy away from most of the international attention. In Myanmar's case, India's reactiveness is modest, for it has faced only modest pressure from NGOs and Western governments.

Thus, the level or degree of reactiveness of China and India in these three cases is a direct response to the level of external pressure on both countries' foreign policies toward Iran, Sudan, and Myanmar – that is, China's and India's high-level reactiveness toward Iran directly results from the intensity of external pressures from the top leaders of the United States and its allies. China's high-level reactiveness toward Sudan and Myanmar is also directly related to the high level of external pressure applied by the top leaders of the Western countries, as well as the intensive social opprobrium heaped on China by non-state actors, including human rights NGOs, the news media in the United States, certain celebrities in the United States, and certain UN officials. In contrast, India's low reactive level toward Sudan and modest reactive level toward Myanmar resulted from the respectively low and modest intensity of pressures it faced with regard to its policies toward those two countries.

Shaping Energy Diplomacy: Two-Level Pressures

China's and India's proactive and reactive energy diplomacy has been mainly shaped by two-level pressures – that is, at the domestic level, both countries seek to maximize their economic wealth, so both have tried to help their NOCs to increase their commercial profits, which will, in turn, benefit governmental revenues and increase domestic employment; at the international level, both China and India seek to enhance their status in the international system, so both countries have been sensitive not only to social opprobrium of their energy diplomacy, but also to their asymmetrical interdependence with the Western countries, especially the United States and its allies.

China and India Seeking Wealth

The proactive feature of China's and India's energy diplomacy in general, and toward Iran, Sudan, and Myanmar in particular, can be explained by both countries' desire to maximize their domestic economic wealth. Specifically, both governments help their NOCs to grow their commercial profits by maximizing their acquisitions of overseas energy assets, which benefits China's and India's domestic economic development in at least three ways.

First and most importantly, both governments employ energy diplomacy to help their NOCs to maximize their oil and gas equities across the world, to facilitate their great efforts to procure energy security. As discussed in Chapter 2, both China's and India's oil supplies have been faced with some potential threats arising from their high dependence on the Middle East. So, diversification of oil import sources is of great importance for both countries' energy security. Thanks to their governments' proactive energy diplomacy, both countries' NOCs have already succeeded in their diversification strategy, by investing in all energy-rich countries and regions across the world. According to the IEA, from January 2002 to December 2010, China's three NOCs signed more than 40 deals to explore overseas oil and gas in Southeast Asia, Latin America, Central Asia, Africa, Russia, and North America, as well as in the Middle East, totaling approximately $83.2 billion.[257]

---

[257] Julie Jiang and Chen Ding, *Update on Overseas Investments by China's National Oil Companies: Achievements and Challenges since 2011*(Paris: OECD/IEA, 2014), 38–39.

However, some of these deals have been unsuccessful, due to domestic instability in the host countries, for example Syria, Venezuela, Ecuador, and Sudan, which has resulted in stressed assets and bad loans in these countries.[258] Learning from the lessons of their previous reckless investments and lending, since 2011, China's three NOCs' targeted investment destinations have been noticeably "moving away from riskier parts of the world towards more politically stable investment climates such as those in Organisation for Economic Co-operation and Development (OECD) member countries," especially North America.[259] As regards India's NOCs' investment diversification, according to the EIA, India's NOCs "have purchased equity stakes in overseas oil and natural gas fields in South America, Africa, Southeast Asia, and the Caspian Sea region to acquire reserves and production capability."[260]

Iran, Sudan, and Myanmar in particular, represent a boon for both the Chinese and Indian governments, as well as their NOCs, because these countries have been regarded as "pariah states" by Western governments who have forbidden their own international oil companies from investing in these countries' energy sectors. This implies that by going-out to these "pariah states," Chinese and Indian NOCs would be able to get a foothold in the international oil industry with little competition from the Western international oil companies, only competition between themselves.[261] It is no surprise that Beijing and New Delhi have done their utmost to ensure their energy security by providing diplomatic support for their NOCs to enter energy-rich countries in general, and Iran, Sudan, and Myanmar in particular. According to Meckling, Kong, and Madan, the Chinese government has played "the role of resource supplier for the going abroad of

---

[258] I owe this point to one of the anonymous reviewers of this book's manuscript.
[259] Jiang and Ding, *Update on Overseas Investments by China's National Oil Companies*, 7.
[260] EIA, "India" (June 14, 2016), 4, available at www.eia.gov/beta/international/analysis_includes/countries_long/India/india.pdf, accessed July 22, 2016.
[261] For more information on the competition and cooperation between the Chinese and Indian NOCs in the period 2000–6, see 伍福佐 [Fuzuo Wu], 《亚洲能源消费国间的能源竞争与合作》 [*Energy Competition and Cooperation among Asian Energy-Consuming Countries*], Chapter 4.

NOCs," and the Indian government is also "slowly emerging as a resource supplier" for its NOCs.[262]

Both Chinese and Indian NOCs' going-out acquisition of oil and gas assets has been successful. According to the IEA's report, Chinese NOCs have invested in 42 countries across the world, and their combined overseas oil and gas production totaled 2.5 mb/d equivalent by the end of 2013.[263] "Of this, oil production reached 2.1 million barrels per day (mb/d), a significant increase from 1.36 mb/d in 2010 and equivalent to Brazil's total oil production in 2013 and to half of Chinese domestic production in the same year."[264] In terms of Indian NOCs' success, according to its 12th FYP, by March 2011 (during its 11th FYP), Indian NOCs had invested $13 billion in the acquisition of assets abroad, mainly in oil-producing assets, which were mainly located in seven countries, that is, Russia, Sudan, Brazil, Syria, Vietnam, Venezuela, and Colombia; and production from overseas oil and gas blocks was then about 10.22 percent of India's domestic production.[265] According to the IEA's *World Energy Outlook 2015*, during India's 12th FYP period, "as of 2014, the overseas production entitlement of Indian companies operating abroad had risen to around 140 kb/d of oil and 6.1 bcm of gas, and that Indian oil and gas companies invested some $3.5 billion outside their home country in 2014."[266]

It is true that both Chinese and Indian NOCs have sold their equity oil on the international market, rather than shipping it back to their domestic consumers as a result of their focus on their own corporate profits. However, in times of crisis on the international energy market, resulting from supply disruption or price spikes, both countries' NOCs' equity oil and gas could help China and India hedge against the crisis by reducing the impact of supply disruption on their energy security and minimizing the negative effects of high oil prices on their economies.[267]

---

[262] Jonas Meckling, Bo Kong and Tanvi Madan, "Oil and State Capitalism: Government–Firm Coopetition in China and India," *Review of International Political Economy* 22, no. 6 (2015), 1160.
[263] Jiang and Ding, *Update on Overseas Investments by China's National Oil Companies*, 7.
[264] Ibid.
[265] Planning Commission, *Twelfth Five-Year Plan (2012–2017)*, Volume II, 172.
[266] IEA, *World Energy Outlook 2015*, 525.
[267] I make this point based on the hedging strategy argument of China's energy security strategy put forward by Øystein Tunsjø. See Tunsjø, *Security and Profit in China's Energy Policy*, 3.

Second, helping their NOCs to maximize their overseas energy assets or their commercial profits is an indirect way of ensuring both countries' own economic wealth, in that the more profits their NOCs make overseas, the higher the contributions they are able to make to their governments' revenues. In China's case, according to Bo Kong, in 2010, the total taxes and fees paid by the three Chinese NOCs to the Chinese government amounted to more than $100 billion, accounting for 9.1 percent of the Chinese government's total fiscal revenues in that year.[268] In 2013 and 2014, CNPC and Sinopec paid the largest amount of taxes to the Chinese government of all the 2,692 Chinese companies listed on the stock markets, with CNPC and Sinopec paying 401 billion yuan and 297 billion yuan respectively in 2013, and 408 billion yuan and 292 billion yuan respectively in 2014.[269]

Similarly, India's NOCs, more especially ONGC, have been making contributions to the Indian government's revenues, but through a different means. ONGC plays an important role in India's policy of subsidization of fuel prices, which has been a historically significant burden on government spending, with around $16 billion spent annually.[270] In 2011, for example, fuel subsidies alone cost 0.8 percent of India's GDP.[271] According to Patey, ONGC has played an important role in helping the Indian government to alleviate this heavy burden, that is:

ONGC provides around one-third of the government's total subsidy payment each year to help cover losses of Indian national refiners, Indian Oil, Bharat Petroleum and Hindustan Petroleum. From 2003 to 2008, ONGC provided $12 billion to help subsidise fuel prices. In 2011, the company gave $8 billion, representing a staggering 57 percent of its total sales. In addition, ONGC transfers vast sums of its earnings annually to the Indian Exchequer in the form of levies, duties, royalties, dividends and taxes. In effect, the subsidy payments and other contributions by ONGC ease the burden on the Indian government brought on by its fuel subsidy programme.[272]

---

[268] Meckling, Kong, and Madan, "Oil and State Capitalism," 1176.
[269] 人民网 [People.cn], "上市公司纳税哪家最多？谁增长最快？ [Which Company Has Paid the Largest Amount of Tax among the Companies on the Stock Market? Which Company's Tax Increases Fastest?]" (May 15, 2015), available at http://finance.people.com.cn/money/n/2015/0515/c218900-27005743.html, accessed May 19, 2016.
[270] Patey, The New Kings of Crude, 130.   [271] Ibid.   [272] Ibid., 130–131.

Chinese and Indian NOCs have also provided significant employment opportunities for both countries' huge labor forces. For instance, according to Bo Kong's calculation, the total number employed by the three Chinese NOCs was 2.6 million people in 2010, which was even greater than the People's Liberation Army.[273] The number of ONGC employees in 2014 and 2015 alone was more than 33,000.[274]

In sum, the proactive feature of Chinese and Indian energy policies toward energy-rich countries in general, and toward Iran, Sudan, and Myanmar in particular, has been motivated by the Chinese and Indian governments' desire to facilitate their NOCs' maximization of their commercial profits and international competitiveness, which has significantly benefited both governments' efforts to maximize their economic wealth, by hedging against potential oil supply disruptions and price spikes, and through their NOCs' contributions to increased governmental revenue in China's case, and by helping to ease the heavy burden of fuel subsidies in India's case, in addition to significant levels of employment for both countries' huge labor forces.

China and India Seeking Status

Aside from domestic pressures to maximize their economic wealth, China and India have faced international pressures in the process of seeking to enhance their status in the international system, that is, social opprobrium of their energy diplomacy behavior due to its breach of internationally accepted norms and principles. As far as the three cases discussed in this chapter are concerned, China's and India's reactive behavior in their dealings with Iran, Sudan, and Myanmar can be largely attributed to their desire to acquire the status of responsible great powers in the international system, meaning that they have had to react to social opprobrium of their proactive energy diplomacy toward these three so-called "pariah states."

In Iran's case, China's and India's desire to enhance their responsible great-power status in the international system has made them very sensitive to any rhetoric or policy measures that might bring shame upon their own identity or status, arising from the mechanism of

[273] Meckling, Kong, and Madan, "Oil and State Capitalism," 1176.
[274] ONGC, *Annual Report 2014–15*, available at www.ongcindia.com/wps/wcm/OngcHTML/Annual_Report_2014_15/images/AR-14-15.pdf, 34, accessed March 7, 2018.

"naming, shaming and sanctioning."[275] For instance, Zoellick's speech mentioned above was an implicit way of blaming China for not having been a responsible stakeholder in the international system, making China feel ashamed, given that it has long desired to be treated as a responsible great power by the international community. Under such circumstances, "naming and shaming" has led China to shift its policy toward Iran. As far as sanctions are concerned, the United States has imposed sanctions on Chinese firms and individuals for their weapons of mass destruction (WMD)-related exports to some sensitive countries, including Iran, since the 1990s. These sanctions, in fact, further indicate the gap between China's energy diplomacy behavior and international non-proliferation norms and rules, another way of naming and shaming China's (especially Chinese firms and individuals) proliferation behavior. Thus, in order to prove to the international audience that it is a responsible great power, the Chinese government has had to take measures to strengthen its export controls toward Iran. India's "nuclear pariah" status in the international system was not shaken off until July 2005, when the United States signed the civil nuclear agreement with India. Against this backdrop, in order to gain the status of or identity as a responsible nuclear weapons state, India has had to side with the international community, led by the United States, to restrain Iran's nuclear weapons ambitions.

With regard to Sudan and Myanmar, the most important mechanism employed by human rights NGOs to force China and India to shift their policies toward these two countries is the "mobilization of shame," using public exposure of their economic, military, and diplomatic support to Sudan and Myanmar, two "pariah states" that have severely violated the human rights norms. The reason that shaming can change China's and India's policy behavior toward Sudan and Myanmar lies in the fact that:

shaming then constructs categories of "us" and "them," that is, in-groups and out-groups, thus reaffirming particular state identities. Some repressive governments might not care. Others, however, feel deeply offended, because they want to belong to the "civilized community" of states. In other words, shaming then implies a process of persuasion, since it

---

[275] H. Richard Friman (ed.), *The Politics of Leverage in International Relations: Name, Shame, and Sanction* (New York: Palgrave Macmillan, 2015).

convinces leaders that their behaviour is inconsistent with an identity to which they aspire.[276]

By using shaming, China's and India's identity was constructed by human rights NGOs as belonging to an in-group with Sudan and Myanmar, two "pariah states," which is obviously inconsistent with their desired identity as responsible great powers in the international system. Faced with potential damage to their desired identity, China and India have had to adjust their proactive policies toward Sudan and Myanmar to various degrees.

China and India Facing Asymmetrical Interdependence[277]

China's and India's high-level reactiveness in terms of their energy diplomacy toward Iran has been shaped by another systemic pressure, that is, China's and India's asymmetrical dependence on the United States and its allies.

First and most importantly, the degree of China's and India's economic interdependence with the United States is much higher than that with Iran, which can be illustrated by both bilateral trade and investment figures. In terms of trade, in 2011, for instance, China and the US's bilateral two-way trade totaled more than $500 billion, with China enjoying a surplus of more than $290 billion,[278] while Sino-Iranian trade was merely $45 billion, which means that China's trade volume with Iran is less than one-tenth of its trade with the United States. Similarly, the Indo-US bilateral two-way trade volume reached $57.7 billion, with India enjoying a surplus of nearly $15.0 billion.[279] In contrast, India and Iran's bilateral trade amounted to only $14 billion, with India suffering a $4.2 billion deficit in 2014.[280] With regard to investment, the United States was China's fifth-largest foreign direct

---

[276] Margaret E. Keck and Kathryn Sikkink, *Activists beyond Borders: Advocacy Networks in International Politics* (Ithaca, NY: Cornell University Press, 1998), 15.

[277] This section largely draws on my article, Wu, "China's Puzzling Energy Diplomacy toward Iran," 61–63.

[278] United States Census Bureau, "Trade in Goods with China" (2011), www.census.gov/foreign-trade/balance/c5700.html, accessed March 2, 2018.

[279] United States Census Bureau, "Trade in Goods with India" (2011), available at www.census.gov/foreign-trade/balance/c5330.html, accessed March 2, 2018.

[280] BBC, "How Iran's Nuclear Deal Affects India" (July 16, 2015), available at www.bbc.com/news/world-asia-india-33547061, accessed March 2, 2018.

investor in the first half of 2012, and China is the largest holder of US treasury securities, having reached $1.258 trillion by September 2015, accounting for 20.6 percent of the US's total foreign holdings.[281] In contrast, China's investment in Iran is "considerably smaller than the $100–$120 billion frequently cited."[282] US investment in India has been rising since 2000; by 2014, US investment in India had reached nearly $28 billion.[283] Against this backdrop, maintaining a healthy trade and investment relationship with the United States is of great importance for China's and India's economic health and growth.

Second, the United States and some of its Western allies, such as Australia and Canada, are rich in energy, so these countries are also important targets for Chinese and Indian NOCs' energy-seeking efforts. Accordingly, Chinese and Indian NOCs have tried to avoid challenging the US's Iranian policy, by restraining their energy-related activities in Iran in order to gain access to the energy sectors of both the USA and its allies. It has turned out that their restraint in developing Iran's energy projects has paid off to a certain extent. For instance, CNOOC, although it had failed to purchase Unocal in 2005, due to strong opposition from the US Congress, succeeded in obtaining energy assets in the United States, together with Sinopec, totaling more than $4.6 billion in 2010.[284] In September 2010, CNPC won a contract to explore Australian gas with the US major Chevron.[285] In 2011, CNPC established a subsidiary in the USA, namely CNPC USA, "to serve as a bridge between North America and CNPC worldwide operations."[286] The Indian government has also tried to help its NOCs

---

[281] Wayne M. Morrison, *China–U.S. Trade Issues* (Congressional Research Service report, December 2015), 14–15, available at www.fas.org/sgp/crs/row/RL33536.pdf, accessed March 2, 2018.

[282] Downs and Maloney, "Getting China to Sanction Iran," 18.

[283] Statista, "Direct Investment Position of the United States in India from 2000 to 2014" (2015), available at www.statista.com/statistics/188633/united-states-direct-investments-in-india-since-2000, accessed September 5, 2015.

[284] Scott Harold and Alireza Nader, "China and Iran: Economic, Political, and Military Relations," Occasional Paper (Santa Monica, CA: RAND, 2012), 16, available at www.rand.org/content/dam/rand/pubs/occasional_papers/2012/RAND_OP351.pdf, accessed January 14, 2013.

[285] Chen Aizhi, "Exclusive: China Slows Iran Oil Work as U.S. Energy Ties Warm," *Reuters* (October 28, 2010), available at www.reuters.com/article/us-china-iran-oil-idUSTRE69R1L120101028, accessed March 2, 2018.

[286] CNPC USA Corporation, "CNPC USA," available at www.cnpc-usa.com, accessed March 2, 2018.

to seek energy resources in countries that are US allies. For instance, in 2008, when the Indian Minister of External Affairs visited Australia, both countries signed five action plans, in the areas of coal, mining, power, petroleum, and natural gas, and new and renewable energy.[287] After negotiating for several years, India eventually succeeded in securing a uranium deal with Canada in April 2015.[288] Simply put, Chinese NOCs' commercial interests in the energy sector of the USA (and its allies), and India's interests in securing energy resources in Western countries allied to the USA, especially Australia and Canada, have forced them to yield to US pressure on the Iranian nuclear issue.

In addition, China's and India's increasingly strengthened energy ties with Saudi Arabia, the world's largest oil producer and the USA's close ally in the Middle East, have also stimulated both countries to react to US and Saudi pressures to curb Iran's nuclear ambitions. Saudi Arabia has already become China's and India's largest oil supplier, accounting for 16 percent of China's total oil imports in 2014, and 20 percent of India's oil imports in 2015,[289] which could hardly be replaced by any other countries' supplies, including those of Iran. Therefore, the contribution that oil supplies from Saudi Arabia make to China's and India's economic development is irreplaceable. In this context, China's and India's participation in international efforts to curb Iran's nuclear ambitions helps Saudi Arabia alleviate its geostrategic concerns about Iran, and therefore facilitates Sino-Saudi and Indo-Saudi energy ties.[290]

Furthermore, in terms of arms supplies, Israel, the USA's important ally in the Middle East, has been of great importance for India. Israel has long surpassed Russia in being India's largest arms supplier. For instance, during the period 2010–14, 46 percent of Israeli arms exports

---

[287] MEA, "India–Australia Relations," available at http://meaindia.nic.in/meaxpsite/foreignrelation/australia.pdf, accessed January 15, 2011.

[288] Sushant Singh, "Explained: Significance of India's Uranium Deal with Canada," *The India Express* (April 17, 2015), available at http://indianexpress.com/article/explained/explained-a-deal-in-canada, accessed March 2, 2018.

[289] EIA, "China" (May 14, 2015) 11; EIA, "India," 11.

[290] Parris H. Chang, "China's Policy toward Iran and the Middle East," *The Journal of East Asian Affairs* 25, no. 1 (Spring/Summer 2011), 1–14; Khalid Rahman, "India-Iran Relations and Current Regional Dynamics," *Policy Perspectives* 7, no. 2 (July–December 2010), 35; Harsh V. Pant, "India's Relations with Iran: Much Ado about Nothing," *The Washington Quarterly* 34, no. 1 (2011), 66–68.

went to India.[291] Among those defense purchases from Israel, surface-to-air missiles, surveillance, and missile-defense technology are part of a high-priority modernization of the Indian armed forces.[292] So Israel is pivotal for India's strategic security.[293]

Conclusion

China and India – given their continued high dependency on overseas energy supplies, especially oil supplies from the Middle East – have proactively used diplomacy to facilitate their NOCs "going-out" to seek overseas energy resources, by strengthening their ties with all energy-rich countries across the world, including the so-called "pariah states," such as Iran, Sudan, and Myanmar. At the same time, however, China's and India's energy diplomacy toward Iran, Sudan, and Myanmar bears reactive features. On the one hand, China and India have been proactively seeking to build their energy ties with Iran, Sudan, and Myanmar, through not only political and economic, but also military and diplomatic means, despite international efforts to isolate and/or to impose sanctions against these three countries. On the other hand, China and India have taken some measures that obviously run counter to their efforts to secure oil supplies from these three countries: China and India have not only voted against Iran at the IAEA and the UNSC, but have also cut their oil imports from Iran in 2012 and 2013 respectively; in addition, China's and India's NOCs have been more cautious about their activities in the Iranian energy sector. In the case of Sudan, China has not only tried to persuade the Sudanese president to agree to peacekeeping troops being stationed in Darfur, but has also supported the peaceful secession between the north and south. In terms of China's and India's reactive measures toward Myanmar, Chinese leaders called on the Myanmar junta government to improve its governance, while India halted its military exports to Myanmar.

---

[291] SIPRI, "Trends in International Arms Transfers, 2014" (March 2015), available at books.sipri.org/files/FS/SIPRIFS1503.pdf, accessed August 20, 2015.
[292] Amol Sharma and Diksha Sahni, "Attacks Test Ties between India, Iran," *Wall Street Journal*, Eastern edn (February 15, 2012), A.9.
[293] P.R. Kumaraswamy, "Israel: The Non-Parallel Player," *Strategic Analysis* 36, no. 6 (2012), 976.

The two-level pressures have shaped China's and India's energy diplomacy in general, and their energy diplomacy toward Iran, Sudan, and Myanmar in particular. At the domestic level, both countries have tried to facilitate their NOCs "going-out" to seek overseas energy resources, so as to maximize their commercial profits and international competitiveness in the international energy market, which is a way for the Chinese and Indian governments to maximize their domestic economic wealth, by not only hedging their energy security against potential supply disruptions and against the negative impacts of high oil prices on their economies, but also by maximizing China's own tax revenues through the taxes paid by its NOCs, and by helping the Indian government ease the heavy financial burden of its fuel subsidies, in addition to providing a lot of employment opportunities for both countries' huge labor forces.

At the same time, however, China and India have faced international pressures, that is, their desire to seek great-power status has been constrained not only by social opprobrium stemming from their breaches of international norms and principles in their energy diplomacy behavior, but also by their asymmetrical interdependence with the United States and some of its allies. More specifically, in Iran's case, the United States and its allies have played an important role in putting pressure on China and India, to force them to side with the international community to restrain Iran's nuclear ambitions. The main reason for China's and India's reactiveness to pressures from the United States and its allies lies in their desire to seek great-power status, as well as in the asymmetrical interdependence between Sino-India and the United States and its allies. That is to say, compared with Iran, the United States and its allies are far more important for China's and India's economic health and development, their energy security, and India's military development.

In terms of Sudan and Myanmar, both non-state and state actors have put pressure on China's and India's energy diplomacy toward these two countries. Human rights NGOs, in particular, have led the social opprobrium of China's and India's ignorance of violations of human rights and poor governance in Sudan and Myanmar. China and India have been sensitive to this social opprobrium, since both countries have aspired to gain "responsible great power" status in the international system. Accordingly, both countries have had to alter their proactive policy toward the two countries. Thus, their energy diplomacy bears proactive and reactive features.

PART III

The Outside-In

5 | Negotiating Climate Change: Proactively Free-Riding and Reactively Burden Sharing[1]

Proposition 3: China's and India's negotiating stances in international climate change negotiations are increasingly shaped by asymmetrical interdependence, climate protection norms, and social opprobrium at the systemic level.

In stark contrast with the inside-out process of China's and India's policies for addressing their energy insecurity, both countries' policies for addressing climate change can be characterized mainly by an outside-in process, that is, the driving forces that have stimulated China and India to take action, domestic and foreign, to address climate change, stem from pressures at the international level.

This chapter explores China's and India's foreign policies on how to address climate change at the international level during the post-Kyoto Protocol period. China's and India's climate change foreign policies during this period can be characterized by both countries taking proactive negotiating strategies in order to continue to free-ride on developed countries' efforts to mitigate GHG emissions. But at the same time, both countries have had to agree to share the burden, by mitigating their increased GHG emissions in response to international pressures. To illustrate this proactive and reactive feature of China's and India's climate diplomacy, and the two-level pressures that have led to this, the chapter is arranged as follows. The first section examines the proactive and reactive feature of Sino-Indian climate diplomacy. The second section explores how domestic pressures on China

---

[1] This chapter is derived from my three articles: Fuzuo Wu, "Shaping China's Climate Diplomacy: Wealth, Status, and Asymmetric Interdependence," *Journal of Chinese Political Science* 21, no. 2 (June 2016), 199–215; Fuzuo Wu, "China's Pragmatic Tactic in International Climate Change Negotiations: Reserving Principles with Compromise," *Asian Survey* 53, no. 4 (2013), 778–800; and Fuzuo Wu, "Sino-Indian Climate Cooperation: Implications for the International Climate Change Regime," *Journal of Contemporary China* 21, no. 77 (2012), 827–843.

and India to seek wealth, and the international pressures they have faced in the process of enhancing their status in the international system, characterized by asymmetrical interdependence, have shaped this proactive and reactive feature. The third section offers a brief conclusion.

Dual-Track Climate Diplomacy

One noticeable feature of China's and India's climate diplomacy is that it has been dual-track. That is to say, both countries have participated in UN climate negotiations and non-UN climate arrangements. The main characteristic of their UN track climate negotiations is that both states have proactively tried to forge new alliances, to facilitate their bargaining power, while maintaining their traditional alliance with the Group of 77 plus China (G-77/China), under the UNFCCC's various negotiating forums, especially the Conference of the Parties (COP). At the same time, China and India have been voluntarily and reactively bandwagoning the US- and the EU-initiated minilateral and bilateral climate arrangements. The outcome of this dual-track climate diplomacy is that China and India have had to make some significant compromises on their traditional negotiating positions. Specifically, China and India have already begun to shift away from their traditional free-riding behavior to reactively burden sharing.

UN Track Climate Negotiations: Building New Coalitions while Maintaining Old Ones

Building coalitions to defend their negotiating stances under the UNFCCC negotiations is nothing new, given the fact that China and India have aligned themselves with the broad coalition of developing countries, that is, the G-77/China, since the inception of ICCN in the early 1990s. Nevertheless, in recent years, a new characteristic of Sino-Indian alignment strategy under the UNFCCC negotiations is that both states have tried to build new coalitions to strengthen their bargaining power.

It is well-known that the central tenet of the Sino-Indian approach to addressing climate change at the global level over the past two decades has been equitable burden sharing, which is guided by the principles of historical responsibility for GHG emissions and "common but differentiated responsibilities (CBDR)."[2] In other words, maintaining the norm of "equity" has been a priority for Sino-Indian climate diplomacy.[3] In practice, both China and India have vehemently opposed taking on binding emissions reduction obligations, emphasizing their own need to grow and reduce poverty, while pushing hard for developed countries to undertake their historical responsibility to address climate change,[4] which is acknowledged as one of the fundamental norms and principles of the UNFCCC and its Kyoto Protocol. In order to maintain this stance in the post-Kyoto Protocol climate change negotiations, China and India have adopted the strategy of building new coalitions to strengthen their bargaining power against developed countries and some of developing countries in the G-77/China, given that both states' increased CO_2 emissions have made it increasingly hard for them to maintain the unity of their traditional coalition with the G-77/China.

As illustrated in Table 5.1, in terms of absolute CO_2 emissions annually, China's emissions were almost equal to the United States and the EU combined in 2011, while India has become the third-largest emitter. As for per capita emissions, China's 6.72 tCO_2 has already surpassed the world average (4.63 tCO_2) while India's emissions (1.52 tCO_2) are much lower. As regards cumulative CO_2 emissions, between 1990 and 2011, China accounted for nearly 19 percent of the world total, second only to the United States. In contrast, India accounted for only 4 percent of global cumulative CO_2 emissions in the same period. However, according to the IEA, India's CO_2 emissions will increase by 34 percent by 2020, and will double by 2030 under its existing policies, and it will overtake the United States in terms of annual CO_2 emissions before 2040.[5] Simply put, China and India have become and will continue to be two of the largest CO_2 emitters in the world. Thus, the traditional dichotomy in the Kyoto Protocol, characterized by Annex I Parties (i.e. developed countries and countries in transition)

---

[2] Wu, "Sino-Indian Climate Cooperation," 832.
[3] Phillip Stalley, "Principled Strategy: The Role of Equity Norms in China's Climate Change Diplomacy," *Global Environmental Politics* 13, no. 1 (February 2013), 1–8; Atteridge et al., "Climate Policy in India," 70.
[4] A. Korppoo and A. Luta (eds), *Towards a New Climate Regime: Views of China, India, Japan, Russia and the United States on the Road to Copenhagen* (Helsinki: The Finnish Institute of International Affairs, 2009), 19.
[5] IEA, *World Energy Outlook 2014*, 89.

Table 5.1 *Current and historical CO_2 emissions, 2011*

	$MtCO_2$	Share of global CO_2 emissions (%)	Per capita emissions (tCO_2)	Cumulative CO_2 emissions as share of global emissions, 1990–2011 (%)*
USA	5,603.82	17.4	17.92	21.6
EU	3,774.27	11.7	7.43	15.9
China	9,034.97	28.0	6.72	18.5
India	1,860.92	5.8	1.52	4.4
World	32,273.73	100.0	4.63	100.0

Source: World Resource Institute (WRI), CAIT 2.0, available at http://cait2.wri.org/wri, accessed December 22, 2014
* *Note:* Figures in this column are calculated by the author, based on the data in WRI's CAIT 2.0. Total CO_2 emissions ($MtCO_2$): world (555454.74); USA (119843.31); China (102628.75); EU (88159.11); India (24325.14).

abiding by binding emissions cut obligations, while non-Annex I Parties, including China and India, having been exempted from such obligations, has failed to capture "observed changes in emissions trajectories over the pre- and post-Kyoto periods."[6]

Against this backdrop, not only developed countries, but also some developing countries in the G-77/China, such as the Alliance of Small Island States (AOSIS) and the least-developed countries (LDCs), have required China and India to undertake emissions cut obligations. Faced with mounting pressures, China and India have adopted a strategy of building new coalitions to facilitate their bargaining power, which had been weakened by the fragmentation of the G-77/China.[7] So far, China and India have built not only a strong bilateral coalition, but also two new mini-multilateral ones.

In terms of the Sino-Indian bilateral climate coalition, in October 2009, both states signed a Memorandum of Agreement on Cooperation

[6] Erick Lachapelle and Matthew Paterson, "Drivers of National Climate Policy," *Climate Policy* 13, no. 5 (2013), 552.
[7] Sjur Kasa, Anne T. Gullberg, and Gørild Heggelund, "The Group of 77 in the International Climate Negotiations: Recent Developments and Future Directions," *International Environmental Agreements: Politics, Law and Economics* 8 (2008), 113–127; Antto Vihma, "Elephant in the Room: The New G77 and China Dynamics in Climate Talks," Briefing Paper (Finnish Institute of International Affairs, 2010).

Table 5.2 *Joint statements with China during public negotiation sessions, December 2009–December 2015*

Partner	Number of joint statements	Joint statements of China's statements (%)	Joint statements of partners' statements (%)
India	63	20.7	28.4
Saudi Arabia	37	12.2	14.3
Brazil	36	11.8	15.1
Philippines	20	6.6	17.2
South Africa	13	4.3	7.1
AOSIS	9	3.0	3.0
G-77/China	8	2.6	2.7

Source: Coding of *Earth Negotiation Bulletin* (*ENB*) reports, following the coding method described in Florian Weiler, "Determinants of Bargaining Success in the Climate Change Negotiations," *Climate Policy* 12, no. 5 (2012), Appendix 1, 572

on Addressing Climate Change, in which they announced that an India–China Working Group on Climate Change would be established to "exchange views on important issues concerning international negotiation on climate change."[8] This official announcement ushered in the Sino-Indian bilateral climate alliance, which has facilitated their coordinated efforts at the UNFCCC's various negotiating forums, especially the COPs.[9] This can best be illustrated by the joint statements they have issued during those negotiations.[10] From Tables 5.2 and 5.3, we can see that China and India made 63 joint statements in the UN climate negotiations between December 2009 and December 2015, accounting for 20.7 percent and 28.4 percent of China's and India's statements respectively. This coordination confirms an observation made by Jairam Ramesh, India's former Environment Minister, who led India's negotiations under the UNFCCC until July 2011, that is, cooperation between India and China on climate change has been

---

[8] Press Information Bureau, GOI, "India and China Sign Agreement on Cooperation on Addressing Climate Change" (October 21, 2009), available at www.pib.nic.in/newsite/erelease.aspx?relid=53317, accessed March 3, 2018.
[9] Wu, "Sino-Indian Climate Cooperation."
[10] I do not analyze the detailed contents of those joint statements here because it is very obvious that all the debates under the UNFCCC negotiations focus on burden sharing between parties related to climate mitigation and adaption.

Table 5.3 *Joint statements with India during public negotiation sessions, December 2009–December 2015*

Partner	Number of joint statements	Joint statements of India's statements (%)	Joint statements of partners' statements (%)
China	63	28.4	20.7
Saudi Arabia	30	13.1	11.6
Brazil	20	9.0	8.4
Philippines	17	7.6	14.7
South Africa	9	4.1	4.9
G-77/China	9	4.1	3.0
AOSIS	2	0.9	0.3

Source: Coding of *ENB* reports, following the coding method described in Weiler, "Determinants of Bargaining Success in the Climate Change Negotiations," 572

"one of the outstanding success stories of this bilateral relationship," despite the huge gap between their respective GHG emissions.[11]

In addition, China and India have built two mini-multilateral coalitions. More specifically, in November 2009, in the run-up to the Copenhagen climate summit (COP-15), China and India formed the BASIC group, together with two other emerging economies – Brazil and South Africa. Tables 5.2 and 5.3 show that in the UN climate negotiations during the period 2009–15, Brazil and South Africa proved to be close allies for China and India, given the fact that Brazil shared 36 and 20 joint statements respectively with China and India, while South Africa shared 13 and 9 Sino-Indian statements. This group significantly boosted China's and India's negotiating power at COP-15, so that the BASIC group's leaders were able to play an important role in shaping the Copenhagen Accord, jointly with US President Barack Obama, following their hard bargaining against Obama in the final days of this summit. Although this Accord was not officially adopted at COP-15, it paved the way for a fundamentally different approach to the future climate regime that succeeded the Kyoto Protocol, that is, "pledge and review" in general, and voluntary GHG

---

[11] Anantha Krishnan, "Climate Cooperation Changing India–China Ties, Says Jairam Ramesh," *The Hindu* (April 9, 2010), available at http://beta.thehindu.com/news/international/article392921.ece, accessed March 3, 2018.

emissions mitigation actions in particular.[12] Even so, as early as the Cancun climate summit (COP-16) in 2010, a divergence of views among the BASIC group had emerged, which became more apparent at the Durban climate summit (COP-17), where both Brazil and South Africa not only agreed to take their own GHG mitigation actions, but also "indicated an openness to the idea" of negotiating a legally binding climate change agreement,[13] which obviously runs counter to China's and India's traditional negotiating stances. Given the divisions among the BASIC group, it seems unlikely that China and India will continue to rely on this group to strengthen their bargaining power.[14] Accordingly, China and India have built a new coalition – the likeminded developing countries (LMDCs) – an informal negotiating bloc, targeted at upholding "the Convention's principles of common but differentiated responsibilities and equity, as well as developed countries' historical responsibility for climate change."[15]

The LMDCs, consisting of Bolivia, China, Cuba, Dominica, Ecuador, Egypt, El Salvador, India, Iran, Iraq, Malaysia, Mali, Nicaragua, Philippines, Saudi Arabia, Sri Lanka, Sudan, and Venezuela, have negotiated together under the UNFCCC since the Bonn session in June 2012, and held their first meeting in Beijing in October 2012.[16] It has turned out that this group has significantly facilitated China's and India's negotiating power. For instance, from Tables 5.2 and 5.3, we can see that Saudi Arabia and the Philippines, both members of the LMDCs, were China's and India's second- and fourth-closest allies respectively, given the number of their joint statements with both countries. Moreover, the effectiveness of the LMDCs is illustrated by the outcome of the negotiations. For example, at the Warsaw climate summit (COP-18) in 2013, thanks to strong support from the LMDCs,

[12] Wu, "Sino-Indian Climate Cooperation," 836–838.
[13] Kathryn Hochstetler and Manjana Milkoreit, "Emerging Powers in the Climate Negotiations: Shifting Identity Conceptions," *Political Research Quarterly* 67, no. 1 (2014), 231.
[14] Kathryn Ann Hochstetler, "The G-77, BASIC, and Global Climate Governance: A New Era in Multilateral Environmental Negotiations," *Revista Brasileira de Política Internacional* 55 (special edn, 2012), 67.
[15] International Institute for Sustainable Development (IISD), "Summary of the Doha Climate Change Conference: 26 November–8 December 2012," *ENB* 12, no. 567 (December 11, 2011), 28.
[16] Karl Hallding, Marie Jürisoo, Marcus Carson, and Aaron Atteridge, "Rising Powers: The Evolving Role of BASIC Countries," *Climate Policy* 13, no. 5 (2013), 621.

India and China succeeded in insisting that the principle of equity remain in the agreement when they negotiated hard against developed countries on this issue. A media article reiterated the importance of the LMDCs for China's and India's negotiation on this occasion: "The support from the LMDC group helped both [China and India] save the day."[17]

Although China and India have tried to build new coalitions following the fragmentation of their traditional alliance, the G-77/China, both states have also tried to maintain unity with the G-77/China in order to avoid a public split. In Tables 5.2 and 5.3, we can see that China and India have also made some joint statements with the G-77/China. This implies that Sino-India and the G-77/China still have shared interests on certain issue areas in ICCN. As a result, in an attempt to maintain the unity of the G-77/China, China and India invited some representatives of other countries from the G-77/China to attend meetings of their new alliances, especially those of the BASIC group;[18] in addition, China and India, along with the BASIC group, repeatedly iterated their desire to strengthen the G-77/China coalition. In April 2010, for instance, the BASIC group, at its third ministerial meeting, stated that "Ministers of the BASIC countries agreed that, remaining anchored in the G77&China."[19] Once again, in the Joint Statement issued at their 11th meeting in July 2012, the BASIC Ministers wrote, "Ministers emphasized that BASIC countries, as part of the G-77 and China, continue to work to maintain the strength and unity of the Group. The Ministers reaffirmed the importance of the unity of the G-77 and China as the common voice of developing countries in the climate change negotiations."[20] In other words, although China

---

[17] "US, EU Nations Try to Split Developing Country Group at Climate Talks," *Business Standard* (August 30, 2014), available at http://smartinvestor.businessstandard.com/market/story-262778-storydet-US_EU_nations_try_to_split_developing_country_group_at_climate_talks.htm#.Wp5bIWdLH4g, accessed March 6, 2018.

[18] Hochstetler, "The G-77, BASIC, and Global Climate Governance," 60.

[19] "Joint Statement Issued at the Conclusion of the Third Meeting of BASIC Ministers," Cape Town (April 25, 2010), available at http://envfor.nic.in/downloads/public-information/BASIC-statement.pdf, accessed March 3, 2018.

[20] "Joint Statement Issued at the Conclusion of the 11th BASIC Ministerial Meeting on Climate Change," Johannesburg (July 12–13, 2012), 4, available at www.indiaenvironmentportal.org.in/files/file/Joint%20Statement%20at%2011th%20BASIC%20Ministerial%20Meeting%20on%20Climate%20Change.pdf, accessed December 15, 2014.

Table 5.4 *China's foreign aid related to addressing climate change in developing countries, 2000–2012*

Type of aid provided	Number of developing countries assisted	Content of the aid
Assisting construction projects	58	64 projects on the utilization of renewable energy resources
Providing materials	13	16 batches of equipment and supplies for environmental protection
	9	More than 500,000 energy-efficient lamps and 10,000 energy-efficient air conditioners
Providing assistance in capacity building	3	Utilization and management of solar power, hydropower, and other clean energy
Training workshops	120 (2010–2012)	150 training sessions on environmental protection and addressing climate change for over 4,000 officials and technical personnel
	- (2000–2009)	50 training workshops for more than 1,400 people on the development and use of renewable resources

Sources: Information Office of the State Council of the PRC, "China's Foreign Aid (2014)", July 10, 2014, available at www.china.org.cn/government/whitepaper/node_7209074.htm, accessed May 20, 2015; Information Office of the State Council of the PRC, "China's Foreign Aid" (April 21, 2011), available at www.china.org.cn/government/whitepaper/node_7116362.htm, accessed March 3, 2018
Note: - represents no data available.

and India have built new coalitions to strengthen their bargaining power, they have tried to remain in the G-77/China.

Aside from the above-mentioned symbolic gestures used to maintain their traditional alliance with the G-77/China, China and India have also deployed foreign aid to help the least-developed countries, small island countries, and African countries, to enhance their capacity to adapt to climate change so as to alleviate pressures from these groups.

Table 5.5 *China's and India's non-UN climate bandwagoning diplomacy*

Minilateral climate arrangements	G8+5; APP; MEF; G20	
Bilateral climate arrangements	China–United States	India–United States
	• Bilateral agreement to phase out HFCs, 2013 • Bilateral climate deal, 2014 • Joint presidential statement, 2015	• Indo–US nuclear deal, 2008 • US–India Partnership to Advance Clean Energy (PACE), 2009 • India agreed to phase out HFCs, 2015 • Fulbright Climate Fellowship program, 2015
	China–EU	India–EU
	• EU–China Joint Declaration on Climate Change, 2005 • EU–China Joint Statement on Coordination and Cooperation on Climate Change, 2010 • EU–China 2020 Strategic Agenda for Cooperation, 2013 • EU–China Joint Statement on Climate Change, 2015	• India–EU Strategic Partnership–Joint Action Plan, 2005 • Joint Work Programme for EU–India Cooperation on Energy, Clean Development and Climate Change, 2008

Source: Author

Table 5.4 shows, in the period 2000–2012, China provided aid to developing countries in order to help them improve their capacity to address climate change, including assisting construction projects on renewable energy, providing equipment and supplies for environmental protection, and providing training workshops on environmental protection and addressing climate change.

In summary, given their increased levels of GHGs, especially CO_2 emissions, it has become increasingly hard for China and India to maintain their traditional stances on how to address climate change under the UNFCCC negotiations. Accordingly, both states have adopted a strategy of building new coalitions, including their bilateral alliance, the BASIC group, and the LMDCs, while maintaining their old coalition – the G-77/China – to strengthen their bargaining power in ICCN.

Non-UN Track: Bandwagoning Climate Arrangements

China and India have also participated in some non-UN track climate arrangements and negotiations (see Table 5.5). More specifically, China and India have been voluntarily and reactively bandwagoning the climate change arrangements initiated by developed countries, especially the United States and the EU. So far, the USA and the EU have initiated both minilateral and bilateral arrangements to engage emerging economies, including China and India, to address climate change outside the UNFCCC process, which are regarded by Keohane and Victor as club-making efforts on the part of developed countries.[21]

Bandwagoning Minilateral Climate Arrangements
The first climate club to have been bandwagoned by China and India is the G8 plus 5 (G8+5). In 2005, China and India, together with Brazil, Mexico, and South Africa, accepted an invitation from Tony Blair, then Britain's prime minister, to attend the G8 summit at Gleneagles to join talks on some topics, including climate change. In 2006, the G8+5 Climate Change Dialogue was launched, and was institutionalized at the G8 summit held in Heiligendamm in 2007. At this summit, Chinese President Hu Jintao's speech signaled a subtle change in Chinese positions on how to tackle climate change at the global level.[22] More specifically, Hu pointed out that:

developing countries' industrialization, urbanization and modernization is far from being completed. So these countries' tasks of economic development and improvement of people's livelihood are arduous. In order to achieve development goals, energy demand in developing countries will increase, which is the basic condition for development in developing countries.

---

[21] Robert O. Keohane and David G. Victor, "The Regime Complex for Climate Change," *Perspectives on Politics* 9, no. 1 (March 2011), 10.
[22] According to a Chinese scholar, in Hu's speech, the Chinese government did not insist on its traditional position that China would not undertake GHG emission mitigation obligations until it reached the level of middle developed countries. See Zhuang, "Post-Kyoto International Climate Governance and China's Strategic Options," 9.

Therefore, at this stage, it is inappropriate to require developing countries to make mandatory emission reductions.[23]

At the same time, however, Hu said that "developing countries should also take measures within their capabilities based on their own situation to make some positive contributions to promote global sustainable development."[24] By the same token, at this summit, India's Prime Minister Manmohan Singh declared that, "We [India] recognize wholeheartedly our responsibilities as a developing country," and India wished "to add weight to global efforts to preserve and protect the environment."[25]

The second climate club once bandwagoned by China and India is the now-defunct APP, initiated by US President George W. Bush on July 28, 2005. China and India, together with the United States and its three allies – Australia, Japan, and South Korea – were the founding members.[26] The potential benefits for China and India included some clean, efficient, and cost-effective technologies transferred from the United States and the other three developed countries in the APP, through their participation in the eight public-private sector task forces established to facilitate the transfer of environmentally sound technologies in energy sectors.[27] Through their participation in these task forces, China and India intended to build capacity to implement climate-friendly policies and projects to mitigate the results stemming

---

[23] 胡锦涛 [Hu Jintao], "在八国集团同发展中国家领导人对话会议上的讲话 [Speech at the Summit Dialogue between G8 and Developing Countries' Leaders]" (June 8, 2007), available at www.fmprc.gov.cn/123/wjdt/zyjh/t328680.htm, accessed March 3, 2018.

[24] Ibid.

[25] Manmohan Singh, "PM's Intervention on Climate Change at the Heiligendamm Meeting" (June 8, 2007), available at http://mea.gov.in/in-focus-article.htm?18822/PMs+intervention+on+Climate+Change+at+the+Heiligendammn+meeting, accessed March 3, 2018.

[26] Canada joined in 2007 as its seventh member; see APP website at www.asiapacificpartnership.org/english/default.aspx, accessed December 16, 2014.

[27] Mike Atkinson, Peter Castellas, and Paul Curnow, *Independent Review of Asia-Pacific Partnership Flagship Projects* (December 2009), 4, available at www.asiapacificpartnership.org/pdf/resources/Final_Flagship_Review_-_English.pdf, accessed December 23, 2014.

from their high dependence on coal – a major source of their CO_2 emissions.[28] However, the APP was not successful, so it was officially dissolved.[29]

The third climate club that China and India have bandwagoned is the US-initiated MEF. In 2007, given the APP's lack of practical consequences due to its limited membership, the US Bush administration initiated the Major Economies Meetings on Energy Security and Climate Change, to include 16 countries plus the EU to address climate change. China and India were both in this club.[30] In 2009, this club was relaunched and rebranded by the Obama administration as the Major Economies Forum on Energy and Climate Change (MEF), with essentially the same 17 participants.[31] The aim of MEF is to set its own rules for a more flexible strategy to reduce emissions among its member states. This group has conducted a series of meetings on how to address climate change.[32] For instance, in July 2009, at the MEF summits in L'Aquila, Italy, all 17 member countries, including China and India, not only accepted the "scientific view" that increases in global average temperature "ought not to exceed 2 degrees C," but also agreed to "identify a global goal for substantially reducing global emissions by 2050."[33] China's and India's participation in this declaration symbolically signaled that they were willing to peak their GHG emissions by the middle of the twenty-first century. In another instance, at the 7th MEF meeting, held in Rome in 2010, Jairam Ramesh reframed the concept of "equity" as "equitable access to

[28] IEA, *World Energy Outlook 2007*.
[29] By the time the APP became defunct in 2011, only 9 out of 100 projects had been fully completed. See APP "project roster" of APP 8 Public-Private Sector Task Forces, available at www.asiapacificpartnership.org/english/task_forces.aspx, accessed December 23, 2014.
[30] MEF's member states are: Australia, Brazil, Canada, China, the EU, France, Germany, India, Indonesia, Italy, Japan, Korea, Mexico, Russia, South Africa, the UK, and the USA. See MEF's website at www.majoreconomiesforum.org, accessed December 24, 2014.
[31] Maximilian Terhalle and Joanna Depledge, "Great-Power Politics, Order Transition, and Climate Governance: Insights from International Relations Theory," *Climate Policy* 13, no. 5 (2013), 572–588.
[32] By the end of 2014, MEF had held 20 meetings of its members. See MEF, "Past Meetings," available at www.majoreconomiesforum.org/past-meetings, accessed December 24, 2014.
[33] MEF, "Declaration of the Leaders of the Major Economies Forum on Energy and Climate," L'Aquila, Italy (July 9, 2009), available at www.majoreconomiesforum.org/resources.html, accessed December 24, 2014.

sustainable development."[34] This is further "clear evidence of new substance in India's stance" on the non-UNFCCC process.[35]

China and India have also been bandwagoning the G20 to address climate change. This club was originally created by Canada and the United States to handle global financial issues in the wake of the Asian Financial Crisis in the late 1990s. It resumed functioning after the 2008 global financial crisis, when the leaders of the G20 countries began to meet regularly and issue communiqués on global financial issues, as well as on measures that might be adopted by its member states to cut their GHG emissions.[36] At the forum, both China and India agreed not only to reduce their fossil fuel subsidies over the medium-term as a way to reduce their GHG emissions, but also to report on their implementation strategies and timelines to the following G20 meeting.[37] So far, one of the most significant achievements of the G20 related to climate change is that in September 2013, its leaders signed an agreement to cooperate on phasing out the use of hydrofluorocarbons (HFCs), gases used in refrigerators, air conditioners, and some industrial equipment, through which as much as 90 billion tonnes of CO_2 equivalent will be reduced by 2050.[38]

Bandwagoning Bilateral Climate Arrangements

At the same time, China and India have also been bandwagoning bilateral climate-related agreements initiated by the EU and the United States. In terms of China and the EU's bilateral agreements on climate change, for instance, in 2005, China and the EU established the EU–China partnership on climate change. In 2009, this partnership was upgraded to ministerial level. Under this partnership, China has

---

[34] Ministry of Environment and Forests, "Statement of Mr Jairam Ramesh, Minister of Environment & Forests (Independent Charge) Government Of India, at the 7th MEF Meeting, Rome, Italy," available at http://moef.nic.in/downloads/public-information/speech-mef.pdf, accessed March 3, 2018.
[35] Atteridge et al., "Climate Policy in India," 70.
[36] Keohane and Victor, "The Regime Complex for Climate Change," 11.
[37] The White House, "The Pittsburgh Summit: Key Accomplishments" (2009), available at www.whitehouse.gov/files/documents/g20/Fact_Sheet_Pittsburgh_Outcomes.pdf, accessed December 16, 2014.
[38] The White House, "United States, China, and Leaders of G-20 Countries Announce Historic Progress toward a Global Phase Down of HFCs" (September 6, 2013), available at www.whitehouse.gov/the-press-office/2013/09/06/united-states-china-and-leaders-g-20-countries-announce-historic-progress, accessed December 24, 2014.

obtained know-how and technology transferred by the EU.[39] Moreover, the EU has not only helped China to draft the legislation related to renewable energy and energy efficiency that is directly contributing to reductions in its GHG emissions, but it has also supported China's participation in the Kyoto Protocol's Clean Development Mechanism, through the CDM Facilitation Project.[40] Moreover, in the September 2012 EU–China Summit, a new series of cooperation projects related to GHG mitigations were launched, including supporting the design and implementation of an emissions trading system (ETS) in China, to assist China in creating sustainable urbanization and achieving environmental sustainability.[41]

Furthermore, during the June 2015 EU–China summit, China and the EU issued a separate joint statement on climate change, in which China and the EU committed to cooperating with each other to address the climate change challenge, and more especially to reach a legally binding agreement at the Paris climate conference in late 2015, as well as an agreement on joint cooperation on 13 different aspects related to addressing climate change.[42] In the meantime, the Chinese government submitted its INDC to the UNFCCC Secretariat on June 30, 2015,[43] only a day after the EU–China summit and Chinese Premier Li Keqiang's meeting with French President Hollande, which can be regarded as one of the most obvious signals of China bandwagoning the EU to address climate change.

Like China, India has also been bandwagoning the EU on climate change. More specifically, in September 2005, at the 6th EU–India

---

[39] Pietro De Matteis, "The EU's and China's Institutional Diplomacy in the Field of Climate Change," Occasional Paper No. 96 (European Union Institute for Security Studies, 2012), 17.
[40] Pietro De Matteis, "EU–China Cooperation in the Field of Energy, Environment and Climate Change," *Journal of Contemporary European Research* 6, no. 4 (2010), 449–477.
[41] European Commission, "The European Union and China Join Forces to Address Environment, Urbanization and Climate Change Challenges," Brussels (September 20, 2012), available at http://europa.eu/rapid/press-release_IP-12-989_en.htm?locale=en, accessed March 3, 2018.
[42] European Council, "EU–China Joint Statement on Climate Change" (June 29, 2015), available at www.consilum.europa.eu/en/press/press-releases/2015/06/29-eu-china-statement, accessed June 1, 2016.
[43] NDRC, "Enhanced Actions on Climate Change: China's Intended Nationally Determined Contributions" (June 30, 2015), 5, available at www4.unfccc.int/submissions/INDC/Published%20Documents/China/1/China's%20INDC%20-%20on%2030%20June%202015.pdf, accessed July 24, 2015.

summit in New Delhi, the India–EU Strategic Partnership–Joint Action Plan was initiated. This Action Plan establishes an EU–India Initiative on Clean Development and Climate Change,[44] "which focuses on cooperation in the area of clean technology and the CDM as well as on adaptation to climate change and the integration of adaptation concerns into sustainable development strategies."[45] At the 2008 EU–India summit, both sides initiated a Joint Work Programme for EU–India Cooperation on Energy, Clean Development and Climate Change.[46]

When it comes to India and China bandwagoning the US-initiated bilateral climate arrangements, in China's case, one of the more significant episodes is its bilateral agreement with the United States to phase out HFCs in September 2013, along with the same agreement within the G20 mentioned above.[47] In addition to this agreement, China and the United States signed a bilateral climate deal – the US–China Joint Announcement on Climate Change – on November 12, 2014.[48] According to a news commentary, the idea for such a bilateral deal on climate change was first initiated by the US Secretary of State John Kerry in early February 2013, when he visited Beijing. After the Chinese leaders seemed "potentially receptive" to his idea, President Obama sent a personal letter to President Xi Jinping, suggesting the two countries start to move to cut CO_2 emissions. Then, after a series of bilateral negotiations, both states eventually reached a deal,[49] which will be discussed in detail in the next section. Furthermore, in the run-up to the Paris climate conference, in September 2015, China and the

[44] Council of the European Union, "The India–EU Strategic Partnership Joint Action Plan," Brussels (September 7, 2005), 14–15, available at http://ec.europa.eu/enterprise/policies/international/files/eu_india_joint_action_plan_en.pdf, accessed December 16, 2014.
[45] European Commission, "Climate Action: India," available at https://ec.europa.eu/clima/policies/international/cooperation/india_en, accessed March 6, 2018.
[46] Ibid.
[47] The White House, "United States, China, and Leaders of G-20 Countries Announce Historic Progress toward a Global Phase Down of HFCs."
[48] The White House, "US–China Joint Announcement on Climate Change" (November 11, 2014), available at www.whitehouse.gov/the-press-office/2014/11/11/us-china-joint-announcement-climate-change, accessed March 3, 2018.
[49] "Secret Talks and a Personal Letter: How the US–China Climate Deal Was Done," *The Guardian* (November 12, 2014), available at www.theguardian.com/environment/2014/nov/12/how-us-china-climate-deal-was-done-secret-talks-personal-letter, accessed March 3, 2018.

United States issued a joint presidential statement to reaffirm their 2014 announcement, and to lay out a joint "vision" for the conference.[50] In this statement, China announced that it would make available ¥20 billion (roughly $3.1 billion) to set up the China South–South Climate Cooperation Fund, to support other developing countries to combat climate change, which matched the US pledge of $3 billion to the Green Climate Fund (GCF).[51]

In India's case, the United States initiated a civil nuclear agreement in 2005, to help India reduce its GHG emissions to the expected level, despite the controversy that surrounded this deal as a result of concerns about its negative impact on the nuclear non-proliferation regime.[52] Since India has never signed up to the NPT, it is ineligible to enter the global nuclear market. However, once the US-Indian civil nuclear deal had been officially signed in October 2008, this taboo was broken, allowing India to gain access to US as well as global civil nuclear technology, to facilitate its efforts to expand nuclear energy.[53] Thus, as Navroz Dubash neatly points out, the US-Indo nuclear deal has pressured India to shift its traditional negotiating stances, so as to "be more closely aligned with broader foreign policy."[54]

Moreover, in November 2009, when US President Obama met with Indian Prime Minister Singh, both sides signed an MOU to Enhance Cooperation on Energy Security, Energy Efficiency, Clean Energy and Climate Change, which established the US–India Partnership to Advance Clean Energy (PACE).[55] PACE is targeted at helping India accelerate its inclusive, low-carbon growth by supporting research and

---

[50] The White House, "U.S.–China Joint Presidential Statement on Climate Change" (September 25, 2015), available at www.whitehouse.gov/the-press-office/2015/09/25/us-chinajoint-presidential-statement-climate-change, accessed January 20, 2016.
[51] Ibid.
[52] Jayshree Bajoria and Esther Pan, "The U.S.-India Nuclear Deal" (Council on Foreign Relations, November 5, 2010), available at www.cfr.org/india/us-india-nuclear-deal/p9663, accessed March 3, 2018.
[53] Mistry, *The US-Indian Nuclear Agreement*.
[54] Navroz K. Dubash, "The Politics of Climate Change in India: Narratives of Equity and Cobenefits," *WIREs Climate Change* 4 (May/June 2013), 194.
[55] The Department of Commerce (DOC), Department of Energy (DOE), Department of State (DOS), Export–Import Bank of the United States (Ex–Im), Overseas Private Investment Corporation (OPIC), US Agency for International Development (USAID), and US Trade and Development Agency (USTDA), *U.S.–India Partnership to Advance Clean Energy (PACE): An Initiative of the U.S.–India Energy Dialogue* (progress report, June 2013), 1, available at

deployment of clean energy technologies. This partnership has turned out to be highly successful, so both governments have committed to strengthening and expanding PACE through a series of policy initiatives.[56] Although India did not sign a bilateral climate deal with Obama in January 2015, during the latter's bilateral meeting with Indian Prime Minister Narendra Modi in New Delhi, it eventually agreed to phase out HFCs in April 2015, after years of opposition.[57] Furthermore, in September 2015, the United States signed another MOU with India on energy security, climate change, and clean energy, in addition to launching a Fulbright Climate Fellowship program to facilitate the exchange of key research information.[58]

Compromises under Dual-Track Climate Diplomacy

As discussed above, China and India have proactively built new coalitions to facilitate their increasingly weakened negotiating stances under the UNFCCC, but they have made some significant compromises on their traditional negotiating positions under external pressures from both their negotiating partners under the UNFCCC and those from the various non-UN track climate arrangements they have been bandwagoning.

Compromises on GHG Mitigation

Objectively speaking, throughout the history of ICCN, China and India have tried their best to avoid undertaking any mitigation commitments. This position can be illustrated clearly by both countries' consistent rejection of "voluntary commitments," a concept put

---

www.seriius.org/pdfs/062013_indo_us_pace_report.pdf, accessed March 3, 2018.

[56] The White House, "Fact Sheet: U.S. and India Climate and Clean Energy Cooperation" (January 25, 2015), available at www.whitehouse.gov/the-press-office/2015/01/25/fact-sheet-us-and-india-climate-and-clean-energy-cooperation, accessed March 3, 2018.

[57] Valerie Volcovici, "India Takes 'Significant Step' in HFC Decision: U.S. Envoy," *Reuters* (April 17, 2015), available at www.reuters.com/article/us-india-climatechange-hfc-idUSKBN0N82CW20150417, accessed March 3, 2018.

[58] "India, US Sign Agreement on Energy Security and Climate Change" (September 23, 2015), available at www.ndtv.com/india-us-sign-agreement-on-energy-security-and-climate-change-1220636, accessed June 1, 2016.

forward by developed countries at COP-3, held in Kyoto, Japan in December 1997. For example, in the general debate session at COP-3, the EU's representative put forward a proposal to "foster voluntary limitation of developing country GHG emissions."[59] This suggestion was strongly rejected by the G-77/China, led by China and India, on the ground that emissions were related to development progress, a top priority for developing countries; besides, the Berlin Mandate did not include any voluntary commitments requirement of developing countries.[60] Despite their opposition, voluntary commitments for non-Annex I parties were still written into the drafted version of the protocol's Article 10. In order to defeat these imposed voluntary commitments, China, together with India, flatly rejected the drafted article and insisted on deleting it in the closing hours of the negotiations, on the ground that although the commitments were voluntary in name, they would determine a level of limitation or reduction of anthropogenic emissions, imposing an obligation that did not apply to developing countries. Moreover, the Article, China contended, endangered the non-Annex I status of parties joining its activities and imposed new commitments on developing countries. Due to China's and India's strong opposition, no consensus was reached on the voluntary commitments, so the original Article 10 in the drafted protocol was deleted at the last minute of COP-3.[61] As a result, the Kyoto Protocol was adopted without any reference to voluntary commitments on the part of developing countries. Thus, Andrew Hurrell and Sandeep Sengupta argue that it was developing countries, rather than developed countries, that had shaped the existing international climate change regime – the UNFCCC and the Kyoto Protocol – given the fact that developing countries had succeeded in defending not only the regime's "'differentiated' architecture, which exempted developing countries from having to undertake any uncompensated mitigation actions," but also "its various principles and provisions that explicitly

---

[59] IISD, "Report of the Third Conference of the Parties to the UNFCCC: 1–11 December 1997," *ENB* 12 no. 76 (December 13, 1997), 5.

[60] Berlin Mandate was reached at COP-1, held in Berlin in 1995, which is a mandate to begin a process toward appropriate action for the period beyond 2000, including the strengthening of the commitments of Annex I Parties in Article 4.2(a) and (b). See IISD, "Summary of the First Conference of the Parties for the Framework Convention on Climate Change: 28 March–7 April 1995," *ENB* 12, no. 21 (April 10, 1995), 11.

[61] IISD, "Report of the Third Conference of the Parties to the UNFCCC," 13.

accepted that their share of global emissions would grow in the future to meet their development needs, and furthermore recognized sustained economic growth and poverty eradication as legitimate national priorities."[62]

In the subsequent negotiations, divergence over the issue of voluntary commitments emerged between Sino-India and some member countries in the G-77/China. At COP-4, held in Buenos Aires, Argentina in 1998, for instance, in the opening plenary, the host country placed on the provisional agenda an item on voluntary commitments for developing countries, which incurred strong opposition from other members within the G-77/China. China put forward some reasons for its opposition, including, among others, that developing country "survival emissions" were different from developed country "luxury emissions,"[63] and the item proposed by Argentina would not promote the FCCC, but would help some countries avoid existing commitments.[64] Given the G-77/China's opposition, Argentina had to drop the item from the agenda, although it still suggested that informal consultations between interested countries should proceed. At COP-5 (Bonn, October 25–November 5, 1999), during the high-level segment, AOSIS called for all countries to participate formally in the effort to reduce GHG emissions, while Argentina and Kazakhstan expressed their willingness to accept the voluntary commitments. In contrast, China, together with India, opposed these positions, reiterating that Annex I countries had the main responsibility, and making it clear that they would not undertake any commitments until they achieved a "medium development level."[65]

---

[62] Andrew Hurrell and Sandeep Sengupta, "Emerging Powers, North–South Relations and Global Climate Politics," *International Affairs* 88, no. 3 (2012), 469.

[63] These two terms were originally put forward by Anil Agarwal and Sunita Narrain, two Indian scholars at the Centre for Science and Environment, in response to the WRI report in 1991, which argued that China and India should share responsibility for global warming. See Anil Agarwal and Sunita Narrain, *Global Warming in an Unequal World: A Case of Environmental Colonialism* (New Delhi: Centre for Science and Environment, 1991), 3.

[64] IISD, "Report of the Fourth Conference of the Parties to the United Nations Framework Convention on Climate Change: 2–13 November 1998," *ENB* 12, no. 97 (November 16, 1998), 3.

[65] IISD, "Summary of the Fifth Conference of the Parties to the UNFCCC: 25 October–5 November 1999," *ENB* 12, no. 123 (November 8, 1999), 13.

However, China and India's position on mitigation commitments or actions has begun to shift since COP-13 (Bali, December 3–15, 2007), where developing countries in general, and China and India in particular, agreed to take on mitigation actions for the first time. At the Conference, during the talks on long-term cooperative action under the Convention, the G-77/China at first strongly resisted the stronger language on developing countries' actions/commitments demanded by the United States and its allied Umbrella Group.[66] Both sides held their own positions and were unwilling to compromise, so the conference was almost on the verge of breakdown until India proposed the concept of "nationally appropriate mitigation actions (NAMAs) by developing country parties in the context of sustainable development," which would be "supported by technology and enabled by finance and capacity building in a measurable, reportable and verifiable manner" on the part of developed countries.[67] In other words, the compromise suggested by India was that developing countries should undertake NAMAs, while developed countries gave their financial and technological support to developing countries under a strict verification system. This compromise text was finally accepted by the G-77/China and developed countries, more especially the United States, which made it possible for the decision on long-term action under the Convention to be adopted by the conference. Thus, developing countries in general, and China and India in particular, finally conceded to the demands of developed countries to undertake actions to mitigate climate change. Even though these were only actions, compared with their historical opposition to GHG mitigation, NAMAs represented a significant compromise by China and India once made in ICCN.

At COP-16 (Cancun, November 29–December 11, 2010), China's and India's concessions to NAMAs began to become more operational. According to the Cancun Agreements adopted at this conference, developing countries would take on NAMAs "aimed at achieving a

---

[66] This group, a loose coalition of non-EU developed countries, is usually made up of Australia, Canada, Iceland, Japan, New Zealand, Norway, Russia, Ukraine, and the United States, and has been formed since the adoption of the Kyoto Protocol.
[67] UNFCCC, "FCCC/CP/2007/6/Add.1: Report of the Conference of the Parties on its Thirteenth Session, Held in Bali from 3 to 15 December 2007" (March 14, 2008), 3, available at http://unfccc.int/resource/docs/2007/cop13/eng/06a01.pdf, accessed March 4, 2018.

deviation in emissions relative to 'business-as-usual' by 2020," and a registry would be set up to record NAMAs requiring international support; and those supported by developing countries themselves would be recorded in a separate section of the registry.[68]

Apart from making compromises under the UN track climate negotiations, China and India have made some further compromises under the non-UN track climate arrangements. For instance, at the G8 summit held in Heiligendamm in 2007, Prime Minister Manmohan Singh announced that India would ensure that its per capita emissions would never exceed those of developed countries.[69] This was regarded as "a subtle (perhaps unintended) shift away from India's emphasis on historical responsibility,"[70] under pressure from the G8+5, although the real intention behind this announcement was to tell developed countries that India was not yet ready to contemplate undertaking mitigation actions. In another example, China and India have also accepted the 2-degree target. As mentioned above, China and India, at the MEF in L'Aquila, Italy, in 2009, agreed with other members that the increase in global average temperatures above pre-industrial levels should not exceed 2 °C, which implied that China and India, as well as other emerging developing countries, would need to undertake mitigation actions in order to reach this target.[71] The target was first written into the Copenhagen Accord, the outcome of COP-15 (Copenhagen, December 7–19, 2009).

Moreover, under the UN track, China and India have already agreed to take actions to mitigate their CO_2 emissions through a unique way, namely reducing the energy intensity of their per capita GDP. Specifically, in the run-up to the Copenhagen climate conference, China set a "binding goal" to cut CO_2 per unit of GDP by 40–45 percent from 2005 levels by 2020, whereas India announced that it would cut its

---

[68] UNFCCC, "FCCC/CP/2010/7/Add.1: Report of the Conference of the Parties on its Sixteenth Session, held in Cancun from 29 November to 10 December 2010" (March 15, 2011), 8–10, available at http://unfccc.int/resource/docs/2010/cop16/eng/07a01.pdf, accessed March 4, 2018.
[69] Peter Foster, "India Snubs West on Climate Change," *The Telegraph* (June 12, 2007), available at www.telegraph.co.uk/news/earth/earthnews/3297214/India-snubs-West-on-climate-change.html, accessed March 4, 2018.
[70] Rajamani, "India and Climate Change," 346.
[71] Vihma, "India and the Global Climate Governance," 15.

CO_2 emissions per unit of GDP by 20–25 percent from 2005 levels by 2020.[72] When it comes to the cost of fulfilling its pledge, according to a Chinese research team, China would have to spend $30 billion as an incremental cost each year within the next ten years, which is equivalent to each Chinese household bearing an additional $64 each year.[73] Simply put, China and India have already made some practical compromises on their obligations to mitigate GHG emissions.

At COP-16, India's willingness to undertake GHG mitigation obligations went further than China was prepared to accede to. More specifically, during the negotiations at Cancun, then Indian Minister of Environment and Forests, Jairam Ramesh, publicly stated that "All countries should take binding commitments in an appropriate legal form."[74] This statement implied that India "was willing to commit to legally binding commitments as part of an international climate deal."[75] Also, India played a constructive role in brokering a solution on the issue of the international monitoring of domestic mitigation efforts at this climate conference.[76] Ramesh's statement and behavior incurred strong criticism from India's domestic audience,[77] which eventually led to Ramesh being replaced by Jayanthi Natarajan in July 2011.[78]

At COP-17 (Durban, November 28–December 11, 2011), China and India agreed, for the first time under the UN track, to negotiate "a protocol, another legal instrument or an agreed outcome" that should be legally binding on all parties, both developed countries and

[72] BBC, "Where Countries Stand on Copenhagen" (May 28, 2011), available at http://news.bbc.co.uk/2/hi/8345343.stm, accessed March 4, 2018.
[73] 陈晔 [Chen Ye], "中国软实力建设中的环境外交因素 [The Factor of Environmental Diplomacy in the Construction of China's Soft Power]," 人民网 [People.com] (July 2010), available at http://theory.people.com.cn/GB/166866/12089481.html, accessed March 4, 2018.
[74] Nitin Sethi, "India Willing to Accept Legally Binding Pact at Cancun," *Times of India* (December 9, 2010), available at http://timesofindia.indiatimes.com/india/India-willing-to-accept-legally-binding-pact-at-Cancun/articleshow/7070583.cms, accessed March 4, 2018.
[75] Ibid. [76] Atteridge et al., "Climate Policy in India," 71.
[77] Kari-Anne Isaksen and Kristian Stokke, "Changing Climate Discourse and Politics in India: Climate Change as Challenge and Opportunity for Diplomacy and Development," *Geoforum* 57 (2014), 110.
[78] Amol Sharma, "Cabinet Reshuffle: India's Environment after Jairam Ramesh," *Wall Street Journal* (July 12, 2011), available at https://blogs.wsj.com/indiarealtime/2011/07/12/indias-environment-after-jairam-ramesh, accessed March 4, 2018.

developing countries, to be written by 2015 and to come into force after 2020, known as the Durban Platform.[79] This compromise implied that developing countries in general, and China and India in particular, would have to undertake some kind of binding emission cut obligations if this new protocol was to be reached as scheduled.

At COP-19 (Warsaw, November 11–23, 2013), China and India, as well as other developing countries, agreed to create national plans to reduce their GHG emissions, or so-called INDCs.[80] INDCs show that, in terms of mitigating GHG emissions, there is no longer a division between Annex I and non-Annex I parties, but that all countries will be required to take measures to mitigate their GHG emissions. This compromise was further reiterated in the Lima Call for Climate Action – the decision adopted at COP-20 (Lima, December 1–14, 2014), which, *"reiterates* its invitation to each Party to communicate to the secretariat its intended nationally determined contribution towards achieving the objective of the Convention as set out in its Article 2."[81] Also, there is no mention in this document of explicit differences in historical obligations between Annex I parties (developed countries) and non-Annex I parties (developing countries), which signifies a subtle shift away from the norm based on a country's historical responsibilities in the UNFCCC – China's and India's traditional negotiation position in ICCN since the early 1990s. Although this document still states the principle of "common but differentiated responsibilities," it adds a new term, that is, "in light of different national circumstances."[82] This term obviously reflects the fact that the distinction between developed and developing countries has been further weakened because all countries are required to take INDCs to

---

[79] UNFCCC, "Draft Decision -/CP.17: Establishment of an Ad Hoc Working Group on the Durban Platform for Enhanced Action" (December 2011), available at http://unfccc.int/files/meetings/durban_nov_2011/decisions/application/pdf/cop17_durbanplatform.pdf, accessed March 4, 2018.

[80] IISD, "Summary of the Bonn Climate Change Conference: 20–25 October 2014," *ENB* 12, no. 605 (October 28, 2014), 5.

[81] UNFCCC, "Report of the Conference of the Parties on its Twentieth Session, Held in Lima from 1 to 14 December 2014 (FCCC/CP/2014/10/Add.1)" (February 2, 2015), 3, available at http://unfccc.int/resource/docs/2014/cop20/eng/10a01.pdf#page=2, accessed March 4, 2018.

[82] UNFCCC, "Decision -/CP.20: Lima Call for Climate Action" (advance unedited version), 1, available at https://unfccc.int/files/meetings/lima_dec_2014/application/pdf/auv_cop20_lima_call_for_climate_action.pdf, accessed March 4, 2018.

mitigate GHG emissions. Moreover, all parties, including China and India, also agreed that INDCs toward achieving the objective of the UNFCCC should represent a progression beyond current mitigation efforts and prevent backsliding.[83]

A more practical step taken by China and India to show their willingness to mitigate their GHG emissions is that both countries submitted their INDCs to the Secretariat of the UNFCCC. More specifically, on June 30, 2015, China submitted its "Enhanced Actions on Climate Change: China's Intended Nationally Determined Contributions," in which China promised to take the following nationally determined actions by 2030:

to achieve the peaking of carbon dioxide emissions around 2030 and making best efforts to peak early; to lower carbon dioxide emissions per unit of GDP by 60% to 65% from the 2005 level; to increase the share of non-fossil fuels in primary energy consumption to around 20%; and to increase the forest stock volume by around 4.5 billion cubic meters on the 2005 level.[84]

Simply put, China reaffirmed its commitment to peaking GHG emissions by no later than 2030, and indicated the means by which it would reach this goal in its INDC.

Four months later, in October 2015, India submitted its INDC, which reiterated its commitment to keeping its per capita emissions below the level of those of industrialized countries in the future, in addition to its pledge to reduce the emissions intensity of the economy by 33–35 percent by 2030, measured against the level in 2005, as well as the scope and coverage of the intensity target and the methodologies for measuring it.[85] According to some experts at the WRI, the information inked in India's INDC "is crucial for monitoring progress towards India's target and for understanding how it contributes to the global goal of limiting temperature rise to 2 degrees C."[86] Even so, when compared

---

[83] Ibid.
[84] Department of Climate Change, NDRC, "Enhanced Actions on Climate Change: China's Intended Nationally Determined Contributions" (June 30, 2015), 5.
[85] GOI, "India's Intended Nationally Determined Contribution: Working Toward Climate Justice" (October 2015), available at www4.unfccc.int/submissions/INDC/Published%20Documents/India/1/INDIA%20INDC%20TO%20UNFCCC.pdf, accessed March 4, 2018.
[86] Apurba Mitra, Thomas Damassa, Taryn Fransen, Fred Stolle, and Kathleen Mogelgaard, "5 Key Takeaways from India's New Climate Plan (INDC)"

with China's INDC, India's obviously lacks any commitment to peaking its GHG emissions by a certain date. In other words, India's compromises under the UNFCCC negotiations have not gone as far as China's, which is a significant difference between both countries' UN track climate diplomacy, although they have had some shared positions under the UNFCCC negotiations since the early 1990s.

Finally, China and India have already agreed to undertake "nationally determined contributions (NDC)" to mitigate their GHG emissions in accordance with the Paris Agreement adopted by all the parties, including China and India, at COP-21 in Paris on December 11, 2015.[87] This Agreement will put China, India, and other large developing countries on an equal footing with developed countries, to combat climate change in terms of both voluntary GHG emissions mitigation (Article 4) targets and the legally binding review process, or "transparency framework" (Article 13), when the Paris Agreement comes into force, because there is no longer any differentiation between Annex I and non-Annex I parties in this Agreement, but only between developing and developed countries.[88] The compromise made by China and India (and developing countries as a whole) in the Paris Agreement has not been reciprocated by developed countries, given that the latter have only pledged to (but are not legally bound to) provide financial assistance to poor developing countries adapting to climate change. Specifically, according to the Copenhagen Accord, developed countries pledged to mobilize no less than $100 billion a year by 2020 to help developing countries address climate change, but this pledge was postponed until 2025 in the Paris Agreement.[89] Although Article 9 of the Paris Agreement is devoted to financial aspects, none of the nine paragraphs of this Article mention the $100 billion pledges.[90] Thus, "The United States and other developed countries succeeded in excluding a reference to the Copenhagen

(WRI, October 2, 2015), available at www.wri.org/blog/2015/10/5-key-takeaways-india%E2%80%99s-new-climate-plan-indc, accessed March 4, 2018.

[87] UNFCCC, "Adoption of the Paris Agreement" (December 12, 2015), available at http://unfccc.int/resource/docs/2015/cop21/eng/l09.pdf, accessed March 4, 2018. INDCs submitted for the Paris Agreement automatically become NDCs when each party ratifies the Agreement.

[88] IISD, "Summary of the Paris Climate Change Conference: 29 November–13 December 2015," *ENB* 12, no. 663 (December 15, 2015), 43.

[89] UNFCCC, "Adoption of the Paris Agreement," 8.   [90] Ibid., Article 9, 25–26.

one-hundred billion dollar per year mobilization goal in the Paris Agreement itself."[91] So, as Daniel Bodansky points out, "the Paris Agreement's provisions on finance are rather modest."[92]

Compromises on Verification

Apart from compromises on their obligations to mitigate GHG emissions, China and India have also made some compromises on the issue of verification of their mitigation actions. More specifically, China and India have agreed to put their NAMAs under either measuring, reporting, and verifying (MRV), or international consultation and analysis (ICA), or even a legally binding review process. In other words, they have agreed to open up their mitigation actions to international scrutiny, or to make their actions transparent.

It is widely recognized that the issue of transparency, or MRV, especially its scope of application and mechanism for implementation, has long been one of the most contentious issues in the UNFCCC negotiations between developed and developing countries. According to the UNFCCC (Article 12) and the Kyoto Protocol (Articles 7 and 8), MRV requirements are mainly applied to Annex I parties rather than non-Annex I parties. More specifically, Annex I parties are required to submit detailed annual reports on emissions of all six GHGs, in accordance with the guidelines and decisions adopted by COPs. Those reports are to be reviewed by expert review teams, who will only review the methodologies used by the statistical authorities in Annex I parties, but who might increase the level of scrutiny once the relevant resources are available in the future. In contrast, non-Annex I parties are only required to submit GHG reports as part of their national communications and do not have to adhere to those guidelines and decisions.[93] Simply put, the UNFCCC and the Kyoto Protocol acknowledge the substantial differentiation in MRV requirements

---

[91] Daniel Bodansky, "The Paris Climate Change Agreement: A New Hope?" *American Journal of International Law* 110, (March 31, 2016), 34, available at https://conferences.asucollegeoflaw.com/workshoponparis/files/2012/08/AJIL-Paris-Agreement-Draft-2016-03-26.pdf, accessed March 4, 2018.
[92] Ibid., 33.
[93] UN, *United Nations Framework Convention on Climate Change* (1992), 12, available at http://unfccc.int/files/essential_background/background_publications_htmlpdf/application/pdf/conveng.pdf, accessed March 4, 2018; UNFCCC, *Kyoto Protocol to the United Nations Framework Convention on*

between Annex I and non-Annex I parties. However, this status quo has already begun to change since COP-13, where the G-77/China agreed to negotiate MRV mitigation actions that can be not only measured and reported, but also verified.[94]

In addition, at COP-15, China and India made further concessions to MRV requirements demanded by developed countries, especially the United States. In fact, at this conference, when the United States and other developed countries pushed for voluntary actions by developing countries to be independently measured, reported, and verified, China and several other countries, especially Brazil, India, and South Africa, or the BASIC group, strongly resisted this demand. Following intense and tough negotiations between US President Obama and Chinese Premier Wen Jiabao, the Chinese leader eventually accepted a less intrusive version of MRV, namely ICA, because ICA would be more flexible in its interpretation of what the process for transparency would include, and it removes references to "verification," an intrusive term for both India and China.[95] This compromise is regarded as having a significant impact on climate regime building.[96] For example, one observer notes: "Subjecting developing country actions to international consultation and analysis erodes the strict asymmetry between Annex I and non-Annex I countries enshrined in the Kyoto Protocol and builds confidence that all major economies are standing behind their commitments."[97]

At COP-16, China's and India's concessions to MRV/ICA became more operational. According to the Cancun Agreements adopted at

Climate Change (1998), 7–8, available at http://unfccc.int/resource/docs/convkp/kpeng.pdf, accessed March 4, 2018.

[94] UNFCCC, "Decision -/CP.13: Bali Action Plan" (2007), available at http://unfccc.int/files/meetings/cop_13/ application/pdf/cp_bali_action.pdf, accessed March 21, 2009; Michael Grubb, "The Bali COP: Plus ça Change," Climate Policy 8, no. 1 (2008), 3–6; Harald Winkler, "Measurable, Reportable and Verifiable: The Keys to Mitigation in the Copenhagen Deal," Climate Policy 8, no. 6 (2008), 534–547; Clare Breidenich and Daniel Bodansky, "Measurement, Reporting and Verification in a Post-2012 Climate Agreement" (Washington, D.C.: Pew Center on Global Climate Change, 2009).

[95] Namrata Patodia Rastogi, "Winds of Change: India's Emerging Climate Strategy," The International Spectator 46, no. 2, (2011), 133.

[96] Wu, "Sino-Indian Climate Cooperation," 827–843.

[97] Trevor Houser, "Copenhagen, the Accord, and the Way Forward," Policy Brief (Washington, D.C.: Peterson Institute for International Economics, 2010), 13–14.

this conference, internationally supported NAMAs would be under both domestic and international MRV, while those on their own would be subject to both domestic MRV and a biennial ICA.[98] In return, developed countries' mitigation actions would be put under international assessment and review (IAR).[99] In other words, the 2010 Cancun Agreements have largely addressed developing countries' concerns about strong reporting and review requirements, by differentiating between their commitments and those of developed countries.[100] However, this differentiation in transparency between developed and developing countries has been largely excluded from the Paris Agreement. Instead, all countries will be equally treated under an "enhanced transparency framework for action and support," under Paris Agreement Article 13.1.[101] Thus, the Paris Agreement has established a single review system applicable to all countries, which reflects the preferences of developed countries rather than those of developing countries, including China and India.[102] Simply put, China, India, and developing countries as a whole have made a significant compromise on the transparency issue in the negotiations. Given the fact that China and India have had a long history of opposing verification, seeing it as a violation of their sovereignty, and therefore rejecting international infringement, both countries' concessions to opening up their future climate mitigation actions to international scrutiny can be regarded as one of the most significant compromises they have made in the climate change negotiations.

In sum, China and India have largely abandoned their traditional positions on how to address climate change at the international level, under both the UNFCCC negotiations and non-UN climate arrangements. In other words, both China and India have made some significant compromises in their negotiations with developed countries, especially the United States and the EU, with both countries agreeing not only to undertake NDCs to mitigate their GHG emissions in a nonbinding manner, but also to put their NDCs under a legally binding transparency framework, all of which were once regarded by China

[98] UNFCCC, "FCCC/CP/2010/7/Add.1: Report of the Conference of the Parties on its Sixteenth Session," 8–10.
[99] Ibid.
[100] Bodansky, "The Paris Climate Change Agreement: A New Hope?" 35.
[101] UNFCCC, "Adoption of the Paris Agreement," Article 13.1.
[102] Bodansky, "The Paris Climate Change Agreement: A New Hope?" 35.

and India as unacceptable and non-negotiable. According to an analyst, with Sino-Indian as well as developing countries' concessions, the "firewall between developed and developing countries" has been outright broken down.[103] With the adoption of the Paris Agreement at COP-21, the fundamental nature of the balance of responsibilities between developed and developing countries stressed by both China and India, as well as the scientific basis of using a country's cumulative emissions, and not just current or future emissions, to apportion responsibilities, as once insisted upon by India, has been cast into history.[104] China and India have already significantly departed from their traditional positions on the responsibilities for climate mitigation in their climate change foreign policy. Their dual-track climate diplomacy behavior has resulted in both countries shifting away from proactively free-riding to reactively burden sharing.

Shaping Climate Diplomacy: Two-Level Pressures

China's and India's climate diplomacy has been shaped by both countries' two-level pressures. At the domestic level, both countries have been proactively seeking to maximize their economic growth and wealth; at the international level, both countries have been faced with international pressures in the process of seeking to enhance their status in the international system, which is characterized by their asymmetrical interdependence on developed countries, especially the EU and the United States.

China and India Seeking Wealth[105]

China's and India's proactive and reactive climate diplomacy in the first 15 years of the twenty-first century has been shaped, first and most importantly, by domestic pressures, namely, to maximize their

[103] Daniel Bodansky, "The Copenhagen Climate Change Conference: A Postmortem," *The American Journal of International Law* 104, no. 2 (April 2010), 240.
[104] Nitin Sethi, "US Dictates the Limits of Paris Climate Change Deal," *Business Standard* (December 21, 2015), available at www.business-standard.com/article/current-affairs/us-dictates-the-limits-of-paris-climate-change-deal-115122101089_1.html, accessed March 4, 2018.
[105] In this section, I only discuss the economic benefits that China and India have gained through their dual-track climate diplomacy, leaving the potential

domestic wealth and to eradicate poverty through their economic development, which has been a national consensus in both countries. Throughout the 1990s and until 2007, the national consensus provided strong domestic support for both countries to resist any international pressures on them to undertake actions to mitigate their increased GHG emissions under the UNFCCC climate negotiations. However, since 2007 there has been a hot debate within both countries on whether or not climate mitigation would negatively impact both countries' economies and, correspondingly, whether or not both countries should undertake actions to mitigate their increased GHG emissions, especially CO_2 emissions.

In terms of China's domestic debate, there were two opposing camps, that is, the pro-climate mitigation camp and the anti-climate mitigation camp. China's pro-climate mitigation camp is composed of some prominent Chinese scholars, including Qin Dahe (a well-known glaciologist and climatologist, and an academician of the Chinese Academy of Sciences) and Hu Angang (an economics professor at Tsinghua University); some environmental NGOs, such as Greenpeace-China; and some Chinese environmentalists, in addition to some Chinese governmental agencies that are in charge of China's environmental affairs. More specifically, in 2007, Qin Dahe used scientific evidence to prove that global warming would negatively challenge not only China's economy, but also its society and sustainable development in the long run. Qin suggested that, at the domestic level, addressing climate change should be written into the Chinese Communist Party's (CCP) and China's national major principles and policies, including the formulation of national strategic plans for climate change adaptation, and the creation of a new model of industrialization characterized by energy efficiency and environmental protection; at the international level, China should actively participate in global efforts to address climate change, even though it was impossible for China to undertake the same climate mitigation commitments as developed countries, since it was still a developing country, with several million people still living in poverty.[106] In contrast to Qin's conservative policy recommendations, in July 2009, several months

---

economic benefits both countries might gain through their participation in the Paris Agreement for future research.
[106] 秦大河 [Qin Dahe], "气候变化对我国经济、社会和可持续发展的挑战 [The Changes of Climate Bring Challenges to Economy, Society and Sustainable

before the Copenhagen climate conference, Hu Angang pointed out that, given the fact that more than one billion people in China had been suffering from air pollution, it was both an opportunity and the responsibility of China to cooperate with the United States and other countries to mitigate GHG emissions at the global level; and he put forward that China should take the lead in reducing GHG emissions and setting emissions reduction targets for 2020, 2030, and 2050, so that China could realize a "green modernization."[107] In short, China's pro-climate mitigation camp regards addressing climate change as an opportunity for the transformation of the Chinese economic growth model to a more sustainable path, meaning that China should take actions to mitigate its GHG emissions.

In stark contrast, China's anti-climate mitigation camp, consisting of academic scholars, such as Wang Jisi (an IR professor at Beijing University), and some of China's powerful governmental agencies, such as the NDRC, and even Chinese local governments, hold China's traditional negotiating stance in the UN climate negotiations, that is, that China should insist on the principle of "common but differentiated responsibilities," and that developed countries are mainly responsible for climate change, so those countries should take the lead in reducing GHG emissions, while China should not undertake mitigation commitments.[108] For NDRC and local governments, China's economic growth is still a national priority, so they oppose China undertaking mitigation actions in ICCN, because climate mitigation actions might hinder economic growth.[109] Chinese local governments' resistance to Chinese central government's climate-related policy will be discussed in more detail in Chapter 6.

Development of Our Country]," 《外交评论》 [Foreign Affairs Review], no. 97 (August 2007), 6–14.

[107] 胡鞍钢 [Hu Angang], "中国气候变化政策：背景、目标与辩论 [China's Climate Change Policy: Background, Target and Debate]," 《国情报告》 [National Situation Report] 12 (2009), 232–249.

[108] 张海滨 [Zhang Haibin], "气候变化与中国的国家战略 ——王缉思教授访谈 [Climate Change and China's National Strategy: An Interview with Professor Wang Jisi]," 《国际政治研究》 [International Politics Quarterly] 46, no. 4 (2009), 79.

[109] 马丽 [Ma Li], "全球气候治理中的中国地方政府:困境、现状与展望 [Chinese Local Government in Global Climate Governance: Problems, Status Quo, and Prospect]," 《马克思主义与现实》 [Marxism and Reality], no. 5 (2015), 176–183.

In comparison, the Chinese anti-climate mitigation camp's stance has been increasingly weakened in recent years, because China's increasingly severe air pollution, such as smog, especially in some large cities, including Beijing, has caused great public concern and led to many complaints against the Chinese government.[110] In this context, Xie Zhenhua, China's Special Representative for Climate Change, said in an interview with *Caixin*, a Chinese news media group, that China uses the climate targets set out in the climate deal reached between China and the United States in November 2014 as a tool to stimulate China's domestic reform.[111]

With regard to India's domestic debate on how to handle economic growth and climate change issues, and India's climate diplomacy in the UN negotiations, there are, according to Navroz K. Dubash, three distinctive camps: "growth-first realists," "sustainable development realists," and "sustainable development internationalists."[112] More specifically, as with China's anti-climate mitigation camp, "growth-first realists" emphasize rapid growth as the top priority for India, so they oppose Indian negotiators committing India to any GHG mitigation obligations that might compromise India's economic growth, and they also emphasize the principle of equity in addressing climate change in the UN negotiations. This camp regards the climate change challenge as "a geopolitical threat rather than environmental one."[113] The most prominent representatives of this camp are Anil Agarwal and Sunita Narrain, two researchers at the Centre for Science and Environment, whose influential report, *Global Warming in an Unequal World: A Case of Environmental Colonialism*, in 1991, set out the fundamental principles for India's official negotiating stance in the UN climate negotiations.[114] According to Agarwal and Narrain, "The idea that developing countries like India and China must share the blame for

---

[110] *Xihuanet* has issued a series of reports about the severity of smog in China: "问计雾霾治理难题 [Asking How to Deal with the Problems of Smog]," *Xinhuanet*, available at www.xinhuanet.com/energy/zt/xzt/07.htm, accessed June 20, 2016.

[111] 胡舒立 [Hu Shuli], 宫靖 [Gong Jing], and 孔令钰 [Kong Lingyu], "解振华：让气候目标倒逼改革 [Xie Zhenhua: Let Climate Targets Force Domestic Reform]," 财新网 [*Caixin.com*] (November 21, 2014), available at http://m.weekly.caixin.com/m/2014-11-21/100753798_6.html, accessed March 7, 2018.

[112] Dubash, "The Politics of Climate Change in India," 197.   [113] Ibid.

[114] Ibid., 192.

heating up the earth and destabilising its climate, as espoused in a recent study published in the United States by the World Resources Institute in collaboration with the United Nations, is an excellent example of environmental colonialism."[115] Both researchers put forward three points to address climate change globally:

contribution to stocks of greenhouse gas emissions, rather than annual flows of emissions, constitutes the appropriate metric for assessing responsibility for causing climate change; a per capita allocation of global sinks is the only morally defensible metric, and that a distinction should appropriately be made between "survival emissions" of the poor and "luxury emissions" of the rich.[116]

The "sustainable development realists" maintain that India should follow a sustainable development model characterized as co-benefits, that is, promotion of "development objectives while also yielding cobenefits for addressing climate change effectively."[117] This camp has also emphasized the equity principle in the UN negotiations, but it has little faith that those negotiations will deliver desirable outcomes for India's domestic development. The third camp, consisting of "sustainable development internationalists," calls for "an internationally effective climate regime," with India making contributions to such a global regime through its domestic actions.[118] This camp is usually dismissed by the other two camps as naive, and has little influence on India's climate diplomacy formation.[119] In other words, it is the former two camps that have played an important role in shaping and altering India's stance in the UN climate change negotiations,[120] that is, to voluntarily mitigate its domestic GHG emissions so as to reach its long-term sustainable development goals.

Simply put, the outcome of China's and India's domestic debates on economic growth and climate change is that both countries have tended to accept the pro-mitigation camp's stance, that is, to use their

---

[115] Agarwal and Narrain, *Global Warming in an Unequal World*, 1.
[116] These three points were summarized by Dubash, "The Politics of Climate Change in India," 192.
[117] Prime Minister's Council on Climate Change, GOI, *National Action Plan on Climate Change* (2008), 2, available at www.moef.nic.in/sites/default/files/Pg01-52_2.pdf, accessed March 4, 2018.
[118] Dubash, "The Politics of Climate Change in India," 197.   [119] Ibid.
[120] Ibid.

policy shifts on climate mitigation as a way of improving their domestic development. We will discuss this topic in more detail in Chapter 6.

In addition to receiving domestic support to alter their free-riding climate policy in the UN negotiations, China and India have accrued significant tangible benefits from their participation in the CDM, which has helped them to maximize their domestic economic growth or wealth to a certain extent. More specifically, the compromises made by China and India, as discussed above, have been largely stimulated by both countries' desires to maximize their domestic economic growth through the profits they have accrued and will continue to accrue through their participation in the CDM, one of the three flexible mechanisms of the Kyoto Protocol during this Protocol's two commitment periods (2008–20).[121] In a Chinese scholar's words, China's and India's goal is "to maximize the economic interest through the implementation of the CDM projects."[122]

CDM is defined in Article 12 of the Protocol. It allows a country with an emissions reduction or emissions limitation commitment under the Kyoto Protocol (Annex B party; that is, Annex I party under the UNFCCC) to implement some emissions reduction projects in developing countries. These projects can earn saleable certified emission reduction (CER) credits, each equivalent to one tonne of CO_2, which can be counted as part of an Annex I party's efforts toward meeting the GHG mitigation targets set by the Kyoto Protocol.[123] According to some researchers, the CDM was offered as a tangible "carrot" to assist developing countries, while at the same time allowing Annex I parties to implement projects that reduce emissions in any developing country and use the resulting CERs to help Annex I parties meet their own targets.[124] Put differently, the CDM allows Annex I parties to achieve their emissions reductions obligations under the Kyoto Protocol at a lower cost, while helping developing countries that host those projects

---

[121] UN, *Kyoto Protocol to the United Nations Framework Convention on Climate Change.*

[122] Zhuang, "Post-Kyoto International Climate Governance and China's Strategic Options," 12.

[123] UNFCCC, "Clean Development Mechanism (CDM)," available at http://unfccc.int/kyoto_protocol/mechanisms/clean_development_mechanism/items/2718.php, accessed March 4, 2018.

[124] Bob Lloyd and Srikanth Subbarao, "Development Challenges under the Clean Development Mechanism (CDM): Can Renewable Energy Initiatives be Put in Place Before Peak Oil?," *Energy Policy* 37 (2009), 240.

to enhance their sustainable development.[125] As to the benefits of CDM for developing countries, as David Victor pointed out in 2001, developing countries, including China and India, could attract new investment into projects which limit emissions of GHGs which stem from other environmental problems, such as urban smog associated with dirty fossil fuels, and which modernize their energy systems.[126] In other words, the CDM provides opportunities for transferring foreign clean technology to developing countries, including China and India, to help them address domestic environmental pollution.[127]

Interestingly, although China and India had initially cast doubt on, and even opposed the inclusion of CDM in the Kyoto Protocol during the negotiations, for fear of having GHG mitigation obligations imposed on them,[128] it turns out that both countries have been the largest beneficiaries of the CDM – that is, since the inception of the CDM in 2004, China and India have been two of the largest host parties.[129] More specifically, between 2004 and 2015, the total number of registered project activities was 7,630.[130] China and India account for nearly 50 percent and more than 20 percent respectively of these projects.[131] Moreover, China and India have the potential to benefit further from CDM projects during the second commitment

[125] Franck Lecocq and Philippe Ambrosi, "The Clean Development Mechanism: History, Status, and Prospects," *Review of Environmental Economics and Policy* 1, no. 1 (Winter 2007), 134–151; Charlotte Streck and Jolene Lin, "Making Markets Work: A Review of CDM Performance and the Need for Reform," *European Journal of International Law* 19, no. 2 (2008), 409–442.
[126] David G. Victor, *The Collapse of the Kyoto Protocol and the Struggle for Slow Global Warming* (Princeton, NJ: Princeton University Press, 2001), 5.
[127] Patrick Bayer, Johannes Urpelainen, and Alice Xu, "Explaining Differences in Sub-National Patterns of Clean Technology Transfer to China and India," *International Environmental Agreements* 16, no. 2 (April 2016), 261.
[128] IISD, "Report of the Third Conference of the Parties to the UNFCCC," 11.
[129] However, there are some differences between both countries' capabilities to make use of the CDM technology transfer for their own domestic development. For instance, according to Bayer, Urpelainen, and Xu, the Chinese government has a greater capacity to direct subnational governance, which allows it to use the CDM for achieving development-related goals to a greater extent than the Indian government. See Bayer, Urpelainen, and Xu, "Explaining Differences in Sub-National Patterns of Clean Technology Transfer to China and India," 261–283.
[130] UNFCCC, "Distribution of Registered Projects by Host Party," available at http://cdm.unfccc.int/Statistics/Public/files/201504/proj_reg_byHost.pdf, accessed May 13, 2015.
[131] Ibid.

period of the Kyoto Protocol (2012–20). According to the UNFCCC's projection, the total potential supply of CERs by China from the end of the Kyoto Protocol's first commitment period (December 31, 2012) to 2020 will rise from 1,000 million to nearly 2,000 million, and those supplied by India will also rise.[132] Moreover, most of the CDM projects will be invested in the host parties' energy sector, especially indigenous gas and renewable energy. The potential supply of CERs by project type from the end of the Kyoto Protocol first commitment period (December 31, 2012) to 2020 will rise significantly, especially those from the projects in gas and renewable energy.[133] This implies that in the second commitment period of the Kyoto Protocol, China's and India's energy sectors will receive a lot of investment from developed countries in their gas and renewable development, which will obviously help both countries to reduce their high dependence on coal for energy to fuel their economies, and thereby reduce their potential CO_2 emissions. Thus, both countries' domestic economic transition to a low-carbon economy will be facilitated. Obviously, it is in China's and India's own identifiable economic interest to support the Kyoto Protocol's existence, so that they will be able to continue to make full use of the CDM to attract more investment from developed countries in their energy sectors. Therefore, it is no surprise that China and India have made compromises with developed countries, because these countries have conditioned their commitment to the Kyoto Protocol's second commitment period on developing countries' willingness to undertake mitigation actions.[134]

Given the existing and potential benefits they can accrue through CDM projects, China and India have tried to insist on the continuation of the Kyoto Protocol and have promoted the establishment of a Kyoto Protocol second commitment period in the post-Kyoto climate negotiations. For instance, during COP-16, China supported Brazil

[132] UNFCCC, "Distribution of Expected CERs from Projects that Have Issued CERs, Adj. by Past Rate of Issuance by Host Country," available at http://cdm.unfccc.int/Statistics/Public/files/201504/exp_cers_byHost.pdf, accessed May 13, 2015.
[133] UNFCCC, "Distribution of Expected CERs from Projects that Have Issued CERs, Adj. by Past Rate of Issuance by Project Type," available at http://cdm.unfccc.int/Statistics/Public/files/201504/exp_cers_byType.pdf, accessed May 13, 2015.
[134] IISD, "Summary of the Durban Climate Change Conference: 28 November–11 December 2011," *ENB* 12, no. 534 (December 13, 2011), 29–31.

"regarding a signal of commitment to the continuation of the CDM," and "highlighted that the CDM cannot continue unless the Kyoto Protocol continues and requires the establishment of a Kyoto Protocol second commitment period."[135] In other words, China and India want to make developed countries continue to take the lead in mitigating GHG emissions. Conversely, some developed countries pointed out plainly during the negotiations that without compromises on behalf of developing countries on mitigating GHG emissions, they would not accept the Kyoto Protocol second commitment. In fact, Japan, Canada, New Zealand, and Russia refused to undertake the second commitment period during the Doha climate negotiations.[136] Against this backdrop, China and India were obliged to make some compromises regarding the EU and other developed countries' demands for mitigation actions, and greater transparency in monitoring and verification of their actions in exchange for the EU's support for the Kyoto Protocol second commitment period.

Aside from the economic benefits China and India have gained and will continue to gain through their UN track climate diplomacy, both countries have already obtained some tangible economic benefits as a result of their non-UN track bandwagoning climate diplomacy, especially as regards financial support for their development of renewable energy. This is especially true for India. For instance, by bandwagoning the EU on climate change, India has won EU-financed projects, including "FOWIND – Facilitating Off-shore Wind Development in India, SCOPE BIG – Scalable CSP Optimized Power Plant Engineered with Biomass Integrated Gasification, Clean Energy Cooperation with India (CECI) and Clean Development and Sustainable Cities (CDSC) to mainstream low-carbon strategies into urban development."[137] Moreover, in February 2014, the European Investment Bank signed an agreement with the Indian Renewable Energy Development Agency (IREDA), to provide the latter with a EUR 200 million loan for renewable energy and energy efficiency in India.[138] As a result of bandwagoning the United States on climate change, a field investment officer was installed in India by the United States Agency for International Development (USAID) in the summer of 2016, backed by a

---

[135] IISD, "Summary of the Cancun Climate Change Conference: 29 November–11 December 2010," *ENB* 12, no. 498 (December 13, 2010), 6.
[136] IISD, "Summary of the Doha Climate Change Conference," 26.
[137] European Commission, "Climate Action: India."   [138] Ibid.

transactions team to help mobilize private capital for India's clean energy sector. In addition, the US Export–Import Bank is exploring potential projects for its MOU with the IREDA for up to $1 billion in clean energy financing. Moreover, the US Overseas Private Investment Corporation (OPIC) plans to build on its existing portfolio of $227 million in renewable energy, and to continue to identify potential projects to support utility-scale growth and off-grid energy access in India.[139]

China and India Seeking Status

Apart from domestic pressures to maximize their wealth, as discussed above, China and India have also been faced with international pressures to adjust their climate diplomacy in the process of seeking great-power status in the international system. Specifically, there are three mechanisms that have pressured China and India to shift their behavior on climate change, from free-riding to burden sharing, namely, the impact of the international climate change norms on China's and India's behavior, social isolation from their traditional identity group, and the social opprobrium China and India suffered after the failure of the Copenhagen climate conference, which had a negative impact on their international image or reputation.[140]

The international climate change regime and its norms and principles have played an important role in shaping China's and India's policy behavior in ICCN. More specifically, the UNFCCC and its Kyoto Protocol, the only legally binding international climate change regimes, have set up some basic rules and norms on how countries should share the burden of addressing climate change in a collective manner. For instance, the primary principles of the UNFCCC –

---

[139] The White House, "Fact Sheet: U.S. and India Climate and Clean Energy Cooperation" (January 25, 2015), available at https:// obamawhitehouse.archives.gov/the-press-office/2015/01/25/fact-sheet-us-and-india-climate-and-clean-energy-cooperation, accessed March 4, 2018.

[140] For instance, a Chinese scholar argues that COPs under the UNFCCC have become the new platforms through which international society constructs China's negative image. See 张丽君 [Zhang Lijun], "联合国气候变化大会与中国国家形象的塑造 [The United Nations' Climate Change Conferences and the Construction of China's National Image]," 《山西师大学报 (社会科学版)》 [Journal of Shanxi Normal University (Social Science Edition)] 41, no. 4 (July 2014), 32–37.

"equity" and "common but differentiated responsibilities and respective capabilities (CBDR-RC)" – are set out in Article 3(1):

> The Parties should protect the climate system for the benefit of present and future generations of humankind, on the basis of equity and in accordance with their common but differentiated responsibilities and respective capabilities. Accordingly, the developed country Parties should take the lead in combating climate change and the adverse effects thereof.[141]

This principle recognizes that all countries are responsible for climate change, so all should endeavor to reduce or limit their emissions. At the same time, however, developed countries should take the lead in mitigating their GHG emissions because they are responsible for historical GHG accumulation. Thus, based on principles of equity and the CBDR-RC, which represent both effectiveness and fairness, the UNFCCC does not require developing countries to reduce their GHGs, but only developed countries. This has become a guiding moral norm in global efforts to address climate change, and is also a norm that China and India have been so earnestly and persistently upholding in their negotiating stances.

However, given the fact that China and India have become the largest and third-largest CO_2 emitters respectively, it would be ineffective and unfair if these two countries were not to take action to mitigate their emissions, since their GHG emissions account for an increased share of the future concentrations of atmospheric GHGs, and thus future climate change. Accordingly, the norm of equity and CBDR-RC has been increasingly applied to China and India, and more especially to China, since both countries' absolute GHG emissions, as well as China's per capita emissions, have already surpassed the world average and been on a par with those of the EU, as mentioned above. Under such circumstances, based on the norm of equity and CBDR-RC, China and India should undertake obligations to mitigate their increased GHG emissions. As some scholars point out, China's "failure to act more robustly now may place it under greater moral scrutiny by future generations."[142] Thus, under pressure from international climate change norms, China and India have already agreed to undertake

---

[141] UN, *United Nations Framework Convention on Climate Change*, 9.
[142] Paul G. Harris and Hongyuan Yu, "Climate Change in Chinese Foreign Policy: Internal and External Responses," in Paul G. Harris (ed.), *Climate Change and Foreign Policy: Case Studies from East to West* (London: Routledge, 2009), 53.

voluntary NDCs and to put their mitigation actions under legally binding verification. In other words, China and India have come to terms with their responsibilities to address the threat of global climate change in order to contribute to the provision of a stabilized climate – a global public good, through which their responsible great-power status will come to be accepted by the international community. As mentioned above, China and India have already agreed to reduce their emissions through cutting their emissions intensity voluntarily under the Paris Agreement, which marks a significant compromise, given that both countries have agreed to undertake mitigation actions without insisting on the precondition of international financial support. This is a clear signal to the international community that China and India have fully understood and acknowledged their own responsibility in abating climate change. This confirms the observation made by Michele M. Betsill that, "International climate change norms are having effects throughout the international system. They are forcing states to redefine what it means to be a legitimate member of the international community."[143]

In addition, fear of social isolation from their own reference group, that is, the great-power group, has also played a role in soliciting both countries' cooperative behavior in the climate negotiations.[144] As noted above, in the post-Kyoto Protocol negotiations, not only developed countries, especially the EU and the United States, but also some developing countries, especially the AOSIS and LDCs, have required the large developing countries, including China and India, to undertake GHG mitigation actions. If China and India had continued to strictly adhere to their traditional bargaining position – that is, rejection of any GHG mitigation obligations – both countries would be reduced to becoming a distinct minority in the negotiations, which would obviously run counter to prosocial behavior, that is, great powers should commit to stabilizing climate change. As Johnston has pointed out, when it was clear that noncommitment would be highly

---

[143] Cited in Michele M. Betsill, "The United States and the Evolution of International Climate Changer Norms," in Paul G. Harris (ed.), *Climate Change and American Foreign Policy* (New York: St Martin's Press, 2000), 224.
[144] This point echoes the observation made by Atteridge et al., who attribute India's announcement of its carbon intensity target in late 2009 to its fear of isolation. See Atteridge et al., "Climate Policy in India," 71.

isolating, then prosocial behavior would follow.[145] China's and India's fear of isolation is vividly illustrated by the reasons laid out by Jairam Ramesh to the Indian Parliament (in the days preceding the Copenhagen climate conference) as to why India needed to make concessions in the climate negotiations: "We are showing some flexibility because we do not want to become isolated."[146] Right before the Copenhagen climate conference, Ramesh stated once again that "We will not exit in isolation," and "will coordinate our exit if any of our nonnegotiable terms are violated. Our entry and exit will be collective."[147] Simply put, Ramesh intended to prevent India from becoming isolated from the majority of other countries in the international community because of its reluctance to take actions to mitigate its GHG emissions.

Furthermore, social opprobrium in the form of "blaming and shaming" has also played a significant role in pressuring China and India to make compromises, since such opprobrium has had a negative impact on their international reputation or image as great powers. In terms of the link between negotiations and reputation, Arild Underdal pointed out back in 1983 that a party entering into negotiations in fact enters "a game that has certain bearings on its own image and reputation."[148] More specifically, he argued that "negotiation is not only a decision-making process, it is also to some extent an unofficial game of performance and reputation."[149] Based on this logic, it can be said that in ICCN, China's and India's reputations and images have also been revealed in the process of their bargaining with developed and developing countries. Throughout the history of ICCN, many developing countries have voluntarily expressed their willingness to commit to mitigating their GHG emissions. For instance, as mentioned above, as early as 1998, Argentina put forward a proposal for voluntary commitments for developing countries, which was defeated outright

---

[145] Johnston, *Social States*, 39.
[146] Sandeep Joshi, "Jairam Ramesh: India Will not Accept Legally Binding Emission Cuts," *The Hindu* (November 5, 2009), available at www.thehindu.com/todays-paper/jairam-ramesh-india-will-not-accept-legally-binding-emission-cuts/article147746.ece, accessed March 4, 2018.
[147] Cited in William Antholis and Strobe Talbott, *Fast Forward: Ethics and Politics in the Age of Global Warming* (Washington, D.C.: Brookings Institution Press), 2010, 57.
[148] Arild Underdal, "Causes of Negotiation 'Failure'," *European Journal of Political Research* 11, no. 2 (1983), 190.
[149] Ibid.

by China and India and other developing countries.[150] Similarly, during the post-Kyoto Protocol negotiations, an increasing number of developing countries, even some large developing countries such as South Africa and Brazil, both members of China's and India's BASIC group, agreed to undertake mitigation commitments so as to boost their status in the state system.[151] Against this backdrop, China's and India's reluctance to undertake actions to mitigate their GHG emissions has been increasingly regarded as free-riding on other countries' efforts, thereby reaping most of the benefits of participation while avoiding the costs of emissions limitations.[152]

China's and India's free-riding behavior has incurred "blaming and shaming" from the international media. For instance, as early as the negotiations for the Bali Road Map at COP-13, in late 2007, India – and, to a lesser degree, China – had gained the reputation of being "obdurate," or, in the words of a commentary in *The Economist*, "an ugly reputation on the global front against climate change."[153] Right after the failure of COP-15 in 2009, China and India were being widely blamed for this failure in the Western media. For instance, *The Guardian* published an article titled, "How Do I Know China Wrecked the Copenhagen Deal? I Was in the Room," which reported how China, supported by India, had insisted on deleting some critical targets in the negotiation texts, which eventually led to the failure of the conference.[154] In the same vein, another commentary appeared in *Spiegel Online International*, titled, "The Copenhagen Protocol: How China and India Sabotaged the UN Climate Summit," also blaming China and India as being fully responsible for the failure of the Copenhagen climate conference.[155] In fact, Ramesh, prior to the Copenhagen

---

[150] IISD, "Report of the Fourth Conference of the Parties to the UNFCCC," 14.
[151] IISD, "Summary of the Durban Climate Change Conference," 30.
[152] Richard B. Stewart and Jonathan B. Wiener, *Reconstructing Climate Policy beyond Kyoto* (Washington, D.C.: American Enterprise Institute for Public Policy Research, 2003), 40.
[153] "China, India and Climate Change: Melting Asia," *The Economist* (June 5, 2008), available at www.economist.com/node/11488548, accessed March 4, 2018.
[154] Mark Lynas, "How Do I know China Wrecked the Copenhagen Deal? I Was in the Room," *The Guardian* (December 22, 2009), available at www.theguardian.com/environment/2009/dec/22/copenhagen-climate-change-mark-lynas, accessed March 4, 2018.
[155] Tobias Rapp, Christian Schwägerl, and Gerald Traufetter, "The Copenhagen Protocol: How China and India Sabotaged the UN Climate Summit," *Spiegel*

climate conference, in his response to domestic criticism that the Indian government had succumbed to developed countries' pressures to change its negotiating stance, said that the main reason for India's policy shift was to avoid being blamed and shamed: "We do not want to earn reputation as deal-breaker ... tomorrow world should not blame us."[156] In another instance, at COP-17, when faced with India's (and, to a lesser extent, China's) persistent rejection of the timetable and legality of an agreement that would bind all countries, including China and India, Karl Hood, Grenada's Foreign Minister, on behalf of the AOSIS, strongly criticized China and India: "If there is no legal instrument by which we can make countries responsible for their actions, then we are relegating countries to the fancies of beautiful words," and "while they develop, we die; and why should we accept this?"[157] Faced with the threat of being "blamed and shamed,"[158] China and India have been driven to make some compromises, as mentioned previously.

The Chinese government has taken some reactive measures to alleviate its already tarnished international image under the UNFCCC negotiations. Specifically, apart from its commitment to peaking its CO_2 emissions around 2030 and the announced ¥20 billion for setting up the China South–South Climate Cooperation Fund, to support other developing countries to adapt to climate change, the government has already begun to engage in the South–South climate change cooperation. For instance, between the period of 2011 and 2015, the Chinese government provided ¥410 million (roughly $60 million) to small island countries, the least developed countries, and African countries, to help them adapt to climate change.[159] During his speech on the opening day of COP-21 on November 30, 2015, Chinese

Online International (May 5, 2010), available at www.spiegel.de/international/world/the-copenhagen-protocol-how-china-and-india-sabotaged-the-un-climate-summit-a-692861.html, accessed March 4, 2018.

[156] Joshi, "Mairam Ramesh: India Will not Accept Legally Binding Emission Cuts."
[157] Richard Black, "UN Climate Talks End with Late Deal," *BBC News* (December 11, 2011), available at www.bbc.com/news/science-environment-16124670, accessed March 4, 2018.
[158] India has been especially criticized and faced with isolation from its traditional developing and under-developed partner countries. See Jagadish Thaker and Anthony Leiserowitz, "Shifting Discourses of Climate Change in India," *Climatic Change* 123, no. 2 (2014),112.
[159] 国家发展和改革委员会 [NDRC], "'十二五'应对气候变化工作成果丰硕 [Fruitful Results of the Work on Addressing Climate Change during the 12th

President Xi Jinping announced a new series of activities aimed at helping other developing countries to adapt to climate change: in 2016, China would launch cooperation projects to set up ten pilot low-carbon industrial parks, start 100 mitigation and adaptation programs in other developing countries, and provide them with 1,000 training opportunities on climate change.[160] China's and India's compromises, and China's efforts to help other developing countries to adapt to climate change, will strengthen their bargaining position in future climate negotiations, by ensuring that they will be seen as legitimate and responsible actors in the international system, and that they will not be labeled as "obstructionist" states on the issue of climate change. In the case of India, its aspirations for a permanent seat on the UN Security Council have also stimulated it to make some significant compromises in ICCN.[161]

In sum, China's and India's desire to enhance their status as responsible great powers has made both countries sensitive not only to pressures stemming from international climate change norms, but also to the potential negative reputational impact resulting from their isolated positions in the negotiations and social opprobrium on their free-riding behavior. As a result, both countries have made compromises on mitigation and transparency.

China and India Facing Asymmetrical Interdependence

Another international pressure that has shifted China's and India's climate policy behavior from free-riding to burden sharing is their asymmetrical dependence on developed countries, especially the United States and the EU, for technology transfers. More specifically, China and India have relied on and will continue to rely on developed countries to transfer certain advanced technologies to help them mitigate their GHG emissions. Other developing countries, especially the

---

Five-Year Plan]" (February 23, 2016), available at www.sdpc.gov.cn/gzdt/201602/t20160223_775165.html, accessed March 4, 2018.

[160] Xi Jinping, "Work Together to Build a Win-Win, Equitable and Balanced Governance Mechanism on Climate Change (November 30, 2015)," *China Daily* (December 1, 2015), available at www.chinadaily.com.cn/world/XiattendsParisclimateconference/2015-12/01/content_22592469.htm, accessed June 3, 2016.

[161] Thaker and Leiserowitz, "Shifting Discourses of Climate Change in India," 112.

LDCs and some poor countries in the AOSIS, have depended on and will continue to depend on developed countries not only for technology transfers, but also for financial assistance to help them adapt to the negative impact of climate change. Therefore, China and India have had to make some concessions to developed countries in order to ensure the progress of the negotiations on issues of technology transfer and financial assistance.

In accordance with the rules set out by the UNFCCC, developed countries have a responsibility to transfer technologies and provide financial assistance to developing countries to adapt to the negative impact caused by GHGs. For instance, Article 4.5 of the UNFCCC requires developed countries to:

take all practicable steps to promote, facilitate and finance, as appropriate, the transfer of, or access to environmentally sound technologies and know-how to other Parties, particularly developing country parties to enable them to implement the provisions of the Convention. In this process, the developed country Parties shall support the development and enhancement of endogenous capacities and technologies of developing country Parties.[162]

In practice, however, during the UNFCCC negotiations, issues related to technology transfer and financial assistance – two out of the four "building blocks," including mitigation and adaptation[163] – have proved to be two of the most controversial issues between developed and developing countries, with developed countries using their capability to transfer technologies and provide financial assistance as leverage to exploit concessions from developing countries on some other issues, especially mitigation and compliance. In other words, developed countries' manipulation of the provision of technologies and financial resources provides them with the "power" to win concessions from developing countries "by means of strategic choice of investment and technology."[164]

---

[162] UN, *United Nations Framework Convention on Climate Change*, 14.
[163] IISD, "Summary of the Thirteenth Conference of Parties to the UN Framework Convention on Climate Change and Third Meeting of Parties to the Kyoto Protocol: 3–15 December 2007," *ENB* 12, no. 354 (December 16, 2007), 19.
[164] Sudhir A. Shah, "A Noncooperative Theory of Emission-Cap Determination," in Michael A. Toman, Ujjayant Chakravorty, and Shreekant Gupta (eds), *India and Global Climate Change: Perspectives on Economics and Policy from a Developing Country* (Washington, D.C.: RFF Press, 2003), 260.

China and India, with their rapid economic growth over the past two decades, already have the financial capacity to adapt to the negative impact of climate change on their own, rather than depending on financial assistance from developed countries. However, both countries will continue to depend on developed countries, especially the EU and the United States, for the transfer of some advanced technologies to improve their efficiency in mitigating increased GHG emissions. This reality can be best illustrated by the fact that China's and India's innovations in climate mitigation-related technologies have been far too limited to allow them to become technologically self-reliant in addressing climate change. To be sure, China's and India's technology research and development has increased significantly over the past decade,[165] but there is still a huge gap between both countries' innovations and those of Western countries, especially the United States.[166] As Table 5.6 shows, both China and India have been way behind the United States and the EU in terms of patents in climate mitigation-related technologies. More specifically, although the number of Chinese and Indian patents in energy generation from renewable and non-fossil sources has increased, both countries' patents in other technologies, such as combustion technologies with mitigation potential (e.g. using fossil fuels, biomass, waste), technologies specific to climate change mitigation, technologies with a potential or indirect contribution to emissions mitigation, emissions abatement, and fuel efficiency in transportation, as well as energy efficiency in buildings and lighting, are dwarfed by the number of patents held by the United States and the EU.

One factor that has complicated technology transfer from developed countries is the issue of intellectual property rights. Given that most of these technologies are being developed by the private sector in developed countries, there should be a price to pay for technology transfer to developing countries. Against this backdrop, the UNFCCC has required developed countries' governments to take initiatives to facilitate such transfers through the means of institutionalized support for technology transfer. Not surprisingly, some scholars point out that "the incentive for transferring technology will critically depend on the

---

[165] Victor, *Global Warming Gridlock*, 158.
[166] China has been far behind the United States in terms of innovation: Beckley, "China's Century?."

Table 5.6 Patents in climate mitigation-related technologies of the world, USA, EU, China, and India

		Energy generation from renewable and non-fossil sources	Combustion technologies with mitigation potential (e.g. using fossil fuels, biomass, waste)	Technologies specific to climate change mitigation	Technologies with potential or indirect contribution to emissions mitigation	Emissions abatement and fuel efficiency in transportation	Energy efficiency in buildings and lighting
2008	World	3155	247	274	1277	2268	834
	USA	742.8342	53.6167	96.9976	261.0869	273.5071	120.3333
	EU	1560.1051	127.969	121.931	404.85	1141.3369	378.9655
	China	38.475	2	0	16.125	11.4	10.7179
	India	15.5833	2.1667	0	3.5	2.95	1.375
2009	World	3711	260	306	1603	2441	935
	USA	831.3797	79.8	93.6924	285.2857	256.48	125.4167
	EU	1752.7914	114.4857	117.1071	513.904	1250.6715	400.2111
	China	68.869	0	1.0833	23.6722	18.4	14.7556
	India	33.6869	6.4857	1	0.3429	12.7222	2.35
2010	World	3838	252	280	1699	2916	985
	USA	776.1552	78.0389	75.0988	248.2798	263.1361	126.8441
	EU	1875.6754	107.5976	114.6512	542.1944	1393.3298	440.6511
	China	97.14	2	1.4167	37.8333	22.4167	30.8857
	India	20.6	3.25	2	4.0714	6.3889	1
2011	World	3529	223	310	1727	2972	932
	USA	609.0135	64.9667	84.5619	262.8305	319.6231	110.3341
	EU	1915.375	93.85	124.0167	584.2063	1457.4828	437.2685
	China	64.2032	2	2.1667	27.3929	17.5095	19.2833
	India	30.7929	5.2333	2.3524	9.869	9.5143	2.35

Source: OECD, *OECD Main Science and Technology Indicators*, available at http://stats.oecd.org/Index.aspx?DataSetCode=MSTI_PUB, accessed January 20, 2015

seriousness of the commitment of the developed country governments."[167] In these circumstances, in order to obtain compromises from the United States and the EU on governmental support to facilitate technology transfers to developing countries in general, and to China and India in particular, in addition to financial assistance to the LDCs, China and India have had to make some concessions to developed countries during the negotiations. As a Chinese researcher points out, one of the main reasons that the United States was able to apply huge pressure on China regarding climate change is China's asymmetrical dependence on US advanced clean energy technology, that is, since China does not have the same level of technological advances in emissions reductions as the United States, if it wants to reach the same level of reduction in its emissions, it will have to import advanced clean energy technologies from the United States.[168] The Chinese government has stated plainly that it would rely on external advanced technology to reduce its emissions. Specifically, in its first-ever National Climate Change Programme (to be discussed in more detail in Chapter 6), issued in June 2007, the Chinese government pointed out that:

As China is now undergoing large-scale infrastructure construction for energy, transportation and buildings, the features of intensive emissions associated with these technologies will exist for the next few decades if advanced and climate-friendly technologies could not be made timely available. This poses severe challenges to China in addressing climate change and mitigating GHG emissions.[169]

Not surprisingly, one of the main motivations for China and India in bandwagoning the United States and the EU to address climate change is to either obtain technology transfers from the United States and the EU, or to have the opportunity to research and develop clean

---

[167] P. G. Babu, K. S. Kavi Kumar, and Bibhas Saha, "The Clean Development Mechanism: Issues and Options," in Toman, Chakravorty, and Gupta (eds), *India and Global Climate Change*, 320.

[168] 王帆 [Wang Fan]: "不对称相互依存与合作型施压 [Asymmetric Interdependence and Oppressive Influence in Cooperation: Analysis on the Tactics Adjustment of US Foreign Strategy toward China]," 《世界经济与政治》 [*World Economics and Politics*], no. 12 (2010), 48.

[169] NDRC, *China's National Climate Change Program* (June 30, 2007), available at www.china.org.cn/english/environment/213624.htm#9, accessed May 31, 2016.

technology jointly with them. In terms of China, in November 2009, Chinese President Hu Jintao and US President Obama announced that a US–China Clean Energy Research Center (CERC) would be established. On November 14, 2009, CERC was officially launched by both sides. In January 2011, moreover, both sides signed five-year Joint Work Plans to research and develop technologies in the fields of advanced coal, buildings and clean vehicles, supported by public and private funding of at least $150 million over five years, divided equally between the two countries, which accounted for 88 projects, supported 1,100 researchers, and had 10 partners.[170] Given CERC's success during its first five years, the Chinese and US governments jointly agreed, in June 2015, at the seventh US–China Strategic and Economic Dialogue, to continue and expand their cooperation under CERC for a second five-year period (from 2016 through 2020), which is to be jointly funded by both countries to the tune of at least $200 million, with such priority research tracks as advanced coal technology (including carbon capture, utilization and storage), clean vehicles, building energy efficiency, and the energy and water nexus.[171]

With concessions on mitigation and verification from China and India, as well as from developing countries as a whole, developed countries have made some limited compromises on technology transfers and financial assistance to developing countries (see Table 5.7). More specifically, at COP-13, given China's, India's, and developing countries' compromise to NAMAs, developed countries conceded to India's proposal to put their technology transfer and financial assistance to developing countries under MRV. At COP-15, following developing countries' voluntary pledges to take NAMAs to reduce or slow the growth of their GHG emissions, with or without the support of developed countries, developed countries made compromises on financial assistance to developing countries. Specifically, developed countries pledged to establish short- and long-term financing. A sum of $30 billion would be provided for the period 2010–12, which would "be prioritized for the most vulnerable developing countries, such as the least developed countries, small island developing States and

---

[170] For information about CERC, see the official website, "About CERC," available at www.us-china-cerc.org/overview, accessed March 5, 2018.

[171] US Department of State, "U.S.–China Strategic and Economic Dialogue Outcomes of the Strategic Track" (June 24, 2015), available at www.state.gov/r/pa/prs/ps/2015/06/244205.htm, accessed June 9, 2016.

Africa," and long-term finance of a further $100 billion a year would be mobilized from a variety of sources by 2020. In addition, developed countries also agreed to establish four new bodies to facilitate financial assistance and technology transfer, that is, a mechanism on REDD-plus, a High-Level Panel under COP to study the implementation of financing provisions, the Copenhagen Green Climate Fund, and a Technology Mechanism.[172] Not surprisingly, progress on short- and long-term financing was regarded as "the most successful part of the [Copenhagen] Accord."[173]

At COP-16, developed countries agreed to establish a Green Climate Fund (GCF), address fast-start and long-term finance, and create a Standing Committee under COP to assist parties to implement a Technology Mechanism consisting of two components: a Technical Executive Committee made up of 20 elected experts, and a Climate Technology Centre and Network (CTCN), which helps developing countries to access information on mitigation and adaptation technologies, and to facilitate enhanced action on technology development and transfer. Requests for support have ranged from renewable energy policies and biodiversity monitoring to saving mangrove forests for coastal protection.[174] At COP-17, in exchange for China's and India's agreement to launch a negotiating process to develop a universal climate change agreement applicable to all parties by 2015 (and to be implemented by 2020), developed countries agreed to the GCF as an operating entity of the financial mechanism of the Convention, in addition to a fund expected to mobilize $100 billion a year by 2020.[175]

Furthermore, at COP-18, developing countries in general, and Sino-India in particular, accepted the notion of "developed" and "developing countries" under the Convention, as opposed to "Annex I" and "non-Annex I Parties," which opened up the possibility of

---

[172] UNFCCC, "FCCC/CP/2009/11/Add.1: Report of the Conference of the Parties on its Fifteenth Session, Held in Copenhagen from 7 to 19 December 2009" (March 30, 2010), 7, available at http://unfccc.int/resource/docs/2009/cop15/eng/11a01.pdf, accessed March 5, 2018.
[173] IISD, "Summary of the Copenhagen Climate Change Conference: 7–19 December 2009," *ENB* 12, no. 459 (December 22, 2009), 29.
[174] UNFCCC, "FCCC/CP/2010/7/Add.1: Report of the Conference of the Parties on its Sixteenth Session," 16–21.
[175] UNFCCC, "Draft decision -/CP.17: Establishment of an Ad Hoc Working Group on the Durban Platform for Enhanced Action."

Table 5.7 *Major compromises between China and India (developing countries) and developed countries*

Conference of Parties (COP)	Compromises made by developing countries, especially China and India	Compromises made by developed countries, especially the USA and the EU
COP-13 Bali Dec. 3–15, 2007	Acceptance of NAMAs	• Technological and financial support and capacity building to developing countries • MRV the above support
COP-15 Copenhagen Dec. 7–19, 2009	Voluntary pledges to take NAMAs to reduce or slow the growth of GHG emissions, with and without the support of developed nations	• US$30 billion for the period 2010–12 • Long-term finance of a further US$100 billion a year by 2020 Four new bodies: • a mechanism on REDD-plus • a High-Level Panel under the COP to study the implementation of financing provisions • the Copenhagen GCF • a Technology Mechanism
COP-16 Cancun Nov. 29– Dec. 11, 2010	Enhanced MRV/ICA of their NAMAs	• GCF • Fast-start and long-term finance • A Standing Committee under the COP to assist parties • A new Technology Mechanism
COP-17 Durban Nov. 28– Dec. 9, 2011	Agreement to launch a process to develop a universal climate change agreement applicable to all parties by 2015 (and to be implemented by 2020)	• GCF as an operating entity of the financial mechanism of the Convention • A fund expected to mobilize US$100 billion a year by 2020

Negotiating Climate Change: Proactive and Reactive 231

Table 5.7 (*cont.*)

Conference of Parties (COP)	Compromises made by developing countries, especially China and India	Compromises made by developed countries, especially the USA and the EU
COP-18 Doha Nov. 26– Dec. 8, 2012	Replacing "Annex I Parties" and "non-Annex I Parties" with "developed countries" and "developing countries"	• Institutional mechanism to address loss and damage • The Standing Committee on Finance • A mechanism on technology • A UNEP-led consortium as the host of the CTC

Source: IISD, *ENB* Volume 12, various years' summary of COPs, available at www.iisd.ca/vol12, accessed January 7, 2015

differentiation according to levels of economic development among developing countries.[176] In return, developed countries agreed to establish: first, an institutional mechanism to address loss and damage in developing countries that are particularly vulnerable to the adverse effects of climate change; second, a Standing Committee on Finance. As regards technology, a mechanism was created, and a UNEP-led consortium as the host of the Climate Technology Centre (CTC) was confirmed.[177]

In sum, two-level pressures have shaped China's and India's proactive and reactive climate diplomacy. Domestically, both countries face pressures to maximize their economic wealth by their sustained economic growth, which is a national consensus, but there is hot debate about how to reach this goal. The result of this debate is that both countries have tended to seek to maximize their economic wealth, while mitigating their increased GHG emissions as a co-benefit. The fact that both countries have significantly benefited from their participation in the CDM has also stimulated them to undertake voluntary

---

[176] IISD, "Summary of the Doha Climate Change Conference," 27.
[177] UNFCCC, "FCCC/CP/2012/8/Add.1: Report of the Conference of the Parties on its Eighteenth Session, Held in Doha from 26 November to 8 December 2012" (February 28, 2013), 21–37, available at http://unfccc.int/resource/docs/2012/cop18/eng/08a01.pdf, accessed March 5, 2018.

mitigation actions. Internationally, in order to seek their great-power status, both China and India have faced pressures stemming from international climate change norms, social isolation, and social opprobrium on their free-riding behavior, as well as their asymmetrical dependence on developed countries, especially the United States and the EU, for certain advanced technologies to help them reduce their GHG emissions, in addition to other developing countries' asymmetrical dependence on developed countries for both technology transfers and financial assistance to facilitate their climate adaptation.

Conclusion

China's and India's climate diplomacy has been characterized by both proactive and reactive behavior. On the one hand, China and India have been proactively building new coalitions, such as their bilateral coalition, the BASIC group, and LMDCs to facilitate their increasingly weakened bargaining power with developed countries and some developing countries under the UNFCCC negotiations. On the other hand, China and India have been reactively bandwagoning the non-UN climate arrangements initiated by developed countries, especially those of the EU and the United States. The result of their proactive and reactive climate diplomacy is that both countries have made some significant compromises, including not only undertaking voluntary NDCs to mitigate their GHG emissions, but also putting their mitigation actions under a legally binding transparent process in the Paris Agreement. These compromises have obviously departed from their traditional negotiation positions, and will have to allow the implementation of the Paris Agreement to impinge upon their sovereignty to a certain extent once it comes into force.

These compromises have been driven by two-level pressures. At the domestic level, China and India face pressures to maximize their economic wealth through sustained, fast economic growth, a national consensus, and although, as we have seen, there is hot debate within both countries on how to reach this goal, China and India have adopted the co-benefits approach, namely to develop their economies in a low-carbon way. Moreover, the significant benefits to their domestic growth of their participation in the CDM have also stimulated China and India to make some compromises in exchange for the maintenance of this mechanism. At the international level, China and

India have faced some external pressures, that is, in the process of seeking great-power status in the international system, both countries have had to react to pressures from the international climate change regime, and to the social isolation and social opprobrium on their free-riding behavior. At the same time, both countries have also had to react to their asymmetrical dependence on developed countries, especially the United States' and the EU's climate mitigation-related technology transfers, as well as other developing countries' asymmetrical dependence on those advanced countries for technology transfer and financial assistance to adapt to the negative impact of climate change.

6 | Domestic Climate Policies: Reactive

Proposition 4: Addressing climate change is not China's and India's domestic policy priority. China and India have only adopted domestic climate change policies when they are faced with increased international pressures.

Energy security and climate change are intricately interlinked with each other. This book temporarily separates China's and India's domestic energy policies from their climate policies on the grounds that their driving forces are different. Specifically, as explored in Chapter 3, both countries' energy policy has been largely stimulated by their domestic challenge of energy insecurity. In marked contrast, both countries' domestic climate policy has been driven by external pressures on them, arising from their dual-track climate diplomacy explored in Chapter 5, in addition to pressures from epistemic communities. Simply put, unlike the "inside-out" process of China's and India's energy policy (domestic and foreign), the domestic climate policy of both countries is the continuation of the process of "outside-in."[1] To illustrate this point, the first section explores the pressures from epistemic communities on China and India to mitigate their increased CO_2 emissions. The second section examines how China and India have reacted to external pressures to take some domestic policy measures to address climate change. The third section shows that China and India have

---

[1] Some Chinese scholars also argue that it is external pressures that have stimulated China to take actions at the domestic level to address climate change. For instance, Yu Hongyuan argues that the international climate change negotiations put a lot of pressure on China's climate change policy: 于宏源 [Yu Hongyuan], "国际制度和中国气候变化软能力建设——基于两次问卷调查的结果分析 [Global Climate Change Institutions and China Soft Capacity Building: A Survey Analysis]," 《世界经济与政治》 [*World Economics and Politics*], no. 8 (2008), 16–23; 郭冬梅 [Guo Dongmei], "气候变化法律应对实证分析——从国际公约到国内法的转化 [Analysis of Legal Strategy against Climate Change: From Outer System to Inner Rules]," 《西南政法大学学报》 [*Journal of Southwest University of Political Science & Law*] 12, no. 3, (June 2010), 41–51.

subordinated climate change to energy security. The fourth section analyzes how two-level pressures have stimulated China and India to take proactive domestic energy policy and reactive domestic climate policy measures; and a brief conclusion is drawn in the final section.

Pressures from Epistemic Communities to Address Climate Change

Apart from the pressures on China and India discussed in Chapter 5, stemming from their dual-track climate diplomacy, both countries have also been under pressure from epistemic communities – "a network of professionals with recognized expertise and competence in a particular domain and authoritative claim to policy-relevant knowledge within that domain or issue-area."[2]

The most significant epistemic community applying pressure on China and India is the Intergovernmental Panel on Climate Change (IPCC) – an intergovernmental organization established in 1988 jointly by the World Meteorological Organization (WMO) and UNEP, "with the mandate to assess scientific information related to climate change, to evaluate the environmental and socio-economic consequences of climate change, and to formulate realistic response strategies."[3] In other words, the IPCC is aimed at "transforming research recommendations into practical policies," through mobilizing science and building consensus.[4] So far, the IPCC has issued a series of Assessment Reports on global climate change, in 1990, 1996, 2001, 2007, and 2013 (and 2014),[5] and has increasingly found stronger evidence of the impact of human activities on global climate change.[6] For instance, in its Fourth Assessment Report (AR4), the IPCC states that "Global

---

[2] Peter M. Hass, "Introduction: Epistemic Communities and International Policy Coordination," *International Organization* 46, no. 1 (Winter 1992), 3.
[3] IPCC, *Climate Change 2007: Synthesis Report* (Geneva: IPCC, 2008), v.
[4] Rolf Lidskog and Göran Sundqvist, "When Does Science Matter? International Relations Meets Science and Technology Studies," *Global Environmental Politics* 15, no. 1 (February 2015), 2.
[5] IPCC, *Climate Change 2013: The Physical Science Basis* (Cambridge and New York: Cambridge University Press, 2014), 123.
[6] Joseph E. Aldy and Robert N. Stavins, "Introduction: International Policy Architecture for Global Climate Change," in Joseph E. Aldy and Robert N. Stavins (eds), *Architectures for Agreement: Addressing Global Climate Change in the Post-Kyoto World* (New York: Cambridge University Press, 2007), 3.

GHG emissions due to human activities have grown since pre-industrial times, with an increase of 70% between 1970 and 2004."[7] Accordingly, AR4 proposes that not only should developed countries take actions to ambitiously reduce their GHG emissions, but that developing countries also need to "deviate below their projected baseline emissions within the next few decades."[8] This implies that all countries, both developed and developing, including China and India, should take actions to mitigate GHG emissions. Moreover, following AR4, doubts about the links between human activity and climate change all but died down.[9] In other words, with the publication of AR4, there is a new awareness of the phenomenon of global warming and its potentially catastrophic effect on the planet.[10]

In its Fifth Assessment Report (AR5), the IPCC further confirmed that:

Human activities are continuing to affect the Earth's energy budget by changing the emissions and resulting atmospheric concentrations of radiatively important gases and aerosols and by changing land surface properties ... Unequivocal evidence ... shows that the atmospheric concentrations of important greenhouse gases such as carbon dioxide (CO_2), methane (CH_4), and nitrous oxide (N_2O) have increased over the last few centuries.[11]

Thus, the IPCC's scientific reports have increasingly reminded the international community of the urgency of addressing climate change, increasing the international community's awareness of the direct causal relationship between human activities, especially CO_2 emissions and

---

[7] IPCC, *Climate Change 2007: Synthesis Report*, 5.
[8] IPCC, *Climate Change 2007: Mitigation*, Contribution of Working Group III to the Fourth Assessment Report of the IPCC (Cambridge and New York: Cambridge University Press, 2008).
[9] Trevor Houser, Rob Bradley, Britt Childs, Jacob Werksman, and Robert Heilmayr, *Leveling the Carbon Playing Field: International Competition and US Climate Policy Design* (Washington, D.C.: Peterson Institute for International Economics/WRI, May 2008), 1, available at http://pdf.wri.org/leveling_the_carbon_playing_field.pdf, accessed March 5, 2018.
[10] Robert Orttung, Jeronim Perovic, and Andreas Wenger, "The Changing International Energy System and its Implications for Cooperation in International Politics," in Andreas Wenger, Robert W. Orttung, and Jeronim Perovic (eds), *Energy and the Transformation of International Relations: Toward a New Producer-Consumer Framework* (Oxford: Oxford University Press, 2009), 4.
[11] IPCC, *Climate Change 2013: The Physical Science Basis*, 121.

global warming, adding scientific knowledge of the risks of climate change, as well as increasing each country's awareness of their own individual vulnerability.[12] Against this backdrop, China and India, the world's largest and third-largest GHG emitters respectively, have undoubtedly been exposed to international pressures to mitigate their increased GHG emissions, especially their CO_2 emissions.

The second source of pressure from an epistemic community on China and India comes from the IEA. In this case, the most influential medium is the IEA's annual *World Energy Outlook* (*WEO*) in general, and its *World Energy Outlook 2007* in particular. Specifically, its *WEO 2007* focuses exclusively on China and India, in which the IEA projects that:

China and India together account for nearly half of the entire growth in world energy demand between 2005 and 2030. China is likely to have overtaken the United States to become the world's largest emitter of energy-related carbon dioxide this year and, by 2015, India will be the third largest emitter. By around 2010, China will overtake the United States to become the world's largest consumer of energy. In 2030, India will be the third largest oil importer in the world.[13]

At the same time, the IEA acknowledges in this *WEO* that, given the fact that there are still 400 million people without access to electricity in India, and people in rural China have rather limited access to clean burning fuels for cooking and space heating, "There can be no moral grounds for expecting China and India selectively to curb their economic growth simply because world energy demand is rising unacceptably, with associated risks of supply interruptions, high prices and damage to the environment. These are global problems to be tackled on a global basis."[14] Even so, this issue of the *WEO* puts China and India under close international scrutiny in terms of their energy demands and CO_2 emissions. Moreover, the IEA devotes half of its *WEO 2015* to India's energy outlook, on the grounds that "the country's growing energy consumption also has broad implications for the regional and global energy outlook."[15] Obviously, this issue of the *WEO* applied particular pressure on India in the run-up to COP-21.

---

[12] Hurrell and Sengupta, "Emerging Powers, North–South Relations and Global Climate Politics," 473.
[13] IEA, *World Energy Outlook 2007*, 3.   [14] Ibid.
[15] IEA, *World Energy Outlook 2015*, 32.

In addition, the Netherlands Environmental Assessment Agency had applied particular pressure on China. In 2007, this agency issued a report entitled, "China Now No. 1 in CO_2 Emissions; USA in Second Position."[16] In this report, the agency pointed out that "China's 2006 CO_2 emissions surpassed those of the USA by 8%. This includes CO_2 emissions from industrial processes (cement production). With this, China tops the list of CO_2 emitting countries for the first time."[17] This report incited a strong reaction from the Chinese Ministry of Foreign Affairs, which officially denied its top CO_2 emitter status.[18] More specifically, at a regular press conference on June 21, 2007, Qin Gang, China's spokesperson at that time, was challenged by a question relating to this report. In his response, Qin said that he did not know on which sources this report was based to be able to draw such a conclusion. Qin further pointed out that, as a developing country, it was natural for China's CO_2 emissions to rise as its economy grew, but China's per capita CO_2 emissions were much lower than those of the Netherlands. Therefore, Qin stated that he wished the international community to judge this issue objectively, calmly, and rationally, and not to target China and other developing countries.[19]

In sum, apart from pressures stemming from their dual-track climate diplomacy, China and India have been faced with pressures from epistemic communities, including the IPCC, the IEA, and the Netherlands Environmental Assessment Agency.

Reactive Domestic Policy Measures to Address Climate Change

Faced with pressures stemming from both their dual-track climate diplomacy and from epistemic communities, China and India have had to take some measures to address climate change at the domestic level. In other words, China's and India's domestic climate policy bears

---

[16] Netherlands Environmental Assessment Agency, "China Now No. 1 in CO_2 Emissions; USA in Second Position" (2007), available at www.pbl.nl/en/dossiers/Climatechange/Chinanowno1inCO2emissionsUSAinsecondposition, accessed March 5, 2018.
[17] Ibid.
[18] 侯利红 [Hou Lihong] and 阮真 [Ran Zhen], "外交部否认中国成头号二氧化碳排放国 [Ministry of Foreign Affairs Denied China Becoming the Number One CO_2 Emitter]," 东方网 [*Dongfangnet*] (June 22, 2007), available at www.oeeee.com/a/20070622/483515.html, accessed November 18, 2015.
[19] Ibid.

a reactive feature. As noted by several Chinese scholars, China and India have had to internalize the international climate change norms, rules, and institutions as a result of external pressures.[20] In reaction to these external pressures, the measures taken by China and India at the domestic level to address climate change can be best illustrated by their efforts to set up new institutions specifically to deal with climate change affairs, as well as the formulation of policies targeted at addressing climate change.

Setting up Institutions in Charge of Climate Change Affairs

China and India began to set up domestic institutions in charge of climate change affairs in 2007 (see Table 6.1). In terms of China's domestic institutional building, it established its first domestic bureaucratic agency – the Climate Change Coordination Group – in 1990, as the leading agency to coordinate the evaluation of climate change, foreign affairs strategies and activities, and to carry out research on the international climate change negotiations.[21] In 1998, it established the National Coordination Committee on Climate Change (NCCCC).[22] Since 2007, the Chinese government has begun to systematically create Chinese domestic institutions in charge of climate change. Specifically, in January 2007, the Experts Committee for Climate Change was established to improve scientific decision-making on climate change, regarded by the Chinese media as China's "climate change thinktank."[23] In June 2007, the National Leading Group to Address

---

[20] 康晓 [Kang Xiao], "利益认知与国际规范的国内化—以中国对国际气候合作规范的内化为例 [Perception of Interests and Internalization of International Norms: A Case Study on China's Internalization of Norms in International Climate Cooperation]," 《世界经济与政治》 [World Economics and Politics] no. 1 (2010), 66–83; 马建英 [Ma Jianying], "国际气候制度在中国的内化 [The Internalization of International Climate Institutions in China]," 《世界经济与政治》 [World Economics and Politics], no. 6 (2011), 91–121.

[21] 卓培荣 [Zuo Peiyong], "国务院决定建立气候变化协调小组 [The State Council Decides to Establish the Climate Change Coordination Group]," 《人民日报》 [People's Daily] (March 1, 1990), 2.

[22] Information Office of the State Council of the PRC (IOSC), China's Policies and Actions for Addressing Climate Change (Beijing: October 29, 2008), available at www.china.org.cn/government/whitepaper/node_7055612.htm, accessed March 5, 2018.

[23] "我国成立'气候智囊团'提高科学应对气候变化能力 [Our Country Establishes a 'Climate Think Tank' to Enhance Capabilities to Address Climate Change

Climate Change and Energy Conservation and Pollution Reduction, headed by the Chinese Premier and made up of more than 20 ministries, was set up to formulate important strategies, policies, and measures related to climate change, and to coordinate solutions to major problems in this regard.[24] In September 2007, a leading group was set up at the Ministry of Foreign Affairs to address climate change foreign affairs, in addition to the Special Representative for Climate Change Negotiations.[25] Obviously, the significance of China's great efforts to build new domestic institutions to address climate change taking place in 2007 rather than other years can be directly attributed to its preparations for the tough negotiations at COP-13 held in Bali in late 2007, as discussed in Chapter 5. Moreover, in 2008, the Department of Climate Change was established under the NDRC. As a result, it is the NDRC rather than China's environmental agencies that has been given the authority to handle climate change affairs, unlike the practice in most other countries, where environmental agencies are the major institutions in charge of climate change affairs.[26]

Faced with similar external pressures, the Indian government has also established some domestic institutions to handle climate change since 2007. For instance, in June 2007, then Prime Minister Singh established a Council on Climate Change under his chairmanship, to coordinate national action for assessment, adaptation, and mitigation of climate change.[27] Although there is no national group directly targeted at climate change, in 2011, ahead of India's 12th FYP, the Indian Planning Commission appointed an Expert Group on Low

Scientifically]," 新华社 [*Xinhua*] (January 22, 2007), available at www.gov.cn/jrzg/2007-01/22/content_503752.htm, accessed March 5, 2018.

[24] 国务院 [The State Council], "国务院关于成立国家应对气候变化及节能减排工作领导小组的通知 [The State Council's Notice on the Establishment of the National Leading Group to Address Climate Change and Energy Conservation and Pollution Reduction]" (June 18, 2007), available at www.gov.cn/zwgk/2007-06/18/content_652460.htm, accessed June 5, 2016.

[25] 外交部 [Ministry of Foreign Affairs], "外交部建立应对气候变化对外工作机制 [Ministry of Foreign Affairs Established Foreign Working Mechanism Addressing Climate Change]" (September 3, 2007), available at www.fmprc.gov.cn/chn/gxh/tyb///wjbxw/t358121.htm, accessed March 5, 2018.

[26] 何香柏 [He Xiangbai], "突破还是保守：评我国适应气候变化的法律框架 [China's Legal Framework on Climate Change Adaptation: Big Progress?]," 《南京大学法律评论》 [*Nanjing University Law Review*] (2015 年秋季卷) (Fall 2015), 237–238.

[27] Planning Commission, *Eleventh Five-Year Plan (2007–2012)*, Volume I, 205.

Carbon Strategies for Inclusive Growth to suggest low-carbon pathways consistent with inclusive growth. This group submitted its interim report and final report, which outline the low-carbon strategy for major carbon-emitting sectors, such as power, transport, industry, buildings, and forestry, and energy efficiency in households, buildings and industry, and carbon sequestration.[28] After Modi became India's Prime Minister in 2014, his government reorganized its Ministry of Environment and Forests, a rather weak bureaucratic agency in India's political system,[29] by adding "climate change" as its core function. Thus, the former Ministry of Environment and Forests under the Singh government has become the Ministry of Environment, Forests and Climate Change (MoEFCC) under the Modi government.[30]

Formulating Domestic Climate Change Policy

Aside from establishing domestic institutions to address climate change, China and India have also begun to formulate domestic climate change policy measures in reaction to external pressures.

As regards China's reactive policy measures to address climate change, first of all, China issued a number of national programs and strategies. For instance, in June 2007, China issued its first-ever program to address climate change, that is, the National Climate Change Programme (NCCP).[31] This program illustrates China's policies to address climate change to the year 2010. Although this program was hailed as the model for developing countries, there were some obvious

---

[28] Planning Commission, GOI, *The Final Report of the Expert Group on Low Carbon Strategies for Inclusive Growth* (April 2014), available at http://planningcommission.nic.in/reports/genrep/rep_carbon2005.pdf, accessed March 5, 2018.

[29] For a discussion of India's institutional weakness on environmental affairs, see Sukumar Ganapati and Liguang Liu, "The Clean Development Mechanism in China and India: A Comparative Institutional Analysis," *Public Administration and Development* 28, no. 5 (December 2008), 351–362.

[30] Vishwa Mohan, "India Invokes 'Right to Grow' to Tell Rich Nations of its Stand on Future Climate Change Negotiations," *Times of India* (June 17, 2014), available at http://timesofindia.indiatimes.com/home/environment/global-warming/India-invokes-right-to-grow-to-tell-rich-nations-of-its-stand-on-future-climate-change-negotiations/articleshow/36724848.cms, accessed March 5, 2018.

[31] NDRC, *China's National Climate Change Programme* (June 2007), available at www.ccchina.gov.cn/WebSite/CCChina/UpFile/File188.pdf, accessed April 9, 2015.

Table 6.1 *China's and India's climate change institutions, 2000s*

China	
Institution	Date of initiation
Experts Committee for Climate Change	January 2007
National Leading Group to Address Climate Change and Energy Conservation and Pollution Reduction	June 2007
Special Representative for Climate Change Negotiations	September 2007
Department of Climate Change under the NDRC	2008

India	
Institution	Date of initiation
Prime Minister's Council on Climate Change	June 2007
Expert Group on Low Carbon Strategies for Inclusive Growth	2011
Ministry of Environment, Forests and Climate Change (MoEFCC)	2014

Source: Author

shortcomings. According to Hu Angang, although this program shows that China formed a relatively integrated climate change policy system, it was only a short-term action plan for the period between 2007 and 2010;[32] in addition, this program did not set any targets, even though it recognized the seriousness of the threat from climate change. Instead, it focused on China's socio-economic agenda, namely economic development and poverty reduction. Thus, it is no surprise that this program fell short of the international community's expectations, especially those of developed countries who want to make China bind with GHG mitigation obligations. In November 2013, the Chinese government issued the National Strategy for Climate Change Adaptation, the first official document aimed at enhancing its adaptation to climate change, which specifies the guiding thoughts and principles in adapting to climate change at the national level by 2020.[33] In September 2014,

---

[32] Hu Angang, "China's Climate Change Policy: Background, Target, and Debate," 235.
[33] 国家发展和改革委员会 [NDRC],《国家适应气候变化战略》[*The National Strategy for Climate Change Adaptation*] (November 2013), available at

the Chinese government issued the National Plan on Climate Change (2014–2020), which "proposed guiding principles and the main target for addressing climate change, and also clarified major tasks regarding control of greenhouse gas emissions and adaptation to impacts of climate change."[34]

Addressing climate change has also been written into China's national development plans. Specifically, China has included the target of CO_2 mitigation in its national economic and social development plans since its 11th FYP, that is, a target of 20 percent energy intensity (see Chapter 3). According to research, this target could be translated into an annual reduction of over 1.5 billion tonnes of CO_2 by 2010.[35] In practice, by 2010, China's energy intensity had cumulatively dropped by 19.1 percent compared with that of 2005, equivalent to a cumulative reduction of 1.46 billion tonnes of CO_2 emissions.[36] In its 12th FYP, China set the target of reducing its economy's carbon intensity (namely CO_2 emissions per capita GDP) by 17 percent for the first time, in addition to a target of 16 percent for its energy intensity.[37] According to the NDRC, by the end of the 12th FYP, in 2015, the real cumulative reduction of carbon intensity in its economy was 20 percent, meaning that China had succeeded in exceeding its originally set target by 3 percent.[38]

Furthermore, China has already embarked on the process of establishing emissions trading markets. More specifically, in 2013–14, China successfully rolled out seven carbon emissions trading pilots or domestic carbon market pilots for permits to discharge CO_2, that is, four

www.gov.cn/gzdt/att/att/site1/20131209/001e3741 a2cc140f6a8701.pdf, accessed June 13, 2016.

[34] NDRC, *China's Policies and Actions on Climate Change (2014)* (November 2014), 45, available at http://en.ccchina.gov.cn/archiver/ccchinaen/UpFile/Files/Default/20141126133727751798.pdf, accessed June 13, 2016.

[35] Jiang Lin, Nan Zhou, Mark Levine, and David Fridley, "Taking Out 1 Billion Tons of CO_2: The Magic of China's 11th Five-Year Plan?," *Energy Policy* 36, no. 3 (March 2008), 954.

[36] 王优玲 [Wang Youlin] and 雷敏 [Lei Min], "国家发改委：中国基本实现'十一五'规划节能降耗目标 [NDRC: China Basically Achieved the Energy Conservation Targets in the 'Eleventh Five-Year Plan']," 新华网 [*Xinhuanet*] (January 6, 2011), available at http://news.xinhuanet.com/politics/2011-01/06/c_12953828.htm, accessed March 3, 2015.

[37] The State Council, "The Twelfth Five-Year Plan Guidelines."

[38] 中华人民共和国国家发展和改革委员会 [NDRC], "'十二五'应对气候变化工作成果丰硕 [The Work of Addressing Climate Change in the Twelfth Five-Year Plan Has Been Fruitful]" (February 23, 2016), www.sdpc.gov.cn/gzdt/201602/t20160223_775165.html, accessed March 5, 2018.

municipalities (Beijing, Tianjin, Shanghai, Chongqing), two provinces (Guangdong and Hubei), and one special economic zone (Shenzhen).[39] By 2015, the total trading volume of carbon dioxide in the carbon emissions trading markets of the seven pilot provinces and cities had reached 48 million tonnes of CO_2, and the turnover was 1.4 billion yuan (roughly $2 billion).[40] To regulate this market, in December 2014, the NDRC issued "the Interim Management Rules on Emissions Trading," and embarked on the process of developing formal "Regulations on carbon emissions trading." NDRC has already introduced national standards for 11 key industries and enterprises on carbon emissions accounting and reporting.[41] China's nationwide ETS is expected to be implemented in 2017.[42] Moreover, China has already begun to create legislation to address climate change. More specifically, in 2009, the Resolution of the Standing Committee of the National People's Congress of China on Actively Responding to Climate Change was adopted, which is the first resolution related to climate change to have been adopted by the Chinese top legislature. Although this resolution itself is not legislation and is non-binding, it initiates the process to establish legislation on climate change. By 2014, the drafted legislation framework for addressing climate change had been established and will be further improved based on public feedback.[43]

In addition to these concrete measures, China also made a symbolic gesture to inform the international community about its efforts to address climate change: since 2008, China has issued successive White Papers on its policies and actions to address climate change, namely in 2015, 2014, 2013, 2012, 2011, and 2008.[44] Thus, by 2015, China had

---

[39] 中国能源网 [China Energy Net], "碳交易试点七省市进行时 [Carbon Trading Pilot in Seven Provinces Is Ongoing]," available at www.china5e.com/subject/show_750.html, accessed June 12, 2016.

[40] NDRC, "The Work of Addressing Climate Change in the Twelfth Five-Year Plan Has Been Fruitful."

[41] Ibid.

[42] 商通社 [Suppliers Newswire], "全国碳排放权交易市场体系明年启动 [National Carbon Emissions Trading Market System Will Start Next Year]" (March 30, 2016), available at www.cnpc.com.cn/cnpc/sycj/201603/53e13c789bd643b19b00a4a30c8da99d.shtml, accessed March 5, 2018.

[43] NDRC, *China's Policies and Actions on Climate Change (2014)*, 43.

[44] These six White Papers are as follows: IOSC, *China's Policies and Actions for Addressing Climate Change* (October 29, 2008), available at www.china.org.cn/government/whitepaper/node_7055612.htm; IOSC, *China's Policies and Actions for Addressing Climate Change* (November 22, 2011), available at

already issued six White Papers on climate change, which serves the purpose of showing the international community the Chinese government's willingness and measures taken to address climate change. In fact, climate change is the only issue-area on which the Chinese government has so frequently issued White Papers since it began to issue them in 1991.[45]

India, like China, as a result of external pressures, has formulated some policy measures to address climate change since 2008. This point echoes Navroz Dubash's observation: "Indeed, Indian domestic policy has undoubtedly been influenced by global pressures to demonstrate a commitment to action. The result has been a flurry of policy activity in the years since the Bali COP of 2007."[46] More specifically, India issued its *National Action Plan on Climate Change* (NAPCC) in June 2008. The NAPCC identified the constraints on India's sustainable economic growth, and detailed the existing and future policies and programs targeted at climate change mitigation and adaptation. It emphasized India's overriding priorities as "economic and social development and poverty eradication,"[47] and set out the concept of "co-benefit" climate policy, that is, to "promote our development objectives while also yielding co-benefits for addressing climate change effectively."[48] In order to realize such co-benefits, the NAPCC set up eight national missions, "targeting energy efficiency and renewable energy, as well as improved research capacity on climate change issues."[49] However, this plan lacked a detailed framework of actions, and did not set higher domestic climate change and development goals. Under the NAPCC,

---

www.china.org.cn/government/whitepaper/node_7142680.htm; IOSC, *China's Policies and Actions for Addressing Climate Change* (November 21, 2012), available at www.china.org.cn/government/whitepaper/node_7172407.htm; IOSC, *China's Policies and Actions for Addressing Climate Change (2013)* (November 5, 2013), available at www.china.org.cn/government/whitepaper/node_7193982.htm; NDRC, *China's Policies and Actions on Climate Change (2014)* (November 2014), available at http://en.ccchina.gov.cn/archiver/ccchinaen/UpFile/Files/Default/20141126133727751798.pdf; NDRC, *China's Policies and Actions on Climate Change (2015)* (November 2015), available at www.cma.gov.cn/en2014/climate/featutes/201511/P020151120633951236905.pdf, accessed June 2, 2016.
[45] The full list of the White Papers issued by the Chinese government since 1991 is available at www.china.org.cn/e-white, accessed March 5, 2018.
[46] Dubash, "The Politics of Climate Change in India," 197. [47] Ibid., 1.
[48] Ibid., 2.
[49] Prime Minister's Council on Climate Change, *National Action Plan on Climate Change*.

two of the missions directly targeted energy security, while only one mission directly targeted climate change. Not surprisingly, some Indian researchers criticized the NAPCC's goals for being merely a rhetorical co-benefits paradigm of Indian climate change policy, which is in fact "an accidental quality" rather than a "proactive framing" approach.[50] In practice, many of those eight missions were implemented at a slow pace.[51] Not surprisingly, Arunabha Ghosh, CEO of the Council on Energy, Environment and Water, India's climate think-tank, points out that "India's energy and development imperatives will largely shape its response to climate change."[52]

Thus, although the NAPCC represents progress in India's policy toward climate change, like China's NCCP, it does not mention a commitment to cut carbon emissions, nor does it set any concrete targets. An Indian expert therefore points out that the NAPCC puts economic development ahead of emission reduction targets, and "makes a case for the right of emerging economies to pursue development and growth to alleviate poverty without having to worry about the volume of atmospheric emissions they generate in the process."[53]

Moreover, in order to enhance its capacity to adapt to the impact of climate change, the Indian government issued its *National Policy on Disaster Management 2009*.[54] The vision of this policy is to build a safe and disaster-resilient India, by developing a holistic, proactive, multi-disaster oriented and technology-driven strategy through a culture of prevention, mitigation, preparedness, and response, specifically

---

[50] Navroz K. Dubash, "Climate Politics in India: Three Narratives," in Navroz K. Dubash (ed.), *Handbook of Climate Change and India: Development, Politics and Governance* (New Delhi: Oxford University Press), 200.

[51] Sujatha Byravan and Sudhir Chella Rajan, "An Evaluation of India's National Action Plan on Climate Change" (July 2012), available at http://ifmrlead.org/wp-content/uploads/2016/05/NAPCC%20Evaluation.pdf, accessed March 5, 2018.

[52] Arunabha Ghosh, "The Big Push for Renewable Energy in India: What Will Drive It?," *Bulletin of the Atomic Scientists* 71, no. 4 (2015), 32.

[53] Sudhirendar Sharma, "Missing the Mountain for the Snow" (July 24, 2008), available at http://indiatogether.org/napcc-environment, accessed March 5, 2018.

[54] Indian Ministry of Home Affairs, GOI, *National Policy on Disaster Management 2009* (October 22, 2009), available at www.ndma.gov.in/images/guidelines/national-dm-policy2009.pdf, accessed May 20, 2015.

to meet challenges such as climate change-induced natural disasters, like cyclones, floods, and droughts in the coming years.[55]

Like China, India has also begun to establish policy measures to address climate change in its national development plans. For instance, in its 11th FYP, targets related to climate change include reducing energy intensity with respect to GHG emissions, by 20 percent from the period 2007–8 to 2016–17, and increasing energy efficiency by 20 percent, as discussed in Chapter 3.[56] Also in this FYP, the Indian government states clearly that, with a share of just 4 percent of global emissions, India is not responsible for climate change, so India will not commit to GHG mitigation, but will only focus on efforts to ensure that its GHG emissions intensity continues to decline.[57] India's 12th FYP recognizes a need to adopt a low-carbon strategy for inclusive growth to improve the sustainability of its growth, with carbon mitigation being an important co-benefit. In order to achieve the co-benefit of both economic development and emission mitigation, this FYP sets a goal of reducing the emission intensity of its GDP in line with the targeted 20 to 25 percent reduction over 2005 levels by 2020, which India pledged before the Copenhagen climate conference (see Chapter 5), and to add 30,000 MW of renewable energy capacity during this FYP period.[58]

In addition, India has been in the process of experimenting cap and trade under the PAT scheme, which is an innovative, market-based trading scheme announced by the Indian government in 2008, under its National Mission on Enhanced Energy Efficiency in NAPCC. It aims to improve energy efficiency in industries by trading in energy efficiency certificates in the eight energy-intensive sectors.[59] Those eight sectors will cover 65 percent of India's total industrial energy consumption. Once the scheme is implemented, expected savings are 19 GW of energy, and emissions reductions of 98 million tonnes

---

[55] Ibid., 7.
[56] Planning Commission, *Eleventh Five-Year Plan (2007–2012)*, Volume I, 207.
[57] Ibid., 206.
[58] Planning Commission, *Twelfth Five-Year Plan (2012–2017)*, Volume I, 9, 35.
[59] Institute for Industry Productivity, "IN-2: Perform Achieve Trade Scheme (PAT Scheme)," available at http://iepd.iipnetwork.org/policy/perform-achieve-trade-scheme-pat-scheme, accessed June 18, 2016.

per year. It is estimated that investment in this scheme will amount to approximately $15 billion.[60]

In sum, faced with external pressures, China and India have already established new institutions and formulated new policies to address climate change at the domestic level. In other words, both China and India have already begun to internalize international climate change norms and principles. Both countries' climate policies are framed as reductions in the emissions intensity of their economies (CO_2 per unit of GDP), but neither country sets specific GHG reduction targets. Apart from these similarities, the degree of internalization in these two countries is different. Table 6.2 shows that the extent of China's internalization of climate protection is obviously higher than that of India, given the fact that China has not only issued a national program and strategies to mitigate GHG emissions and adapt to climate change, and drafted climate change legislation, but has also been in the process of establishing a nationwide ETS based on the experience of the emissions trading pilot. Simply put, China has set up a more comprehensive domestic climate policy framework than India.

Subordinating Climate Change to Energy Security

Having explored China's and India's respective domestic policies on energy security and climate change through the process of inside-out and outside-in, it is worthwhile comparing the profiles of these two issue-areas, namely energy security and climate change, in both countries.

As discussed above, China and India have already taken some significant measures to address climate change at the domestic level, but the climate institutions and policy measures they have established do not necessarily guarantee that both countries have put the issue of climate change above their energy security. In practice, both countries still do not treat the climate change issue as a priority for them, which can be best illustrated by comparing the profile of energy security and climate change in both countries (see Table 6.3).

As discussed in Chapter 3, both China and India consider energy security to be a national security issue of great importance for their

---

[60] IEA, "Perform, Achieve, Trade (PAT) Scheme," available at www.iea.org/policiesandmeasures/pams/india/name-30373-en.php?, accessed March 5, 2018.

Table 6.2 *China's and India's domestic climate change policies*

	China
Policies	Brief contents
China's National Climate Change Programme (June 2007)	A strategy to address climate change and sustainable development for the period 2007–10
National Five-Year Plans (11th and 12th)	Set the energy-intensity target of 20% in the 11th FYP and the target of 17% reduction of carbon intensity in its economy in the 12th FYP
Carbon emissions trading pilot markets (2013–14)	Beijing, Tianjin, Shanghai, Guangdong, Hubei, and Shenzhen
Emissions trading scheme (ETS)	To be implemented nationwide from 2017
National Strategy for Climate Change Adaptation (2013)	Specifies the guiding thoughts and principles in adapting to climate change at the national level by 2020
National Plan on Climate Change (2014–2020) (2014)	Proposes guiding principles and the main targets for addressing climate change, and also clarifies major tasks regarding control of GHG emissions and adaptation to impacts of climate change
Legislation on climate change (2014)	Drafted legislation framework for addressing climate change
China's Policies and Actions for Addressing Climate Change	White Papers on climate change issued annually since 2008 to communicate China's policies and actions to address climate change to the international community
	India
Policies	Brief contents
National Action Plan on Climate Change (2008)	Sets up eight national missions, targeting energy efficiency and renewable energy, as well as improved research capacity on climate change issues
National Policy on Disaster Management (2009)	Addresses the challenges of climate change-induced natural disasters like cyclones, floods, and droughts in the coming years
National Five-Year Plans (11th and 12th)	Planning for longer-term economic growth, emphasizing meeting development targets, with climate goals as co-benefits

Table 6.2 (cont.)

	India
Policies	Brief contents
Perform, Achieve, Trade (PAT) Scheme (2008)	A market-based trading scheme that aims to improve energy efficiency in industries by trading in energy efficiency certificates in India's eight most energy-intensive sectors

Source: Author

Table 6.3 *The profile of energy security and climate change in China and India*

	Energy security	Climate change
National security issue?	Yes	No
Strategically imperative issue?	Yes	No
Priority?	High	Low
Independent issue?	Yes	No
Independent policy actions?	Yes	No
Sources for policy actions?	Internal/domestic	External/international
Characteristic of Sino-Indian policy actions?	Proactive	Reactive
Effectiveness of policies?	High	Low

Source: Author

economic growth, and therefore it has been typically granted high priority on their policy-making agenda. In contrast, neither China nor India regards climate change as a national security issue, which explains both countries' persistently strong opposition to having climate change recognized as an international security threat by the UNSC.[61] More specifically, as early as April 2007, when the UNSC held its first-ever debate, initiated by the UK, on the impact of climate

---

[61] Ed King, "China and Russia Block UN Security Council Climate Change Action" (February 19, 2013), available at www.climatechangenews.com/2013/02/18/china-and-russia-block-un-security-council-climate-change-action, accessed March 5, 2018.

change on peace and security,[62] China, India, and the majority of developing countries, with the exception of small island countries, strongly opposed this idea of bringing climate change into the UNSC. During this debate, China's representative stated clearly that "The developing countries believe that the Security Council has neither the professional competence in handling climate change – nor is it the right decision-making place for extensive participation leading up to widely acceptable proposals."[63] India's representative concurred that "the appropriate forum for discussing issues relating to climate change was the United Nations Framework Convention on Climate Change," rather than the UNSC, because "Nothing in the greenhouse gas profile of developing countries even remotely reflected a threat to international peace and security."[64] In February 2013, China and India, together with some other developing countries, once again opposed the proposition of climate change being a UNSC issue, when the Council met to discuss the potential effects of global warming.[65]

Rhetorically, neither country's top leaders have ever treated climate change as a priority, which contrasts strikingly with their frequent emphasis on energy security as their top priority, as discussed in Chapter 3. In other words, for both China and India, climate change is not an independent issue-area, but only an element in both countries' development. For instance, in his speech to the UN climate summit in 2009, Hu Jintao, then Chinese President, clearly stated that climate change was fundamentally a development issue, so neither stopping development to deal with climate change, nor developing the economy without considering climate change was acceptable.[66] By the same token, Xi Jinping, the incumbent Chinese President, stressed green development in his speech at the plenary session at the COP-21 climate summit in Paris in late 2015: "The agreement should ... set up incentive mechanisms to encourage countries to pursue green, circular and low-carbon development featuring both economic growth and an

---

[62] UNSC, "Security Council Holds First-Ever Debate on Impact of Climate Change on Peace, Security, Hearing over 50 Speakers" (April 17, 2007), available at www.un.org/press/en/2007/sc9000.doc.htm, accessed March 5, 2018.
[63] Ibid.   [64] Ibid.
[65] King, "China and Russia Block UN Security Council Climate Change Action."
[66] Hu Jintao, "Join Hands to Address Climate Challenge" (September 23, 2009), available at www.china-un.org/eng/gdxw/t606111.htm, accessed March 5, 2018.

effective response to climate change."[67] His explanation for the way in which China has addressed the issue of climate change is that, "We have integrated our climate change efforts into China's medium- and long-term program of economic and social development."[68] Thus, Xi does not regard climate change as an independent issue-area, but only an element of China's transformation from a high-carbon to low-carbon economy, its green development, or as an element of China's economic and social development.

In terms of India's non-prioritization of climate policy, at a meeting of the Prime Minister's Council on Climate Change, chaired by Indian Prime Minister Shri Narendra Modi, he called for a paradigm shift in global attitudes toward climate change, from "carbon credit" to "green credit," and "instead of focusing on emissions and cuts alone, focus should shift on what we have done for clean energy generation, energy conservation and energy efficiency, and what more can be done in these areas."[69] Moreover, in his speech at the plenary session at COP-21, Modi reiterated India's traditional negotiating stances, especially the historical responsibility of developed countries, and their responsibility to provide financial and technological support to developing countries. He also pointed out that India has 300 million people without access to energy, so he stated clearly that "developing countries should have enough room to grow."[70] Obviously, like his Chinese counterpart, Modi's policy is preoccupied with energy as a focal point or priority, rather than climate change. In other words, like China, India does not regard climate change as an independent issue-area, but only an element of its energy security.

Another piece of evidence to show that climate change is not a priority issue for India is that its newly established institution in charge of climate change has not held regular meetings. Specifically, as mentioned in the previous section, India established a new Council on

[67] Xi, "Work Together to Build a Win-Win, Equitable and Balanced Governance Mechanism on Climate Change."
[68] Ibid.
[69] PM India, "PM Chairs Meeting of the Council on Climate Change" (January 19, 2015), available at http://pmindia.gov.in/en/news_updates/pm-chairs-meeting-of-the-council-on-climate-change, accessed March 5, 2018.
[70] "Full Text: PM Narendra Modi's Speech at the Plenary Session at COP 21 Summit in Paris," *First Post* (December 1, 2015), available at www.firstpost.com/world/full-text-pm-narendra-modis-speech-at-the-plenary-session-at-cop-21-summit-in-paris-2527862.html, accessed March 5, 2018.

Climate Change in 2007, under the Indian Prime Minister, "to coordinate national action plans for assessment, adaptation and mitigation of climate change."[71] However, it was reported that this Council had not met for over three years until November 2014, when it was reconstituted.[72] This fact suggests that there had been no coordination between India's different ministries on climate change for at least three years (2011–14), which in turn indicates that climate change is not a priority for India's ministries.

Moreover, neither China nor India has addressed climate change directly, but only in the shadow of their energy security. For instance, some research points out clearly that, on the domestic front, India has not addressed climate change directly, but indirectly by "domestic legislation for energy conservation and increased use of renewable energy,"[73] so "in India, concerns about energy security are key, not climate change."[74] Simply put, climate change mitigation is just one element of their focus on energy security. In his explanation of why the Indian government issued the NAPCC, India's Finance Minister said outright: "It is because we recognize the linkages between climate change and energy security that we have adopted a National Action Plan on Climate Change."[75] In the NAPCC itself, moreover, the term "energy security" is mentioned eight times,[76] which is sufficient to prove that India officially regards the issue of climate change not as an independent issue-area, but only as an integrated element of its energy security. Thus, it is no surprise that Indian domestic climate change policy has in fact been derived from its energy security policy.

Similarly, when it comes to China's NCCP, according to a Chinese scholar, although it is a climate change program in name, in practice it is not a climate change policy per se, but merely a series of policy

---

[71] Sharma, "Missing the Mountain for the Snow."
[72] Meena Menon, "PM's Climate Change Council Recast," *The Hindu* (November 6, 2014), available at www.thehindu.com/sci-tech/energy-and-environment/govt-reconstitutes-pms-council-on-climate-change/article6567187.ece, accessed March 5, 2018.
[73] Thaker and Leiserowitz, "Shifting Discourses of Climate Change in India," 109.
[74] Ingrid Boas, "Where Is the South in Security Discourse on Climate Change? An Analysis of India," *Critical Studies on Security* 2, no. 2 (2014), 157.
[75] P. Mukherjee, "India and Global Challenges: Climate Change and Energy Security," cited in Dubash, "The Politics of Climate Change in India," 197.
[76] Prime Minister's Council on Climate Change, *National Action Plan on Climate Change*.

combinations being carried out through energy reform, economic development, and transition.[77] Moreover, the fact that it is the NDRC – the most powerful bureaucratic agency in charge of China's overall long-term economic and social planning – that has been tasked with leading China's climate change policy-making is sufficient to show that the Chinese government's priority is development rather than climate change.

Simply put, neither China nor India has regarded climate change as an independent issue. Rather, both countries treat it as an element of their energy security. As a result, they have both taken some proactive actions to procure their energy security while mitigating their CO_2 emissions, a so-called "co-benefit" approach, which can be attributed to the fact that the forces that have driven China and India to take action to procure their energy security stem from their increased energy demand at the domestic level, while their policies to address climate change have been largely driven by international pressures applied on them by epistemic communities, as discussed earlier in this chapter, as well as those arising from their dual-track climate diplomacy discussed in Chapter 5. This echoes other scholars' observations about the driving forces behind China's climate policy, that is: "China's answer to climate change has been very limited nationally as well as internationally. External pressures generated by international negotiations on climate change have placed the issue of global warming on China's domestic agenda and forced the country to respond."[78]

Given the proactive features of China's and India's energy security policies, and the reactive features of both countries' climate policies, it is no surprise that their energy security policies have turned out to be quite effective, witnessed by both countries' enhanced energy efficiency and their successful expansion of renewable energy, especially wind and solar, as discussed in Chapter 3. In stark contrast, both countries' policies on climate change have been not nearly so effective. It is true that both countries have adopted the method of reducing their economy's carbon intensity as a means of mitigating their CO_2 emissions, as discussed in Chapter 3. However, according to research, carbon intensity targets "can be used to obfuscate the fact

---

[77] He Xiangbai, "China's Legal Framework on Climate Change Adaptation: Big Progress?," 239.
[78] Carmen Richerzhgen and Imme Scholz, "China's Capacities for Mitigating Climate Change," *World Development* 36, no. 2 (2008), 314.

that a targeted reduction in intensity can mean a continued increase in absolute levels," although "they have valuable properties in managing economic uncertainty and focus the target formulation on structural and technological change, rather than GDP growth, which itself is not a policy variable."[79] In other words, even if China and India have pledged to cut their carbon intensity, both countries' absolute emissions are likely to continue to grow. According to the IEA's *WEO 2016*, China's energy-related CO_2 emissions will not peak until around 2030, while India's CO_2 emissions will continue to rise through 2040.[80]

In summary, the issue of climate change is not a national priority for either China or India, so both have subordinated their climate change policy to their development and energy security. Accordingly, both China's and India's climate policies lack independence and are, to a large extent, the co-beneficiaries of both countries' policies targeted at energy security. Put differently, addressing climate change is, in practice, a co-benefit from China's and India's efforts to address the challenges posed by their energy insecurity. This reality has been succinctly pointed out by a Chinese scholar, that is, the Chinese government (as well as the Indian government) has treated climate change mainly as an economic and energy security problem, rather than an environmental problem.[81] Against this backdrop, the slower growth rate of China's CO_2 emissions in recent years can be attributed mainly to the "side-effect" of China's other economic policies, including those targeted at improving its energy efficiency, as discussed in Chapter 3.

Shaping Domestic Energy and Climate Policy: Two-Level Pressures

As explored in Chapter 3, although China and India have been taking some proactive policy measures to address their energy insecurity at the domestic level, neither country had started to take concrete actions to address climate change until 2007, despite the fact that global

---

[79] David I. Stern and Frank Jotzo, "How Ambitious Are China and India's Emissions Intensity Targets?," *Energy Policy* 38, no. 11 (November 2010), 6776.
[80] IEA, *World Energy Outlook 2016*, 324.
[81] He Xiangbai, "China's Legal Framework on Climate Change Adaptation: Big Progress?," 241.

efforts to address this challenge had been initiated in the early 1990s. This section explores the reasons for China's and India's proactive domestic energy policies and reactive domestic climate change policies, which can also be attributed to both countries' two-level pressures. At the domestic level, both countries have faced pressures to maximize their wealth through sustained economic development. At the international level, both countries have reacted to external pressures on their policy behavior in the process of enhancing their status in the international system, in which they have been asymmetrically dependent on Western countries' markets, as well as on their advanced clean technology.

China and India Seeking Wealth

China's and India's proactive energy policies and reactive climate policies have been driven by their "growth imperative" rather than a "green imperative,"[82] or by their domestic pressures to maximize their energy security and economic growth. This argument can be tested by the following aspects.

First, both countries have been reluctant to put restraints on their total energy consumption because energy consumption is fundamental to their sustained economic growth. As mentioned earlier, in both countries' 11th and 12th FYPs, the Chinese and Indian governments set some ambitious targets on energy efficiency and the reduction of energy intensity and carbon intensity, which reflects their strong commitments to sustainable development. At the same time, however, there are no specific targets set for the total energy consumption in these plans. In China's case, according to a news report by Xinhua, China's official news agency, in the originally drafted 12th FYP, the Chinese government set a hard target for its overall energy consumption, that is, capping it at four billion tce by 2015, which would serve as "a mandatory ceiling."[83] In practice, however, this mandatory cap on its

---

[82] Sufang Zhang, Philip Andrews-Speed, Xiaoli Zhao, and Yongxiu He, "Interactions between Renewable Energy Policy and Renewable Energy Industrial Policy: A Critical Analysis of China's Policy Approach to Renewable Energies," *Energy Policy* 62 (November 2013), 349.
[83] "全国政协委员张国宝详解'十二五'能源发展战略 [CPPCC Member Zhang Guobao Explains the Strategy for Energy Development during the Twelfth Five-Year Plan]," 新华网 [*Xinhuanet*] (March 5, 2011), available at

energy consumption was not able to be inked into the final text of its 12th FYP, which only states that "China is devoted to controlling the total amount of energy consumption rationally."[84] Such an ambiguous policy statement is obviously intended to provide leeway for the potential growth of its total energy consumption. Without setting a limit to cap its total amount of energy consumption, China's total CO_2 emissions might still increase as usual, even if it achieved its energy efficiency and carbon intensity targets.[85] This explains why China has not set a nationwide absolute emissions cap, which, according to some researchers, poses some challenges for its efforts to develop its piloted ETS and a nationwide ETS, as discussed above.[86]

Similarly, India has failed to set any limits on its total amount of energy consumption in its latest two FYPs. In its 11th FYP, for example, the Indian government states clearly that in order to eradicate poverty, India would "accelerate the pace of growth while also making it more inclusive. The growth objective is to achieve an average growth rate of 9% per annum for the Plan period."[87] This plan recognizes that "the availability or affordability of energy" is of critical importance for India's rapid growth and the challenges India faces in its energy sector, including the sharp increase in oil prices on the world market, and the need to increase energy efficiency and reduce carbon emissions. This plan further points out that all of these challenges "must be met in a manner which does not come in the way of our objective of achieving faster and more inclusive growth."[88] Although the plan admits "the emerging threat of climate change ... over the longer term," it states clearly that "the burden for reducing energy emissions must fall largely on the industrialized countries," because India's per capita emissions are much lower than those of the industrialized countries.[89] Simply

---

http://news.xinhuanet.com/politics/2011-03/05/c_121152018_2.htm, accessed May 20, 2015.

[84] The State Council, *The Twelfth Five-Year Plan*.

[85] Xueliang Yuan and Jian Zuo, "Transition to Low Carbon Energy Policies in China – from the Five-Year Plan Perspective," *Energy Policy* 39, no. 6 (June 2011), 3858.

[86] Tao Pang and Maosheng Duan, "Cap Setting and Allowance Allocation in China's Emissions Trading Pilot Programmes: Special Issues and Innovative Solutions," *Climate Policy* 16, no. 7 (2016), 816–817; Alex Y. Lo, "Challenges to the Development of Carbon Markets in China," *Climate Policy* 16, no. 1 (2016), 120.

[87] Planning Commission, *Eleventh Five-Year Plan (2007–2012)*, Volume I, vii.

[88] Ibid., x.   [89] Ibid., x.

put, in its 11th FYP, the Indian government determines its policy to seek economic development with energy security as its top priority, while climate change should be left for developed countries to address.

In its 12th FYP, the Indian government further emphasizes economic growth by stating that "our first priority must be to bring the economy back to rapid growth," although this plan adds that such rapid growth should be "inclusive and sustainable."[90] However, the plan also emphasizes the importance of increasing energy supplies for its economic growth, that is, "A growth rate of 8 per cent in GDP requires a growth rate of about 6 per cent in total energy use from all sources."[91] In other words, India's 12th FYP sets a target of a 6 percent increase in its total energy consumption to sustain its target of 8 percent in its GDP growth.

China's short-lived "Green GDP" is another piece of evidence that proves China's growth imperative.[92] More specifically, the concept of "Green GDP" was first put forth by Chinese President Hu Jintao, in a speech at a symposium on population, resources, and environment held by the Chinese central government in March 2004. After this symposium, the process of calculation of green GDP was initiated. Two years later, in 2006, the results of the calculation were published in a document entitled, *China Green National Accounting Study Report 2004*. Although the resulting "Green GDP" estimates were highly conservative, they still served as a wake-up call for the Chinese leadership, by proving that the annual costs incurred by its environmental pollution were $51 billion (at the official exchange rate at the time of the study in 2004), equaling 3.05 percent of China's GDP in 2004. However, the concept of green GDP was rejected and blocked so forcefully by local governments that the Chinese central government was forced to postpone it indefinitely. Not surprisingly, some Western

---

[90] Planning Commission, *Twelfth Five-Year Plan (2012–2017)*, Volume I, v.
[91] Ibid., 33.
[92] Content in this paragraph draws mainly on my work on China's environmental governance: Fuzuo Wu, "China's Environmental Governance: Evolution and Limitations," in Jing Huang and Shreekant Gupta (eds), *Environmental Politics in Asia: Perspectives from Seven Asian Countries* (Singapore: World Scientific, 2014), 100.

Domestic Climate Policies: Reactive 259

scholars point out that "locally driven economic growth will never easily be trumped by environmental considerations."[93]

China's and India's "growth imperative" can be further demonstrated by some of the environmental problems that resulted from their renewable energy development.[94] As discussed in Chapter 3, both the Chinese and Indian governments have adopted the policy of encouraging the development of renewable energy as one of the major measures to assure their energy security while reducing CO_2 emissions. This policy measure has brought considerable economic wealth to both countries. For instance, according to a report by Chinese Greenpeace, in 2010 China's solar photovoltaic (PV) production chain's annual output value was more than 300 billion yuan (roughly $45 billion), its import and export value was $22 billion, and it created 30 million jobs.[95] According to a report published by the Natural Resources Defense Council (NRDC) and the Council on Energy, Environment and Water (CEEW), grid-connected solar and wind energy was estimated to have created nearly 70,000 full-time jobs in India by February 2015.[96] However, the processes involved in their renewable energy expansion have turned out to be not so "green," leading to environmental problems.

In China, environmental regulations in the renewable energy production chain are lax. Chinese local governments typically compete with each other to attract firms to invest in their local areas, so they tend to offer firms policy incentives, such as privileged use of land and tax breaks, while lacking stringent environmental regulations. Not

---

[93] Coraline Goron and Cyril Cassisa, "Regulatory Institutions and Market-Based Climate Policy in China," *Global Environmental Politics* 17, no. 1 (February 2017), 116.

[94] China's and India's "growth imperative" is also reflected in the fact that both countries' environmental performance has been rather poor. For instance, China and India rank 118th and 155th respectively (out of 178 countries) on the comprehensive Environmental Performance Index, and 176th and 174th respectively on air quality. "Country Rankings," *Environmental Performance Index* (New Haven, CT: Yale University, 2014), available at www.epi.yale.edu/epi/country-rankings, accessed March 4, 2017.

[95] 绿色和平 [Greenpeace], "风光无限 – 中国风电发展报告2011 [Unlimited Wind and Sunshine: China Wind Power Outlook 2011]," available at www.greenpeace.org/china/Global/china/publications/campaigns/climate-energy/2011/windpower-briefing-2011.pdf, accessed March 5, 2018.

[96] CEEW/NRDC, "Clean Energy Powers Local Job Growth in India," Interim Report (New Delhi: Natural Resources Defense Council and Council on Energy, Environment and Water, 2015).

surprisingly, firms usually do not pay much attention to the environmental impacts of their investment in renewable energy. This is especially the case in the solar sector, given that some solar facilities have already caused local air and water pollution. For instance, in September 2011, Jinko Solar factory in Haining township, Zhejiang province, released fluoride into a local creek, exceeding the national standard by a factor of 10, resulting in its contamination, which led to the death of the local fish stock. This incident incurred protests from local villagers. To solve this problem, the local government fined the factory 470,000 yuan ($70,000), but still allowed it to continue production after it paid the fine and cleaned up the pollutant.[97]

Similarly, India's expansion of its renewable energy industry has also led to some environmental and even social problems, which have largely resulted from a specific policy targeted at facilitating the development of wind and solar PV plants (up to a plant size of 50 hectares), and other renewable energy projects that are exempted from environmental impact assessment.[98] This policy has led to pressure on water and land usage. In terms of water pressure, solar thermal plants require large quantities of water for cooling the steam used to power the electric turbines, which puts further pressure on water resources in some states located in dry and arid areas, such as Gujarat and Rajasthan.[99] When it comes to the land use pressure, in Karnataka, for example, "many of the high-wind potential sites fall in forest lands and the non-availability of non-forest land contiguous to forest land for compensatory afforestation," which has been regarded as one of the major problems in that state.[100]

China's and India's ambitions to acquire wealth or economic growth have led to the two countries passively enduring the heavy losses caused by extreme weather events stemming from the negative impact of climate change. For instance, according to the National Strategy for Climate Change Adaptation, since the 1990s, China's average annual direct economic and human losses due to extreme weather and climate

---

[97] For further details about this pollution, see a Chinese website's series of reports, 世纪新能源网 [21st Century New Energy net], "笼罩在阳光下的阴影：关注晶科能源污染事件 [Shrouded in the Shadow of the Sun: Attention to the Jinko Pollution Incident]" (September 2011), available at www.ne21.com/special/show-10.html, accessed May 20, 2015.

[98] Krithika and Mahajan, "Governance of Renewable Energy in India," 27.

[99] Ibid.   [100] Ibid.

events have been more than 200 billion yuan ($30 billion), and more than 2,000 deaths.[101] India's official documents have not comprehensively recorded the details of the economic and human losses caused by extreme weather events. But according to *Down to Earth*, an Indian magazine with environmental issues as its focus, "India is extremely vulnerable to the impact of climate events. Every year, it faces extreme weather events in the form of floods and cyclones, which take lives, destroy homes and agricultural yields, and result in huge revenue losses."[102] It recorded the total economic and human losses, caused by five typical extreme weather events, such as floods and cyclones, between 2005 and 2014, being 132,204 crore rupees (roughly $24.48 billion), and 10,297 deaths.[103] Not surprisingly, a Western scholar points out that "in a country where the overriding goal is economic growth, an enhanced sense of vulnerability is insufficient to offset concerns about the cost of restricting emissions."[104]

Simply put, domestic pressure on China and India to seek wealth has stimulated both countries to proactively procure energy supplies, especially renewable energy, to fuel their domestic economic growth, despite the negative environmental impacts arising from their weak environmental regulations on firms, and the economic and human losses caused by extreme weather events derived from the negative impact of climate change.

China and India Seeking Status

China's and India's proactive energy policies and reactive climate policies at the domestic level have also been shaped by pressure to enhance their great-power status in the international system, through becoming the most advanced technological leaders in the global transition to low-carbon energy and a low-carbon economy. According to Yu Hongyuan, a Chinese scholar at the Shanghai Institute for

---

[101] NDRC, *The National Strategy for Climate Change Adaptation*, 3.
[102] "Extreme Weather Events in India in the Past 10 Years," *Down to Earth* (September 19, 2014), available at www.downtoearth.org.in/news/extreme-weather-events-in-india-in-the-past-10-years-46450, accessed March 5, 2018.
[103] Ibid. The figures are calculated by the author, based on the data available at this source.
[104] Stalley, "Principled Strategy: The Role of Equity Norms in China's Climate Change Diplomacy," 2.

International Studies, to become a responsible great developing country, China should make some breakthroughs in new energy,[105] because new energy will play an important role in determining China's status in the international system, given the fact that the emergence of new energy has been closely related to global power transition in the history of the world.[106]

In this regard, both Chinese and Indian leaders have explicitly linked their countries' capacity to develop low-carbon energy and green technology with increased competitiveness against developed countries. For instance, in 2010, Li Keqiang, then Chinese Vice Premier, pointed out that:

at the global level, green economy, low-carbon technologies and so forth are emerging so to seize the high ground of the future development of competitions in those sectors is becoming increasingly fierce. In some areas, the gap between emerging economies and developed countries is relatively small. In this context, as long as we grasp the trend and respond appropriately, it is possible for us to seize the initiative, gain advantages and promote the realization of leapfrog development. Otherwise, we will miss the opportunity and it will be difficult for us to catch up, and we may even lose the initiative and fall behind.[107]

Similarly, former Indian Prime Minister Manmohan Singh also encouraged clean energy growth. For example, Singh announced goals in 2013 to double renewable energy capacity by 2017, and established clean energy subsidies; and in 2009, the National Solar Mission was initiated, with the goal of deploying 20,000 MW of solar panels that would be connected to the grid by 2022, which aims to enable India to

[105] 于宏源 [Yu Hongyuan], "权力转移中的能源链及其挑战 [The Energy Chain in the Power Transition and its Challenges]," 《世界经济研究》 [World Economy Study], no. 2 (2008), 29.
[106] 于宏源 [Yu Hongyuan], "初析全球清洁能源治理的趋势 [Preliminary Analysis of the Trends of the Global Clean Energy Governance]," 《联合国研究》 [UN Studies], no. 1 (2014), 41–42; 于宏源 [Yu Hongyuan], "全球能源治理的功利主义和全球主义 [Globalism and Utilitarianism in Global Energy Governance]," 《国际安全研究》 [Journal of International Security Studies] 31, no. 5 (September/October 2013), 77–78.
[107] 李克强 [Li Keqiang], "深刻理解《建议》主题主线 促进经济社会全面协调可持续发展 [Deeply Understand the Main Themes and Main Threads of the Proposal to Promote Comprehensive, Coordinated and Sustainable Economic and Social Development]," 新华网 [Xinhuanet] (November 14, 2010), available at http://news.xinhuanet.com/2010-11/14/c_12773751.htm, accessed May 17, 2015.

"take a global leadership role in solar manufacturing (across the value chain) of leading edge solar technologies."[108]

China's and India's desire to build great-power status through playing a leadership role in the development of renewable energy and enhancing their competitiveness in the international market can best be illustrated by their development of wind power. According to Joanna Lewis, although it was Europe and the United States that originated modern wind-power technology, China and India have already caught up with Europe and the United States to become "the centre of the global wind power industry," thanks to their strong government subsidies.[109] India first overtook Europe and the United States, and later China overtook India, becoming the largest wind energy market in the world in 2009.[110] Policy support within both countries has played an important role in this significant development of wind power in the two countries. Hence, China's and India's wind power companies have already become important players in the international wind-power market. For example, Goldwind, the leading Chinese wind turbine manufacturer, has already embarked on an ambitious strategy to establish its presence by exporting its goods to the United States and Cuba, in addition to its contracts with Australia, Pakistan, Ethiopia, Chile, Ecuador, Cyprus, and Scotland.[111] The Indian wind turbine company Suzlon is now "operating in 20 countries around the world and supplying turbines to projects in Asia, North and South America and Europe."[112]

China's and India's ambitions to become world leaders in the development of renewable energy by providing significant subsidies to their domestic enterprises have obviously challenged the United States and the EU – the existing leaders in renewable energy. As a result, the United States and the EU have taken steps to frustrate China's and India's ambitions. This can be illustrated by the PV trade disputes

---

[108] Ministry of New and Renewable Energy, GOI, *Jawaharlal Nehru National Solar Mission: Towards Building Solar India* (New Delhi: Ministry of New and Renewable Energy, GOI, 2009), 9.

[109] Joanna I. Lewis, "Building a National Wind Turbine Industry: Experiences from China, India and South Korea," *International Journal of Technology and Globalisation* 5, no. 3/4 (2011), 281.

[110] Ibid.

[111] For details about China's wind power industry in general, and its company Goldwind in particular, see Lewis, *Green Innovation in China*, Chapter 5.

[112] Lewis, "Building a National Wind Turbine Industry," 289.

between China and the United States and the EU, as well as those between India and the United States, on how countries should best foster the development of renewable energy.[113] More specifically, since 2011, both the United States and the EU have launched anti-dumping and – in the case of the United States, also anti-subsidy – investigations into Chinese solar imports.[114] For example, in October 2012, following a year-long investigation, the US government decided to impose anti-dumping and anti-subsidy duties on imported Chinese solar products, ranging from 18.32 percent and 249.96 percent in the case of the former, and 15.78 to 15.97 percent for the latter.[115] Once again, in January 2015, the US International Trade Commission voted on a motion that imports of solar panels from the Chinese mainland and Taiwan hurt the US solar industry. This vote enabled the US government to impose heavy duties on solar products imported from China and Taiwan.[116] In retaliation, the Chinese government has also imposed anti-dumping duties on solar imports from the United States and South Korea.[117] Similarly, in September 2012, the European Commission (EC) launched investigations into allegations of Chinese solar panel manufacturers benefiting from illegal subsidies and dumping their products onto the EU market. In response, the Chinese Ministry of Commerce announced that it was launching its own anti-dumping and anti-subsidy investigations, focusing on solar-grade polysilicon, a key component of solar panels, from the EU.[118] Following a

[113] Joanna I. Lewis, "The Rise of Renewable Energy Protectionism: Emerging Trade Conflicts and Implications for Low Carbon Development," *Global Environmental Politics* 14, no. 4 (November 2014), 10–35.
[114] Ministry of Commerce of the PRC, "Timeline: China–EU, China–US Tug of War on Solar Duties" (June 19, 2013), available at http://english.mofcom.gov.cn/article/zt_solar/column2/201307/20130700218603.shtml, accessed March 5, 2018.
[115] International Center for Trade and Sustainable Development, "China Launches Solar Case against EU at WTO" (November 7, 2012), available at www.ictsd.org/bridges-news/bridges/news/china-launches-solar-case-against-eu-at-wto, accessed March 5, 2018.
[116] Amy He, "Trade Panel Clears Solar Panel Tariffs," *China Daily* (January 22, 2015), available at www.chinadaily.com.cn/world/2015-01/22/content_19377660.htm, accessed March 5, 2018.
[117] Leslie Hook, "China Imposes Tariffs on Polysilicon Exports from US and S Korea," *Financial Times* (July 18, 2013), available at www.ft.com/content/a82b8294-ef9b-11e2-8229-00144feabdc0, accessed May 25, 2017.
[118] International Center for Trade and Sustainable Development, "China Launches Solar Case Against EU at WTO."

year's negotiations, China and the EU reached a settlement, with the solar module trade dispute being solved by the two sides.[119] Even so, on June 5, 2013, the EC imposed provisional anti-dumping duties on EU imports of solar panels from China.[120] At the time of writing, the solar trade disputes between China and the United States continue.[121]

In terms of India's trade disputes with the United States, in 2013 and 2014, the United States issued two cases on India's domestic content requirements for solar cells and solar modules under the National Solar Mission mentioned above, violating the World Trade Organization (WTO) rules on national treatment.[122] It took the WTO nearly three years to resolve the dispute between these two countries. In February 2016, the WTO ruled in favor of the United States regarding its complaint about India's local content requirements for solar equipment.[123] Three months later, in May 2016, the Indian government announced that it would file 16 solar cases against the United States under the WTO dispute.[124] Not surprisingly, Joanna Lewis points out that protectionism has grown in the development of renewable energy, which has led to some trade conflicts between countries such as China, India, the United States, and the EU.[125]

---

[119] Herman K. Trabish, "China, EU Reach Settlement in Solar Module Trade Dispute" (June 29, 2013), available at www.greentechmedia.com/articles/read/china-eu-reach-solar-module-trade-dispute-settlement, accessed March 5, 2018.

[120] Tancrède Voituriez and Xin Wang, "Real Challenges behind the EU–China PV Trade Dispute Settlement," *Climate Policy* 15, no. 5 (2015), 670.

[121] IHS, "Global Solar-PV Polysilicon Prices Rise, as China–US Trade Dispute Continues, IHS Says" (March 22, 2016), available at http://press.ihs.com/press-release/technology/global-solar-pv-polysilicon-prices-rise-china-us-trade-dispute-continues-ih, accessed March 5, 2018.

[122] International Center for Trade and Sustainable Development, "US Launches New WTO Challenge against India Solar Incentives" (February 13, 2014), available at www.ictsd.org/bridges-news/biores/news/us-launches-new-wto-challenge-against-india-solar-incentives, accessed March 5, 2018.

[123] Tom Kenning, "WTO Rules in Favour of US in Local Content Solar Dispute with India," *PVTECH* (February 24, 2016), available at www.pv-tech.org/news/wto-rules-in-favour-of-us-in-local-content-solar-dispute-with-india, accessed March 5, 2018.

[124] Ian Clover, "India Confirms it Will File 16 Solar Cases against US under WTO Dispute," *PV Magazine* (May 13, 2016), available at www.pv-magazine.com/news/details/beitrag/india-confirms-it-will-file-16-solar-cases-against-us-under-wto-dispute_100024597/#axzz4CCvvpF71, accessed March 5, 2018.

[125] Lewis, "The Rise of Renewable Energy Protectionism."

In sum, China's and India's domestic proactive energy policies and reactive climate policies have also been shaped by their desire to obtain great-power status in the international system, through mastering advanced, new energy technology. However, both countries' ambition to become world leaders in the development of renewable energy has challenged the existing leaders in renewable energy – the United States and the EU – who intend "to preserve their economic dominance and advantage at all cost."[126] Accordingly, both the United States and the EU have engaged in some renewable energy-related trade disputes with China and India.

China and India Facing Asymmetrical Interdependence

China's and India's proactive energy policies and reactive climate policies at the domestic level have also been shaped by another systemic pressure, that is, both countries have been asymmetrically dependent on Western countries for advanced technologies (apart from those explored in Chapter 5), as well as on their markets for the development of renewable energy.

In terms of China's asymmetrical dependence on imported technologies, according to *China Human Development Report 2009/10*, issued by the United Nations Development Program (UNDP), in order to mitigate its increased CO_2 emissions, 62 key specialized and general technologies are needed in its six sectors, including power generation, transportation, buildings, steel, cement, chemicals, and petrochemicals. However, China currently lacks 43 of these core technologies.[127] This means that China needs to import nearly 70 percent of its core mitigation technologies. This dilemma has also been highlighted by UNEP in its report about China's green evolution. According to UNEP, Chinese renewable energy companies still lag behind the global technology frontier, although those companies have already been able to compete internationally with some Western competitors. For example, Chinese firms lag behind Western firms in the development of thin film

---

[126] Hurrell and Sengupta, "Emerging Powers, North–South Relations and Global Climate Politics," 473.
[127] UNDP China, *China Human Development Report 2009/10* (Beijing: China Translation and Publishing, 2010), 58–59.

solar panels.[128] Moreover, Chinese manufacturers still rely on foreign imports for several high-end components, such as control systems, hydraulic systems, and main shaft bearings.[129] Accordingly, UNEP points out that "China is a follower, not a leader, in the technology of the solar sector."[130]

This observation is echoed in a report by some Chinese researchers on China's wind-power development, that is, 50 percent of the high-added-value critical parts and components in China's wind-power sector were imported.[131] Another Chinese research study explores the comparative advantages of the United States, the EU, and China in the PV production chain, and concludes that Chinese PV enterprises have to spend a lot of money on importing large quantities of some core raw materials and production equipment, particularly when it comes to manufacturing photovoltaic panels, for which China needs to import not only polycrystalline silicon from the EU and the United States, but also machinery from Germany and France.[132] Thus, Xie Zhenhua, when asked what is the toughest challenge for China in developing new energy, answers that it is the core technologies because those technologies are controlled by "others," meaning that China has had to pay a lot to buy overseas patents, which leads to an increase in the costs of its development of renewable energy.[133]

Apart from its asymmetrical dependence on Western countries' advanced renewable energy technology, China is also asymmetrically dependent on their markets for its development of renewable energy. This is especially the case for China's PV industry. More specifically, China's domestic market for PV products is so small that its PV products have been highly reliant on overseas markets, especially those of the United States and the EU. Nearly 90 percent of China's PV

---

[128] UNEP, *China's Green Long March* (2013), 14, available at www.unep.org/greeneconomy/Portals/88/Research%20Products/China%20synthesis%20report_FINAL_low%20res_22nov.pdf, accessed February 10, 2016.
[129] Ibid., 19.   [130] Ibid., 15.
[131] 李俊峰编 [Li Junfeng (et al.)], 《中国风电发展报告 2012》 [*China Wind Power Outlook 2012*] (Beijing: China Environmental Science Press, 2012), 66.
[132] 张良福 [Zhang Liangfu], "中国,替代美国守卫霍尔木兹海峡?—悄然变动的国际能源格局 [China to Replace the United States to Guard the Strait of Hormuz? Quietly Changed the International Energy Structure]," 《世界知识》 [*World Affairs*], no. 24 (2012), available at www.aisixiang.com/data/60879-2.html, accessed February 9, 2016.
[133] Hu, Gong, and Kong, "Xie Zhenhua: Let Climate Targets Force Domestic Reform."

products are exported to the United States and the EU, leaving Chinese PV enterprises extremely vulnerable to US and EU policy changes related to solar energy. Thus, the anti-dumping investigation by the United States and the EU into Chinese PV manufacturers, and their punitive rulings on Chinese PV manufacturers discussed above, were a devastating blow to Chinese PV module manufacturing. In 2012, for example, 80 percent of Chinese polysilicon enterprises ceased production; 80 percent of the employees engaged in PV manufacturing were laid off or expected to be laid off; the cumulative debt of China's ten largest PV companies reached 111 billion yuan ($16 billion); and the overall debt ratio exceeded 70 percent.[134]

Like China, India's development of renewable energy has also depended, to a certain extent, on imported technologies from developed countries. For instance, according to a 2016 news report, in order to meet the target of 100 GW electricity from renewable energy sources during its 12th FYP, the Indian government needed to ease its rules on the importation of technology equipment.[135] According to some Indian researchers, India's renewable technology lags behind that of some developed countries, and India's manufacturing facilities "are only focused on replicating the existing technologies and are limited to small processing units."[136] In marked contrast, Western renewable firms and government-backed research projects "are engaged in advanced R&D and are continuously setting up bigger, more advanced manufacturing facilities."[137] This observation is echoed by a more recent research study on India's solar energy development, that is, due to the Indian government's local content requirements policy, India's solar energy industry's innovation potential is low, geared toward low-cost assembly, which implies that Indian solar energy development will have to

---

[134] Zhang, "China to Replace the United States to Guard the Strait of Hormuz?"
[135] "India Should Ease Tech Equipment Import Rules to Meet Renewable Energy Target: LM Wind Power," *Business Standard* (April 20, 2016), available at www.business-standard.com/article/companies/india-should-ease-tech-equipment-import-rules-to-meet-renewable-energy-target-lm-wind-power-116042000862_1.html, accessed March 5, 2018.
[136] Vikas Khare, Savita Nema, and Prashant Baredar, "Status of Solar Wind Renewable Energy in India," *Renewable and Sustainable Energy Reviews* 27 (November 2013), 9.
[137] Ibid.

rely on imports of some of the important elements of the value chain, such as balance of systems components and project management.[138]

China's and India's asymmetrical dependence on developed countries for core clean energy technologies has two policy implications. On the one hand, both governments need to adopt policy measures to encourage more research and development investment in clean energy sectors, so that their own renewable energy firms can move into the higher bracket of the value chains.[139] In other words, Chinese and Indian governments have had to adopt proactive policy measures on renewable energy, as discussed in Chapter 3. On the other hand, Chinese and Indian governments have had to take some climate change policy measures to mitigate their increased CO_2 emissions, as a result of pressures from developed countries, in order to obtain the necessary technology transfers from these countries.

Conclusion

China's and India's domestic energy policies have proactive features, while their domestic climate change policies have reactive features, because both countries' domestic energy policies have been formulated out of their own domestic needs. In contrast, both countries' domestic climate change policies and institutions have been created as a result of external pressures stemming from their dual-track climate diplomacy, as discussed in Chapter 5, as well as pressures from epistemic communities.

As we have seen, China's and India's domestic proactive energy policies and reactive climate policies have been mainly shaped by both countries' two-level pressures. At the domestic level, maximizing their domestic economic wealth and reducing poverty through sustained economic development has been a national consensus in China and India. To reach this goal, both countries need to procure sufficient energy supplies to fuel their economies, so they have both undertaken some proactive policy measures, such as the enhancement of their energy efficiency, the expansion of renewable and nuclear energy, and the establishment of the SPRs. In stark contrast, there is no national

---

[138] Oliver Johnson, "Promoting Green Industrial Development through Local Content Requirements: India's National Solar Mission," *Climate Policy* 16, no. 2 (2016), 188, 191.

[139] UNEP, *China's Green Long March*, 14.

consensus in either country on how to address climate change. Faced with pressures from both anti- and pro-climate mitigation camps, both the Chinese and Indian governments adopt the approach of co-benefits, that is, to procure their energy security and economic development while mitigating their GHG emissions.

At the systemic level, China and India are keen to enhance their great-power status by proving themselves to be global leaders in advanced renewable energy technologies, so both countries have taken domestic measures to subsidize their renewable energy development, which has challenged the existing leading powers in the renewable energy industry – the United States and the EU. As a result, both China and India have engaged in some renewable energy-related trade disputes with the United States and the EU. At the same time, both China and India have also depended asymmetrically on the United States and the EU for advanced renewable technologies and markets, in order to expand their renewable energy. So both countries have had to react to pressures from the United States and the EU to mitigate their GHG emissions.

In sum, my main argument in this chapter is that China's and India's domestic climate change policies have been shaped by external pressures, which adds empirical evidence to the "second image reversed" analytical framework. To be sure, there is a counterargument that says it is China's and India's domestic conditions, especially their enormous pollution problems in general and air pollution in particular, that have driven them to introduce domestic policies to address climate change.[140] This view can be refuted by one simple fact and one related question: that is, it is well-known that both China's and India's pollution problems, especially air pollution, have existed and have become far worse for over two decades, since the 1990s in fact, when both countries' economies began to grow fast,[141] so why was it that both countries suddenly started to take domestic action to address climate change in 2007? Obviously, to answer this question, we can only resort to the factors stemming from the systemic level explored in this chapter and Chapter 5.

---

[140] This counterargument was put forward by Professor Robert Keohane in his comments on an earlier version of my manuscript.

[141] A. Hsu et al., *Environmental Performance Index 2016* (New Haven, CT: Yale University, 2016).

PART IV

Implications and Conclusion

7 | Implications for Global Energy and Climate Governance

China's and India's inside-out efforts to address energy insecurity, as well as their outside-in efforts to address climate change, have significant implications for global energy and climate governance. Thus, building on the preceding chapters, this chapter is devoted to assessing those implications. To do so, I first examine the characteristics of current global energy and climate governance. I then explore the implications for global energy and climate governance of these two countries' policy measures to address energy insecurity and climate change. The chapter concludes with a brief summary.

Global Energy and Climate Governance

As reviewed in Chapter 1, there has been a growing literature related to global governance on energy security and climate change in recent decades, which has resulted from the international community's efforts to govern these two issue-areas at the global level. To supplement the existing literature on global energy and climate governance, this section explores the competition and cooperation between the institutions, regimes, forums, and/or clubs that have integrated the two issue-areas of energy security and climate change into their mandates (see Table 7.1).[1] The rationale for these institutions, regimes, forums, and/or clubs for integrating global energy governance with global climate governance lies in the fact that "greenhouse-gas emissions from the energy sector represent roughly two-thirds of all anthropogenic greenhouse-gas emissions and CO_2 emissions from the sector have risen over the past century to ever higher levels."[2] In other words,

---

[1] This section will not discuss those institutions that have energy as their only mandate, such as OPEC and International Energy Forum, although they are important players in global energy governance.
[2] IEA, *Energy and Climate: World Energy Outlook Special Report* (Paris: OECD/IEA, 2015), 20.

Table 7.1 *Integration of energy and climate change by major institutions, regimes, forums, and clubs*

Integration of energy and climate change	Institutions, regimes, forums, or clubs
Governance of energy first and integration of climate change later	IEA
Governance of energy and climate change simultaneously	UNFCCC and Kyoto Protocol; APP; MEF; IRENA; G8+5
Governance of global economy and financial issues first, and integration of energy and climate change into their mandates later	G8/G7; G20

Source: Author

the energy sector is largely responsible for climate change. Against this backdrop, over the past decade, there has been a tendency for some of the existing international institutions, regimes, forums, and/or clubs to expand their mandates to integrate energy and climate change, in order to enhance their significance in governing both issue-areas at the global level.

Objectively speaking, there is no lack of institutions to govern energy security and climate change. In fact, "there are arguably too many."[3] For instance, according to Keohane and Victor, there are more than a dozen major institutions, clubs, and initiatives governing climate change.[4] Similarly, Sovacool and Florini identify more than 50 institutions that have played a role in governing global energy.[5] Under such circumstances, some of the existing institutions, regimes, and clubs have already integrated both energy security and climate change into their mandates, to enable themselves to play a bigger role in global energy and climate governance, including the IEA, the UNFCCC and its Kyoto Protocol, the APP, the MEF, IRENA, G8+5, G8/G7, and G20.

---

[3] Hirst and Froggatt, "The Reform of Global Energy Governance," 7.
[4] Keohane and Victor, "The Regime Complex for Climate Change," 7–23.
[5] Sovacool and Florini, "Examining the Complications of Global Energy Governance," 239–251.

There has been cooperation between these fundamentally fragmented and decentralized institutions. For instance, cooperation has been an ongoing process between the IEA and G8/G7 and G20. Both G8/G7 and G20 have resorted to the IEA for assistance in terms of energy data. So the IEA is sometimes portrayed as the de facto energy secretariat of G8/G7.[6] Thus, the IEA "has today become a leading voice in the energy-climate debate."[7] In another example, in order to help its member countries to phase out energy subsidies, G20 leaders requested the IEA, Organization of the Petroleum Exporting Countries (OPEC), OECD, and the World Bank to "provide an analysis of the scope of energy subsidies, and suggestions for the implementation of this initiative," after their 2009 Pittsburgh summit. In response to this request, these four organizations issued a report which was presented at the G20 summit in Toronto in June 2010.[8] Furthermore, after their November 2010 summit in Seoul, where the leaders committed to rationalizing and phasing out "over the medium term inefficient fossil fuel subsidies that encourage wasteful consumption, with timing based on national circumstances, while providing targeted support for the poorest,"[9] the G20 leaders once again requested the IEA, World Bank, OECD, and OPEC to "further assess and review the progress made in implementing the Pittsburgh and Toronto commitments and report back to the 2011 Summit in France."[10] In response to this further request, these four organizations issued their second joint report in 2011, on trends in support to fossil fuel and other energy sources, and the management of subsidy reform in the context of sustainable development.[11]

A further example of cooperation between the existing fragmented institutions, regimes, and forums on global energy and climate

[6] Dries Lesage, Thijs Van de Graaf, and Kirsten Westphal, "The G8's Role in Global Energy Governance since the 2005 Gleneagles Summit," *Global Governance* 15, no. 2 (April–June 2009), 271.
[7] Van de Graaf, "Obsolete or Resurgent?," 236.
[8] IEA, OPEC, OECD, and World Bank, "Analysis of the Scope of Energy Subsidies and Suggestions for the G-20 Initiative," joint report prepared for submission to the G-20 Leaders' Summit, Toronto, June 2010 (June 16, 2010), available at www.oecd.org/env/45575666.pdf, accessed November 18, 2015.
[9] Cited in IEA, OPEC, OECD and World Bank, "Joint Report by IEA, OPEC, OECD and World Bank on Fossil-Fuel and Other Energy Subsidies: An Update on the G20 Pittsburgh and Toronto Commitments" (2011), 2, available at www.oecd.org/env/49090716.pdf, accessed November 18, 2015.
[10] Ibid.   [11] Ibid.

governance is the Joint Organizations Data Initiative (JODI), which has been established since 2001 under the leadership of the International Energy Forum, bringing together such institutions as the IEA, OPEC, Eurostat (the EU official statistics body), Asia-Pacific Economic Cooperation (APEC), Latin American Energy Organization (OLADE), and the United Nations Statistics Division (UNSD).[12] JODI obviously represents significant progress in cooperation between existing institutions, regimes, and forums, given the decades-long opposition of interests between consuming countries represented by the IEA, and exporting countries represented by OPEC.[13] This cooperation is regarded by some scholars as "bringing more transparency to oil markets by providing data on oil production and trade."[14] Accordingly, such cooperation is depicted as "cooperative fragmentation."[15]

At the same time, however, there is competition between these institutions, regimes, forums, and clubs. The relationship between the IEA and IRENA is a good case in point. More specifically, in January 2009, three of the IEA's member states, namely Germany, Denmark, and Spain, initiated the establishment of IRENA, despite strong opposition by the IEA secretariat and some of the IEA's member states, such as the United States.[16] IRENA, which aims at the comprehensive provision and dissemination of knowledge on renewable energy technology and policy, as well as enhancing international cooperation and partnership in the realm of renewable energy technology and policy,[17] has obviously challenged the IEA's authority and legitimacy in the realm of renewable energy, given that the IEA has been engaged in the governance of renewable energy for more than two decades.[18] Not

---

[12] Rafael Leal-Arcas and Andrew Filis, "The Fragmented Governance of the Global Energy Economy: A Legal-Institutional Analysis," *Journal of World Energy Law and Business* 6, no. 4 (2013), 393.

[13] Ibid., 348.

[14] David G. Victor and Linda Yueh, "The New Energy Order: Managing Insecurities in the Twenty-first Century," *Foreign Affairs* 89, no. 1 (January/February 2010), 67.

[15] Michael Zürn and Benjamin Faude, "Commentary: On Fragmentation, Differentiation, and Coordination," *Global Environmental Politics* 13, no. 3 (August 2013), 128.

[16] Van de Graaf, "Obsolete or Resurgent?" 237.

[17] IRENA, "Vision and Mission," available at www.irena.org/menu/index.aspx?mnu=cat&PriMenuID=13&CatID=9, accessed November 18, 2015.

[18] Regarding the challenges posed by IRENA to the IEA, see Van de Graaf, "Obsolete or Resurgent?," 237.

Implications for Global Energy and Climate Governance 277

surprisingly, "The creation of IRENA presented a serious blow to the IEA, which had been working on renewable energy for more than two decades."[19] Nevertheless, both agencies have shown signs of cooperation. For example, in January 2012, they signed a partnership agreement to cooperate on such issues as data collection.[20]

Certainly, there is both competition and cooperation between these regimes, institutions, forums, and clubs. The relationship between the UNFCCC and some of the clubs is a case in point. More specifically, the UNFCCC process provides all the countries in the international community, big or small, developed and developing, with an equal platform for addressing climate change at the global level. In other words, the UNFCCC represents universality and legitimacy in global climate governance. However, membership of clubs such as APP, G8/G7, G8+5, G20, and MEF is only open to those large countries with a good deal of weight in the world's economic and political affairs. Therefore, the existence of these clubs has posed a challenge, to a certain extent, to the legitimacy of the UNFCCC regime in terms of global climate governance. At the same time, however, the clubs have facilitated the UNFCCC process. For instance, in the run-up to COP-21, G20 leaders reaffirmed their support of the UNFCCC process by stating that:

We reaffirm that UNFCCC is the primary international intergovernmental body for negotiating climate change. We welcome that over 160 Parties including all G20 countries have submitted their INDCs to the UNFCCC, and encourage others to do so in advance of the Paris Conference. We are prepared to implement our INDCs. We will instruct our negotiators to engage constructively and flexibly in the coming days to discuss key issues, among other things, mitigation, adaptation, finance, technology development and transfer and transparency in order to arrive at Paris with a way forward. We commit to work together for a successful outcome of the COP21.[21]

---

[19] Ibid.
[20] Thijs Van de Graaf, "Fragmentation in Global Energy Governance: Explaining the Creation of IRENA," *Global Environmental Politics* 13, no. 3 (August 2013), 30.
[21] "G20 Leaders' Communiqué Antalya Summit, 15–16 November 2015" (Turkey G20, 2015), 6, available at www.mofa.go.jp/files/000111117.pdf, accessed March 7, 2018.

With support from the G20, as well as the support of Sino-US bilateral climate cooperation, in particular the Sino-US bilateral climate deal in 2014 and the bilateral presidential joint statement in 2015, COP-21 succeeded in reaching and adopting the Paris Agreement. Thus, thanks to the efforts of and support from the clubs, as well as Sino-US cooperation, the UNFCCC's legitimacy in global climate governance has been restored, after this was widely cast into doubt after the fiasco at COP-15 in Copenhagen in 2009. Not surprisingly, some scholars point out that although there is contestation between the UNFCCC and those clubs or minilateral forums, results reached at those clubs or minilateral forums in fact "have flowed back into the UNFCCC process."[22]

In sum, given the close linkage between the energy sector and climate change, some of the existing institutions, regimes, forums, and clubs have intentionally integrated these two issue-areas into their mandates, with the IEA, UNFCCC and its Kyoto Protocol, MEF, IRENA, G8/G7, G8+5, and G20 being the most noticeable in this regard. However, these institutions, regimes, forums, and clubs are not hierarchical.[23] So there exists both cooperation and competition between them. As a result, these regimes, institutions, clubs, and forums can only provide the international community with rather weak global energy and climate governance.

Implications of Addressing Energy Insecurity and Climate Change

China and India, the world's largest and third-largest energy consumers and GHG emitters respectively, have adopted policy measures to address their energy insecurity and climate change at both domestic and international levels, which has significant implications for global energy and climate governance in general, and for the UNFCCC process in particular.

---

[22] Sylvia I. Karlsson-Vinkhuyzen and Jeffrey McGee, "Legitimacy in an Era of Fragmentation: The Case of Global Climate Governance," *Global Environmental Politics* 13, no. 3 (August 2013), 74.

[23] Dubash and Florini, "Mapping Global Energy Governance," 15.

Implications for Global Energy and Climate Governance

First, Sino-Indian national and international policy actions/activities to address energy insecurity and climate change have broadened energy and climate governance beyond the realms of the existing global energy and climate institutions, regimes, forums, and clubs discussed above. More specifically, China and India have handled the two issues – that is, energy security and climate change – as their own internal affairs, rather than global affairs that should be addressed by international or multilateral energy and climate institutions. In other words, China and India have maintained their traditional emphasis on their sovereignty on these two issue-areas, so both countries' domestic political and economic concerns have dominated their climate policies, while national energy security looms large as the driver of their energy policies.[24] At the same time, however, with energy security and climate change becoming two globalized issue-areas which have been increasingly put under the governance of those aforementioned institutions, regimes, forums, and clubs at the global level, and with China's and India's increasing engagement with these organizations,[25] both countries' decision-making on these two issue-areas has been increasingly influenced, to various degrees, by global energy and climate institutions, regimes, forums, and clubs, although such influence has been "filtered through national politics and circumstances."[26] Thus, China's and India's efforts to address their energy insecurity and climate change have obviously supplemented the existing global energy and climate governance through exceeding the realms of those existing global energy and climate institutions, regimes, forums, and clubs.

---

[24] Dubash, "From Norm Taker to Norm Maker?," 76.
[25] Julia Xuantong Zhu, *China's Engagement in Global Energy Governance* (Paris: OECD/IEA, 2016); Energy Research Institute, NDRC and Grantham Institute for Climate Change, Imperial College London, *Global Energy Governance Reform and China's Participation: Consultation Draft Report* (February 2014), available at https://workspace.imperial.ac.uk/climatechange/Public/pdfs/Global%20Energy%20Governance%20Reform%20and%20China's%20Participation.pdf, accessed June 23, 2016; 赵庆寺 [Zhao Qingsi], "金砖国家与全球能源治理:角色、责任与路径 [BRICS and Global Energy Governance: Roles, Responsibilities and Means]," 《当代世界与社会主义》 [*Contemporary World and Socialism*], no. 1 (2014), 145–150.
[26] Dubash, "From Norm Taker to Norm Maker?," 76.

Second, China and India may be able to achieve energy security and emission reductions independently of the performance of the existing energy and climate regimes, based on their own national and international policy rules, principles, norms, and procedures, which calls into question the effectiveness, legitimacy, and appropriateness of the existing global energy and climate governance.

As discussed above, China and India have mainly relied on their domestic energy policies and energy diplomacy to address their energy insecurity, rather than relying on global energy institutions. China's and India's energy diplomacy has been overwhelmingly devoted to fostering their bilateral energy ties with any energy-resource countries – regardless of the reputation of those countries in the international system – rather than working through either multilateral forums or global institutions. For instance, China and India, despite being two of the largest energy consumers in the world, have remained outside of the IEA, the world energy consumers' institution, which has obviously significantly undermined the IEA's efficiency and legitimacy. The explicit reason for China's and India's exclusion from the IEA is that neither country is a member of the OECD. The implicit reason for this is that current members of the IEA want to uphold their own interests in this institution: their voting rights at the IEA are based on each country's share of global oil consumption in 1974; this rule would have to be changed in order to admit China and India into the IEA, and the interests of current members would be influenced accordingly, because the share of current members' oil consumption has been declining, while that of China and India has been growing at the global level.[27]

By the same token, China's and India's dual-track climate diplomacy has had significant implications for global climate governance that has "clustered around" the UNFCCC and its Kyoto Protocol, the only legal instrument with legally binding constraints on GHG emissions at the international level.[28]

China's and India's flexible coalition diplomacy under the UN track has successfully watered down the binding level of any potential climate agreements. This can be illustrated by three outcomes reached at the COPs. The first is the Copenhagen Accord reached at COP-15, held

---

[27] Van de Graaf, "Obsolete or Resurgent?," 237; Stewart Patrick, "Irresponsible Stakeholders?: The Difficulty of Integrating Rising Powers," *Foreign Affairs* 89, no. 6 (November/December 2010), 49.

[28] Keohane and Victor, "The Regime Complex for Climate Change," 9.

Implications for Global Energy and Climate Governance 281

at Copenhagen in late 2009. As mentioned in Chapter 5, China and India formed the BASIC group to facilitate their bargaining power in the run-up to this conference. During the negotiations, this group, led by Chinese leader Wen Jiabao, clearly shunning the EU, bargained with US President Obama, and succeeded in its attempt to have some key numbers and words on mitigation targets removed from the text of what became the Copenhagen Accord, in exchange for their acceptance of a special provision on transparency for developing countries insisted on by the United States.[29]

The second outcome that was largely shaped by China and India and their coalition are the provisions reached at COP-17 at Durban in late 2011. More specifically, in the final negotiations, the EU and its allies, namely, the AOSIS and the LDCs, were strongly insistent that all the Parties to the UNFCCC should commit to "a protocol or other legal instrument" as the ultimate goal for a comprehensive global treaty in 2015. In other words, the EU and some developing countries in the G77/China intended to create a legally binding agreement, like the Kyoto Protocol, to assign binding emissions reduction targets on all countries, including China and India. Not surprisingly, China and India strongly opposed this language and sought to water it down by insisting on the inclusion of a new phrase, that is, "or an agreed outcome with legal force."[30] Faced with Sino-Indian intransigence on this issue, the EU had to make a concession. As a result, the final text adopted by COP-17 read: "to develop a protocol, another legal instrument or an agreed outcome with legal force under the Convention applicable to all Parties."[31] In addition to this concession to China and India, the EU also conceded to the BASIC group's further insistence on the year of enforcement of such a new instrument being 2020, which incited fierce criticism of the EU by AOSIS because, according to AOSIS, 2020 would be "too little too late."[32]

The third outcome shaped by China and India and their alliances is one of the decisions reached at COP-20, held in Lima in December 2014. Specifically, China and India, supported by the LMDCs and BASIC, succeeded in watering down the level of the parties' so-called

[29] Hallding et al., "Rising Powers: The Evolving Role of BASIC Countries," 615.
[30] IISD, "Summary of the Durban Climate Change Conference," 30.
[31] UNFCCC, "Draft decision -/CP.17: Establishment of an Ad Hoc Working Group on the Durban Platform for Enhanced Action," 1.
[32] IISD, "Summary of the Durban Climate Change Conference," 30.

"intended nationally determined contributions (INDCs)." To put it in more detail, during the negotiations, the parties disagreed on how INDCs would be communicated and what their possible ex ante consideration or review might look like. China, India, and their allies insisted that the negotiations should only focus on the process of communication, while the United States preferred a "consultative" process or period. But the EU and AOSIS demanded a strong review that would assess the aggregate effect of INDCs against the latest climate science findings and what was deemed necessary to avoid dangerous climate change, which was strongly opposed by the LMDCs, so the final decision text adopted by COP-20 simply requested that the Secretariat publish the communicated INDCs on the UNFCCC website and prepare, by November 1, 2015, a synthesis report on their aggregate effect. This result implies that there would not be any kind of ex ante review of individual countries' contributions in 2015. Thus, this decision text on INDCs is regarded as "the weakest link of the Lima outcome."[33] INDCs have been officially written into the Paris Agreement, which describes them as "nationally determined contributions (NDCs)" (Article 4),[34] in the final official text adopted at COP-21 in Paris in December 2015. Thus, China's and India's (as well as the United States') preferred voluntary mitigation contributions have replaced the Kyoto Protocol's top-down imposed mitigation targets as the new norm in global climate governance.

Moreover, China's and India's non-UN track bandwagoning climate diplomacy has helped the United States to restore its leadership role in ICCN, which can best be illustrated by COP-21. As discussed in Chapter 5, during COP-21, it was not China and India, but the United States, supported by the EU, which played the most important role in shaping the Paris Agreement.[35] In contrast, China and India made some significant compromises, faced with external pressures under their dual-track climate diplomacy. Even so, the Paris Agreement still reflects China's and India's fundamental stance, as well as that of the United States, namely no top-down binding targets on emissions cuts,

---

[33] IISD, "Summary of the Lima Climate Change Conference: 1–14 December 2014," *ENB* 12, no. 619 (December 16, 2014), 44.
[34] UNFCCC, "Adoption of the Paris Agreement," 21. In practice, countries' INDCs submitted to the UNFCCC Secretariat automatically become their NDCs under the Paris Agreement when they ratify this Agreement.
[35] Sethi, "US Dictates the Limits of Paris Climate Change Deal."

although the review process will be legally binding. Thus, the Paris Agreement, a mixture of bottom-up and top-down measures, can only be regarded as a partial success in global climate governance under the UNFCCC. This result confirms Keohane and Victor's observation that "efforts to create an integrated, comprehensive regime are unlikely to be successful."[36]

Implications for the UNFCCC Process

China's and India's non-UN track bandwagoning climate diplomacy has significant implications for the UNFCCC process in several ways. First, by bandwagoning the US- and EU-initiated climate arrangements, China and India have further undermined the prospects of reaching a top-down climate treaty through the UNFCCC negotiations. As noted above, all of the US- and EU-initiated climate arrangements take a bottom-up approach. For instance, the APP focuses on "voluntary practical measures," and on "national strategies, experience-sharing, and technology development and deployment."[37] The MEF, the US effort to build on the APP, has an unidentified "long-term goal" to reduce GHGs, and the path toward this goal is clean energy technology coupled with national strategies.[38] Such a voluntary, technology-focused approach is in stark contrast to the top-down approach embedded in the UNFCCC and its Kyoto Protocol, which groups countries into Annex I and Annex II, based on their responsibility in contributing to global warming, and which prescribed for Annex I (developed countries and countries in transition) their respective quantitative and economy-wide emissions reduction commitments, as well as other obligations, such as reporting, for both developed and developing-countries.[39] For another instance, in their bilateral climate deal, the United States announced that it "intends to achieve an economy-wide target of reducing its emissions by 26%–28% below its 2005 level in 2025 and to make best efforts to reduce its emissions by 28%," while China announced that it "intends to achieve the peaking of CO_2 emissions around 2030 and to make best efforts to peak early and intends to increase the share of non-fossil fuels in

---

[36] Keohane and Victor, "The Regime Complex for Climate Change," 8.
[37] See APP website.   [38] See MEF website.
[39] UNFCCC, *Kyoto Protocol to the United Nations Framework Convention on Climate Change*.

primary energy consumption to around 20% by 2030."[40] A crucial word in this deal – "intends" – implies that this deal is by no means meant to create any legally binding obligations. In other words, both China and the United States prefer a bottom-up approach, rather than a top-down approach, to mitigate their emissions. Also, it implies that neither state will accept any legally binding targets assigned by any international climate agreement reached under the UNFCCC negotiations. The non-binding or voluntary nature of NDCs inked in the Paris Agreement reflects this preference on behalf of China and the United States, as well as India.

Second, the bandwagoning by China and India of the US climate arrangements would make it impossible for the international community to reach the targets for emissions cuts set by the UNFCCC negotiations. According to the Cancun Agreements adopted at COP-16, the parties not only reached the agreement on the need for deep cuts in global emissions in order to limit the global average temperature rise to 2 °C above pre-industrial levels, but also agreed to consider strengthening the global long-term goal during a review by 2015, including in relation to a proposed 1.5 °C target.[41] Similarly, the Paris Agreement sets the goal of "holding the increase in the global average temperature to well below 2 °C above pre-industrial levels and to pursue efforts to limit the temperature increase to 1.5 °C above pre-industrial levels."[42] In practice, however, the result embodied in the Sino-US climate deal, and the INDCs submitted to the UNFCCC Secretariat, make such an ambitious goal unachievable. More specifically, in the Sino-US deal, China has failed to indicate clearly at what level its CO_2 emissions, not to mention its GHG emissions, would peak by 2030. According to a research study, in recent years the annual increase rate of China's CO_2 emissions was 3 percent.[43] If this rate of

[40] The White House, "U.S.–China Joint Announcement on Climate Change" (Beijing, November 12, 2014), available at www.whitehouse.gov/the-press-office/2014/11/11/us-china-joint-announcement-climate-change, accessed March 5, 2018.
[41] UNFCCC, "FCCC/CP/2010/7/Add.1: Report of the Conference of the Parties on its Sixteenth Session," 3.
[42] UNFCCC, "Adoption of the Paris Agreement," 21.
[43] PBL Netherlands Environmental Assessment Agency, *Trends in Global CO_2 Emissions: 2013 Report* (The Hague, 2013), 9, available at http://edgar.jrc.ec.europa.eu/news_docs/pbl-2013-trends-in-global-co2-emissions-2013-report-1148.pdf, accessed March 5, 2018.

increase were to continue, China's emissions would reach 16 gigatons by 2030. In the United States' case, its CO_2 emissions were equivalent to 7.26 gigatons in 2005. If the United States fulfills its announcement to cut its emissions by 28 percent by 2025, it will emit 5.23 gigatons by then, which is equivalent to the amount that it emitted in 1992. Thus, even if China's emissions do peak in 2030, it would emit three times more than the United States by then.[44] According to India's Centre for Science and Environment, the targets for emissions cuts announced by the United States and China mean that both countries per capita emissions would converge at 12 tons by 2030, surpassing the current global average more than twofold.[45] These efforts would lead the global temperature to rise by about 3.7 °C to 4.8 °C, a result projected by the IPCC,[46] which is far above the 2 °C target set in the Cancun Agreements and the Paris Agreement. Not surprisingly, Chandra Bhushan, the Centre's deputy director general, made this comment on the Sino-US climate deal in an interview by Bloomberg: "If this is the benchmark set by the world's two biggest economies – and two biggest polluters – we are on a completely catastrophic path."[47] Glen Peters, a researcher at Norway's Centre for International Climate and Environmental Research and the Global Carbon Project, echoed this concern: "Overall, the Chinese and US emissions targets represent a political step forward, but are broadly not consistent with a likely chance of keeping temperatures below 2 °C. The targets themselves are not so different from the continuation of existing trends."[48] In other words, the targets and measures announced by

---

[44] Ronald Bailey, "The Inconvenient Truth about the U.S.–China Emissions Deal: It's Meaningless," *Time* (November 13, 2014), available at http://time.com/3583621/the-inconvenient-truth-about-the-u-s-china-emissions-deal-its-meaningless, accessed March 5, 2018.

[45] Alex Morales, "China's Move with U.S. on Pollution Spurs Climate Deal," *Bloomberg* (November 12, 2014), available at www.bloomberg.com/news/2014-11-12/china-s-move-with-u-s-on-pollution-spurs-work-on-climate-deal.html, accessed March 5, 2018.

[46] IPCC, *Climate Change 2014: Synthesis Report* 21, available at www.ipcc.ch/pdf/assessment-report/ar5/syr/SYR_AR5_LONGERREPORT.pdf, accessed December 28, 2014.

[47] Morales, "China's Move with U.S. on Pollution Spurs Climate Deal."

[48] Leigh Phillips, "How Big a Deal is the US–China Climate Deal?," available at http://roadtoparis.info/2014/11/16/big-deal-us-china-climate-deal, accessed March 5, 2018.

the United States and China fall well short of what is needed to avoid dangerous climate change at the global level.

Similarly, there is a huge gap between the target of 2 °C and the INDCs submitted to the UNFCCC Secretariat. According to *The Emissions Gap Report 2015* issued by UNEP in November 2015, a total of 119 INDCs had been submitted to the UNFCCC Secretariat by October 1, 2015, which covered 146 countries and 85–88 percent of global GHG emissions in 2012.[49] UNEP points out that, "The emissions gap between what the full implementation of the unconditional INDCs contribute and the least-cost emission level for a pathway to stay below 2 °C, is estimated to be 14 GtCO$_2$e (range: 12–17) in 2030 and 7 GtCO$_2$e (range: 5–10) in 2025."[50] Under such circumstances, even if those INDCs are fully implemented, according to UNEP, emission levels in 2030 have more than a 66 percent chance of limiting the global average temperature increase to below 3.5 °C until 2100.[51] Thus, UNEP straightforwardly states that "the submitted contributions are far from enough and the emissions gap in both 2025 and 2030 will be very significant."[52] UNEP further confirmed this finding in its *Emissions Gap Report 2016*: that is, even with full implementation of the 160 INDCs submitted by all countries, representing 187 out of 195 parties to the UNFCCC, it is "only consistent with staying below an increase in temperature of 3.2 °C by 2100 and 3.0 °C, if conditional Intended Nationally Determined Contributions are included."[53] This finding is also echoed by the IEA's *Energy Technology Perspectives 2017*, in which the IEA finds that "significant changes in policy and technologies in the period to 2060 as well as substantial additional cuts in emissions thereafter ... would result in an average temperature increase of 2.7 °C by 2100, at which point temperatures are unlikely to have stabilised and would continue to rise."[54]

---

[49] UNEP, *The Emissions Gap Report 2015*, xviii, available at http://uneplive.unep.org/media/docs/theme/13/EGR_2015_301115_lores.pdf, accessed March 5, 2018.
[50] Ibid.   [51] Ibid.   [52] Ibid.
[53] UNEP, *The Emissions Gap Report 2016*, xvii, available at www.unep.org/emissionsgap, accessed June 15, 2017.
[54] IEA, *Energy Technology Perspectives 2017: Catalysing Energy Technology Transformations* (Paris: OECD/IEA, 2017), 23.

Implications for Global Energy and Climate Governance   287

Simply put, the Paris Agreement and all of its INDCs are not sufficient to ensure a safe and stable climate.[55]

Furthermore, another consequence resulting from China's and India's non-UN track bandwagoning diplomacy is that the scale and scope of fragmentation in global climate governance has increased. More precisely, the UNFCCC/Kyoto Protocol is universal in scope, whereas all the non-UN climate arrangements are based on small-group negotiations between their limited parties; the UNFCCC/Kyoto Protocol is legally binding, whereas all the non-UN climate arrangements stress voluntary measures; the UNFCCC/Kyoto Protocol focuses on GHG emissions reduction, whereas the non-UN climate arrangements are mainly focused on fostering technological innovation and energy efficiency and cleaner energy; the UNFCCC negotiations set the goal of limiting the global average temperature rise to 2 °C above pre-industrial levels, whereas the non-UN climate arrangements make such an ambitious goal unachievable. Moreover, regarding the compliance mechanisms, according to the Cancun Agreements, mitigation actions taken by developed and developing countries would be subject to a top-down evaluation process. Specifically, "Nationally appropriate mitigation commitments or actions by developed country Parties" are required to be subject to "reporting and review guidelines, processes and experiences,"[56] while "Nationally appropriate mitigation actions by developing country Parties" would be subject to either international consultations and analysis or measurement, reporting and verification.[57] Also, according to the Paris Agreement, all parties' NDCs would be put under "an enhanced transparency framework," legally binding transparency arrangements that include "national communications, biennial reports and biennial update reports, international assessment and review and international consultation and analysis."[58] In stark contrast, none of the US- or EU-initiated climate mini-multilateral or bilateral arrangements has any transparency arrangements (see Table 7.2), except for the "U.S.–China Joint Presidential Statement on Climate Change," issued by the two countries in

---

[55] Jorge E. Viñuales, Joanna Depledge, David M. Reiner, and Emma Lees, "Climate Policy after the Paris 2015 Climate Conference," *Climate Policy* 17, no. 1 (2017), 7.

[56] UNFCCC, "FCCC/CP/2010/7/Add.1: Report of the Conference of the Parties on its Sixteenth Session," 7–8.

[57] Ibid., 10–11.   [58] UNFCCC, "Adoption of the Paris Agreement," 27.

Table 7.2 *Comparison of the UNFCCC/Kyoto Protocol, Sino-India climate bandwagoning, and the Paris Agreement*

	UNFCCC/ Kyoto Protocol	Sino-India climate bandwagoning	Paris Agreement
Approach	Top-down	Bottom-up	Top-down and bottom-up
Membership	Universal	Limited	Universal
Goal	Limitation to 2 °C above pre-industrial levels	Limitation to 2 °C above pre-industrial levels	Limitation to 2 °C above pre-industrial levels, even to 1.5 °C
Legal status	Rules-based, legally binding	Voluntary pledges	Voluntary pledges and legally binding
Transparency	*Developed countries*: reporting and review guidelines, processes, and experiences *Developing countries*: monitoring, reporting and verification/ international consultation and analysis	No	Each party: national communications, biennial reports and biennial update reports, international assessment and review, international consultation and analysis

Source: Author

September 2015, in which the transparency issue was mentioned in a very vague way: "Both sides support the inclusion in the Paris outcome of an enhanced transparency system to build mutual trust and confidence and promote effective implementation including through reporting and review of action and support in an appropriate manner."[59]

[59] The White House, "U.S.–China Joint Presidential Statement on Climate Change."

Furthermore, China's and India's bandwagoning climate diplomacy has undercut the UNFCCC process because both countries have preferred to announce their readiness for climate actions under the non-UN track, rather than as part of the UNFCCC process, in order to enhance their great-power status. As mentioned in Chapter 5, India announced its willingness to cap its per capita emissions at a rate not exceeding those of developed countries at G8+5, while China used the platform of its bilateral climate deal with the United States in November 2014 to announce, for the first time, its intention to peak its CO_2 emissions around 2030. Moreover, according to a media commentary, Chinese Vice Premier Zhang Gaoli plainly told US President Obama that China wanted to move ahead quickly on a separate climate deal, during his meeting with Obama on the sidelines of the United Nations climate summit in September 2014.[60] Not surprisingly, then Indian Environment Minister Ramesh made the observation that China and India have "delinked emissions control actions from the international negotiations."[61] Against this backdrop, Sino-Indian actions have called into question the following three aspects of the UNFCCC: the effectiveness of the UNFCCC process, the image of its COPs as the most appropriate forum to conduct negotiations, and, thus, its legitimacy, regardless of its universal membership.[62]

In sum, Sino-Indian dynamic coalition strategy under the UNFCCC process, and their bandwagoning of the US- and EU-initiated climate arrangements, has made it unlikely that a top-down, comprehensive global architecture to address climate change will be forged collectively. As we have seen, Sino-Indian climate diplomacy has led to the formation of a mixture of bottom-up and top-down norms inked in the Paris Agreement, a result that is regarded by Tom Hale as "a turning point for the climate regime."[63] Moreover, Sino-India bandwagoning climate diplomacy has significantly undercut the UNFCCC process by not only increasing the scale and scope of fragmentation in global

---

[60] "Secret Talks and a Personal Letter," *The Guardian*.
[61] Deborah Seligsohn, "India–China Climate Cooperation Thrives with the 'Spirit of Copenhagen'" (May 10, 2010), available at www.chinafaqs.org/blog-posts/india-china-climate-cooperation-thrives-spirit-copenhagen, accessed March 5, 2018.
[62] De Matteis, "The EU's and China's Institutional Diplomacy in the Field of Climate Change," 37.
[63] Thomas Hale, "'All Hands on Deck': The Paris Agreement and Nonstate Climate Action," *Global Environmental Politics* 16, no. 3 (August 2016), 12.

climate governance under the UNFCCC/Kyoto Protocol, but also by challenging its effectiveness, appropriateness, and legitimacy, even though the latter's centrality in global climate governance was restored, to a certain extent, by the Paris Agreement reached at COP-21 in December 2015.

Restructuring Global Energy and Climate Governance to Integrate China and India

A practical policy implication is that restructuring the existing global energy and climate institutions and regimes to fully integrate both China and India is the only way to achieve the ultimate goals of global energy and climate governance, that is, globalized energy security and a stabilized climate. As discussed in previous chapters, the dramatic growth of China's and India's energy consumption and CO_2 emissions has already transformed the global energy and climate landscape, which has posed serious challenges to those existing institutions, regimes, forums, and clubs discussed above. As Patrick and Dubash note, China and India – yesterday's rule takers – are becoming tomorrow's rule makers.[64] In the coming decades, accordingly, a major challenge for global energy and climate governance will be how to fully integrate China and India into international energy and climate regimes, institutions, forums, and clubs.

So far, the international community has succeeded, to a limited degree, in integrating China and India into global efforts to address climate change through the Paris Agreement. To be sure, the most daunting challenge in fully integrating China and India into global climate governance will be to alleviate both countries' worries that their acceptance of international mitigation commitments could jeopardize their domestic development, which has been one of the main reasons for both countries, and other emerging economies, rejecting any binding GHG emissions cuts obligations, while insisting that developed countries take the lead in GHG mitigation. In this process, China and India have been wrestling with conflicting identities. On the one hand, China and India have been seeking great-power identity or status in the international system, which would require both countries

[64] Patrick, "Irresponsible Stakeholders?," 44; Dubash, "From Norm Taker to Norm Maker?," 66–79.

Implications for Global Energy and Climate Governance 291

to make contributions to provide global public goods, including global energy security and climate protection. On the other hand, however, both countries still maintain their traditional identity as developing countries, so both remain committed to alleviating poverty. To accommodate these two identities, both countries maintain sustained economic growth as their top priority, while accepting their voluntary GHG mitigation obligations under the Paris Agreement.

In terms of the necessity for China and India to be integrated into the existing international energy institutions, some scholars point out that accommodating new consumers such as China and India – heavyweights in the existing energy clubs – is "imperative not only with a view to protect against market failure but also to prevent further negative side-effects on weaker nations and potential regional or global contagion. Consequently, the existing modes of cooperation among consumers must be reviewed and adjusted."[65] Obviously, China's and India's cooperation with the IEA, among other institutions, will be the most crucial factor in dealing with any future oil supply disruptions in the international oil market, because with China's and India's shares in international energy trade increasingly rising while those of the OECD progressively decline, China and India will play an important role in influencing the international energy market.[66] Take China's and India's SPRs as an example. As discussed in Chapter 3, at the domestic level, China and India have already begun to establish their own SPRs, which might have a significant impact on the IEA's existing emergency system. It is beyond doubt that the IEA's existing mechanisms of emergency supply management require substantial adjustments to deal with the increased presence of China and India in the international energy market, which is still potentially at risk of market failure.[67] If China's and India's SPRs were to be integrated into the IEA's existing emergency system, should oil supply disruptions happen in the international oil market, there could be coordination in the use of the IEA's SPRs with those of China and India, which would benefit all energy-consuming countries.[68] On the other hand, if China and India are not fully integrated in the IEA's mechanisms for short-term emergency management, in the event of a disruption, major oil-consuming nations

---

[65] Goldthau and Witte, "Back to the Future or Forward to the Past?," 386.
[66] Ibid.   [67] Ibid., 385.
[68] Hirst and Froggatt, "The Reform of Global Energy Governance," 10.

would not be able to coordinate their policies, which might render the IEA's existing buffer mechanisms ineffective.[69]

In practice, as early as the mid-1990s, the IEA had already begun to take measures to reach out to China and India in an attempt to integrate both countries into its global energy governance. In the 2000s, the IEA stepped up efforts to engage with China and India. For instance, the IEA has treated China and India, together with Brazil, Indonesia, Russia, and South Africa, as its "partner countries," and has invited both countries to attend its ministerial meetings since 2009. At its 2009 ministerial meeting, the IEA issued a joint statement with China, India, and Russia, which detailed plans for collaboration in a variety of energy-related areas, such as energy security, energy efficiency, renewable, cleaner coal, and technology collaboration.[70] At its 2013 ministerial meeting, the IEA and these six partner countries issued a joint declaration of their intention to pursue closer multilateral cooperation on energy.[71] Moreover, in April 2013, the IEA initiated an "Association" between itself and non-members, especially its partner countries, aimed at "creating a closer alliance on energy security, environmental sustainability and data sharing," so as to deepen its partnerships with these countries.[72] In November 2015, the IEA activated China's association status,[73] which, together with China's observer status under the International Energy Charter in May 2015, was hailed by a Chinese scholar as a milestone of China's participation in

[69] Goldthau and Witte, "Back to the Future or Forward to the Past?," 385.
[70] IEA, "IEA Ministers Confirm Commitment to Stabilise CO_2 Emissions and Ensure Transition to Low-Carbon Economy; Welcome Closer Cooperation with China, India and Russia" (October 15, 2009), available at www.iea.org/newsroomandevents/news/2009/october/2009-10-15-.html, accessed March 6, 2018.
[71] IEA, "Joint Declaration by the IEA and Brazil, China, India, Indonesia, Russia and South Africa on the Occasion of the 2013 IEA Ministerial Meeting Expressing Mutual Interest in Pursuing an Association" (November 20, 2013), available at www.iea.org/media/ministerialpublic/2013/jointdeclaration.pdf, accessed November 17, 2015.
[72] Ron Bousso, "International Energy Agency to Offer China Room in Strategic Talks," *Reuters* (April 4, 2013), available at www.reuters.com/article/2013/04/04/iea-china-idUSL2N0CR1IU20130404#PLZiOOZV3Jc1TGoR.97, accessed March 6, 2018.
[73] IEA, "China, People's Republic of (Association Country)," 2016, available at https://www.iea.org/countries/non-membercountries/chinapeoplesrepublicof, accessed June 20, 2016.

global energy governance.[74] Furthermore, the IEA has already conducted an emergency response exercise with India in May 2012, and with China in January 2015, which can be regarded as a breakthrough in the relations between the IEA and these two countries, given the fact that such an emergency response exercise involves sharing sensitive information concerning how the two governments would act in the event of a supply crisis.[75]

Conclusion

Some of the existing regimes, institutions, forums, and clubs have intentionally integrated energy and climate change into their mandates so as to enhance their relevance in global energy and climate governance. The IEA, the UNFCCC and its Kyoto Protocol, MEF, IRENA, G8/G7, G8+5, and G20 have been the most significant of these. However, they do not constitute one hierarchy, so there is both competition and cooperation between them in the process of governing global energy and climate change.

China and India, the world's largest and third-largest energy consumers and GHG emitters respectively, have undertaken their own policy measures to seek energy security while addressing climate change, which has largely remained outside of the realm of global energy and climate governance, which has, in turn, had significant implications. On the one hand, Sino-Indian national and international policy actions/activities to address energy insecurity and climate change have broadened energy and climate governance beyond the realms of those existing global energy and climate institutions, regimes, forums, and clubs. On the other hand, China and India may achieve energy security and emissions reductions independently of the performance of the existing energy and climate regimes, as a result of their own national and international policy

---

[74] 康晓文 [Kang Xiaowen], "签署国际能源宪章 加入国际能源署联盟国 中国开启全球能源治理体系协同进化新进程 [Signing the International Energy Charter and Joining the International Energy Agency's Association Country, China Starts the New Process of Co-Evolution with the Global Energy Governance System]," 《国际石油经济》 [*International Petroleum Economics*] 24, no. 1 (2016), 18–20.

[75] Andrew B. Kennedy, "China and the Free-Rider Problem: Exploring the Case of Energy Security," *Political Science Quarterly* 130, no. 1 (2015), 44; Zhu, *China's Engagement in Global Energy Governance*, 56.

rules, principles, norms, and procedures, which calls into question the effectiveness, legitimacy, and appropriateness of existing global energy and climate governance. Furthermore, restructuring those existing global energy and climate institutions, regimes, forums, and clubs so as to fully integrate both China and India is the only way to achieve the ultimate goals of global energy and climate governance, that is, globalized energy security and a stabilized climate.

8 | Conclusion

Three tasks remain for this concluding chapter. The first is to highlight the principal findings of this book. Second, this chapter discusses the implications of this book's two-level pressures analytical framework for broader international relations research. The final section of this conclusion is a discussion on the basis of the impact of the findings of this book on our understanding of China's and India's foreign policy behavior in the context of their rise in the international system.

Principal Findings

The preceding chapters of this book offer six principal findings. The first is that China and India, two countries with widely differing domestic attributes, have behaved almost identically in the issue-areas of energy insecurity and climate change. In terms of their behavior in addressing energy insecurity domestically, both countries have adopted both demand and supply-side policy measures, such as improving energy efficiency and the development of renewable energy resources and nuclear energy, in addition to establishing SPRs to hedge against any potential supply emergencies. Internationally, both countries have encouraged their NOCs to go out and seek equity oil and natural gas across the world, without any discrimination toward the resource countries. In order to facilitate their NOCs' going-out, both governments have employed a variety of policy tools to boost their energy ties with energy-rich countries, such as high-level official visit exchanges, enhanced trade, the provision of aid, the sale of arms, as well as the provision of diplomatic support to those resource countries at the international level. As regards their behavior in addressing climate change on the domestic front, both countries' climate change policy is characterized by so-called "co-benefits," that is, both countries intend to mitigate their increased GHG emissions by transforming their energy mix from a high dependence on fossil fuels to renewable energy and nuclear energy, so as to transform their economic development path from high-carbon to low-carbon. Internationally, both countries have made some significant compromises to their negotiating partners in ICCN, by accepting voluntary mitigation obligations and a legally binding review process of their mitigation efforts. Simply put, China's and India's efforts to address energy insecurity and climate change at both domestic and international levels share a lot of similarities.

The second finding is that the once widely received view of the first decade of the twenty-first century that potential threats, and even conflicts, were likely to arise from China's and India's mercantilist way of struggling for overseas energy resources has turned out to be exaggerated and empirically unsupported. On the contrary, China's and India's national and international policy measures to procure energy supplies have contributed to global energy security, either directly or indirectly. China's and India's domestic policies to enhance their energy efficiency and develop renewable energy, such as wind and solar power, have contributed to the growth of their own capacity to meet their increased energy demand, thus alleviating the huge pressure they would otherwise put on the international energy market. Their domestic efforts to establish their own SPRs would go on to indirectly contribute to enhancing the global capacity to handle any potential supply disruptions. At the international level, by exploring and developing oil and natural gas reserves in regions and countries barred to Western energy companies, Chinese and Indian NOCs have increased the world's proven reserves of energy available to all consumers. Moreover, China's and India's equity oil production from their overseas operations has increased significantly, and more than half of it was sold to consumers outside the two countries. Thus, both countries' international efforts to develop overseas oil and natural gas have helped to enhance international energy security rather than harm it. This finding echoes the findings by Kong,[1] Downs,[2] and the US Department of Energy (DOE).[3]

---

[1] Kong finds that worries about China locking up international oil are exaggerated and misplaced: Kong, *China's International Petroleum Policy*, 151.
[2] According to Downs, CNPC, China's largest NOC, has historically shipped back home most of its Sudanese equity oil. In 2006, however, CNPC sold most of its Sudanese production on the international market for profits and the consideration of transport costs. See Erica S. Downs, "Fact and Fiction of Sino-African Energy Relations," *China Security* 3, no. 3, (2007), 42–68.
[3] According to the DOE, "Even if China's equity oil investments 'remove' assets from the global market, in the sense that they are not subsequently available for resale, these actions merely displace what the Chinese would have otherwise

The third finding is that China and India have altered the trajectory of their foreign policy behavior in the issue-areas of energy security and climate change, in reaction to external pressures applied by both state and non-state actors. In terms of both countries' energy diplomacy, as noted in Chapter 4, although both China and India have significantly strengthened their energy ties with Iran, Myanmar, and Sudan, so-called "pariah states," both countries have had to modify their proactive policy measures toward these countries. Specifically, under pressure from the United States and its allies, both countries voted against Iran at the IAEA on the Iranian nuclear issue, and agreed to refer this issue to the UN Security Council. Moreover, China consistently sided with the United States and other P-5 members to vote for a series of UNSC Resolutions regarding the Iranian nuclear issue, which made it possible for the Security Council to impose four rounds of incremental sanctions on Iran. As regards Myanmar and Sudan, China has been the specific target of international pressure, not only from the United States, but also from non-state actors, such as human rights NGOs. Under external pressures, China had to shift its traditional non-interference policy to tacit interference in both Myanmar's and Sudan's domestic affairs. Specifically, in Sudan's case, China interfered not only in the resolution of the Sudan Darfur crisis, but also in the referendum and peaceful secession of South Sudan.

With regard to China's and India's climate diplomacy, as a result of pressures stemming from the international climate change negotiations under the UNFCCC, and their bandwagoning with the United States and the EU to address climate change, not only China and India, but developing countries as a whole have agreed to undertake voluntary GHG mitigation actions and to subject their actions to a binding review process. They have agreed to this without their traditional insistence on reciprocity from developed countries in terms of providing funding and transferring technology to help developing countries, especially the LDCs and AOSIS, to adapt to climate change, which is obviously not in the best interests of developing countries, especially those of the LDCs and AOSIS.

> bought on the open market. Regardless of whether China secures its oil through equity investments or purchases on the global market, its increasing demand for these resources will continue to play a role in world oil markets (as will rising demand from other areas, such as the U.S. and India)." DOE, "Energy Policy Act of 2005 Section 1837: National Security Review of International Energy Requirements" (February 2006), 3, available at www.hsdl.org/?view&did=469287, accessed March 6, 2018.

The fourth finding is that China's and India's domestic measures to address climate change have been driven largely by their desire to alleviate external pressures on them. As noted in Chapter 6, China and India only began to take serious measures to address the climate change challenge in 2007, when both countries faced mounting pressure from their dual-track climate diplomacy, as well as from some epistemic communities.

A further finding of this book is that, contrary to the conventional wisdom that multilateral efforts to address climate change, such as MEF, APP, G8+5, G8/G7, and G20, as well as the US bilateral agreement with China and India, would undermine the legitimacy of the UNFCCC process, these non-UN efforts – especially G20 and the US bilateral deal with China – have played an important role in bringing back the centrality and legitimacy of the UNFCCC in global climate governance, which can be evidenced by the Paris Agreement adopted at COP-21.

Finally, subject to two-level pressures, China and India have shifted away from their traditional free-riding of developed countries' efforts to govern energy security and climate change, and have begun to cooperate with developed countries to make contributions to ensuring global energy security and a stabilized climate – two global public goods. At the same time, however, both countries' contributions to global energy and climate governance remain limited as a result of domestic pressures, that is, economic growth and poverty reduction – two of China's and India's top priorities at the domestic level.

Implications for International Relations Research[4]

The two-level pressures analytical framework developed in this book has broader implications for IR research. First, this book offers a new explanation not only for why states' policy behavior at both domestic and international levels on certain issue-areas has been both proactive and reactive, but also when to expect proactive and when to expect reactive policy behavior in international affairs. More specifically, the two-level pressures framework shows that in the process of formulating their domestic and foreign policies, states are faced with both

[4] This section mainly draws on the insightful comments and critiques raised by one of four anonymous reviewers during the first-round review.

domestic pressures – stemming mainly from economic growth – and international pressures – stemming from asymmetrical power/capability distribution at the systemic level – as well as from international norms and principles embedded in global governance. As a result, the behavior of states bears proactive and reactive features. When domestic pressures are strong enough to overwhelm pressures at the systemic level, states' policy behaviors are expected to be proactive, that is, to take whatever policy measures they have at their disposal to procure their economic growth. On the contrary, when the pressures stemming from the systemic level are so strong that states have to take those pressures into account when they adjust their proactive policy measures, then their policy behaviors are expected to be reactive. Thus, this book is part of a growing body of literature on states' foreign policy being shaped by factors at both domestic and systemic levels, that is, Putnam's seminal two-level games,[5] and other scholars' application of Putnam's model to empirically analyze various situations,[6] apart from neoclassical realism.[7] However, what distinguishes this book from existing studies lies in its identification of the proactive and reactive features of states' domestic and foreign policies using a two-level analysis, and in adopting a different aspect of a state's domestic structure. Specifically, rather than focusing on a state's domestic bureaucratic politics, interest groups, and decision-makers, as adopted in the existing literature, this book emphasizes a state's level of development as the most relevant aspect in explaining its policy behavior.

At the same time, the book accords with the rational choice model. More specifically, utilitarian and rational assumptions are used to identify the specific interests driving China's and India's behavior in the two issue-areas, namely, energy security and climate change.

[5] Putnam, "Diplomacy and Domestic Politics."
[6] For instance, Dinshaw Mistry uses the two-level game model to analyze the US–India Nuclear Agreement: Mistry, *The US–Indian Nuclear Agreement*; Fudan University's scholar Bo Yan employs the two-level games model to study the United States' climate diplomacy: 薄燕 [Bo Yan], 《国际谈判与国内政治: 美国与<京都议定书>谈判的实例》 [*International Negotiations and Domestic Politics: Case Study of the United States and the Negotiations of the Kyoto Protocol*], (Shanghai: Shanghai Joint Publishing Company, 2007); Robert G. Kaufman extended this model to a two-level interaction, to study the United States' grand strategy: Robert G. Kaufman, "A Two-Level Interaction: Structure, Stable Liberal Democracy, and U.S. Grand Strategy," *Security Studies* 3, no. 4 (Summer 1994), 678–717.
[7] Rose, "Neoclassical Realism and Theories of Foreign Policy," 144–172.

In other words, China and India have not adjusted their proactive policy measures in seeking overseas energy resources in "pariah states" in response to external pressures because they have suddenly embraced international norms such as non-proliferation, good governance, democracy, and human rights; and China and India have not agreed to accept the voluntary mitigation obligation and legally binding review of their mitigation efforts in response to external pressures because both countries have suddenly embraced the concept of climate protection. Rather, both countries have had to react to external pressures in both issue-areas because in many instances the expected benefits for China's and India's immediate and long-term economic and ideational interests outweigh the costs.

In addition, the "two-level pressures" model offers an analytical framework for China's and India's behavior in global governance on other policy issue-areas that entail global public goods (or bads), such as global trade and health. In other words, China's and India's policy behavior on energy security and climate change is in line with their behavior in other policy issue-areas that need to be governed at the global level. This is especially the case for global trade and health governance in the same timeframe explored in this book, namely, the first decade and a half of the twenty-first century. In terms of global trade governance, in the world trade negotiations, especially the Doha Round of negotiations in the WTO between 2001 and 2008, China and India, in order to boost their domestic development in general, and agriculture in particular, proactively resisted the obligations of reducing trade barriers to further open their markets, which led to deadlock in the negotiations.[8] However, faced with increased pressures from the United States and the EU, China and India eventually dropped their resistance and accepted the agreements reached in Nairobi in December 2015, including the ban on agricultural export subsidies, which was acclaimed by Roberto Azevêdo, the WTO's director general, as the

---

[8] Heather Stewart, "Tariffs: WTO Talks Collapse after India and China Clash with America over Farm Products," *The Guardian* (July 30, 2008), available at www.theguardian.com/world/2008/jul/30/wto.india, accessed March 6, 2018; Susan C. Schwab, "After Doha: Why the Negotiations Are Doomed and What We Should Do About It," *Foreign Affairs* 90, no. 3 (May/June 2011), 104–117.

Conclusion

"most significant" achievement on agriculture in the WTO's history.[9] Similarly, according to Yanzhong Huang, a Senior Fellow for Global Health at the Council on Foreign Relations, China and India have become increasingly active in participating in global health governance, but both countries have failed to shoulder significantly more responsibilities in that issue-area; neither have they offered a viable, sustainable alternative to the existing governance paradigm, which has been mainly shaped by pressures stemming from both countries' domestic health challenges on the one hand, and the ongoing power shift at the global level on the other.[10]

Moreover, the two-level pressures analytical framework can be applied to explain the energy and climate policy behavior of some other emerging powers, such as Brazil and South Africa, which share similar characteristics, including, among others, undergoing rapid economic growth and industrialization, but having not yet reached developed level, being latecomers, and facing asymmetrical power distribution in the international system. On the one hand, these emerging powers have taken proactive measures to seek their energy security based on their own domestic development priorities. On the other hand, they have had to undertake measures to mitigate GHG emissions under increased external pressures stemming from international climate change negotiations under the UNFCCC, as well as some non-UN climate institutions such as G20, in addition to their desire to be recognized as great powers by other state and non-state actors in the international system.

Understanding the Rise of China and India

China's and India's efforts to address their energy insecurity and climate change at both domestic and international levels, as explored in this book, provide us with a supplementary perspective to understanding the

---

[9] Shawn Donnan, "Trade Talks Lead to 'Death of Doha and Birth of New WTO'," *Financial Times* (December 20, 2015), available at www.ft.com/cms/s/0/97e8525e-a740-11e5-9700-2b669a5aeb83.html#axzz4DpGehWTw, accessed July 8, 2016.

[10] Yanzhong Huang, "Enter the Dragon and the Elephant: China's and India's Participation in Global Health Governance" (Council on Foreign Relations, working paper, April 2013), available at www.cfr.org/china/enter-dragon-elephant-chinas-indias-participation-global-health-governance/p30332, accessed March 6, 2018.

rise of both countries in the international system – aside from the ongoing hot debate about the nature of China's rise,[11] and, to a lesser extent, India's rise in both academic and policy circles.[12] The proactive and reactive features of China's and India's policy behavior in the two issue-areas, namely energy security and climate change, as explored in previous chapters – with due regard for caveats about representativeness – suggests that China and India will rise peacefully in the international system. This observation echoes the views of some Chinese and Indian scholars about the emerging powers' collectively peaceful rise in the international system,[13] and some Western

[11] Some argue that China's rise will not be peaceful and will challenge the United States' hegemony: Thomas J. Christensen, *The China Challenge: Shaping the Choices of a Rising Power* (New York: W.W. Norton & Company, 2015); Thomas J. Christensen, "Fostering Stability or Creating a Monster? The Rise of China and U.S. Policy toward East Asia," *International Security* 31, no. 1 (Summer 2006), 81–126; Aaron L. Friedberg, "The Future of U.S.–China Relations: Is Conflict Inevitable?," *International Security* 30, no. 2 (Fall 2005), 7–45. Others contend that China is a status quo state and its rise will be peaceful: Zheng Bijian, "China's 'Peaceful Rise' to Great-Power Status," *Foreign Affairs* 89, no. 2 (March/April 2010), 76–91; Alastair Iain Johnston, "China: A Status Quo Power?," *International Security* 27, no. 4 (Spring 2003), 5–56; Avery Goldstein, "Expectations: Interpreting China's Arrival," *International Security* 22, no. 3 (Winter 1997–1998), 36–73; David C. Kang, "Why China's Rise Will Be Peaceful: Hierarchy and Stability in the East Asian Region," *Perspectives on Politics* 3, no. 3 (September 2005), 551–554; Suijian Guo, *China's "Peaceful Rise" in the Twenty-First Century: Domestic and International Conditions* (Burlington: Ashgate, 2006).

[12] As to the rise of India: Amrita Narlikar, "All That Glitters Is Not Gold: India's Rise to Power," *Third World Quarterly* 28, no. 5 (2007), 983–996; Evan A. Feigenbaum, "India's Rise, America's Interest: The Fate of the U.S.–Indian Partnership," *Foreign Affairs* 89, no. 2 (March/April 2010), 76–91; Harsh V. Pant, *Contemporary Debates in Indian Foreign and Security Policy: India Negotiates its Rise in the International System* (New York: Palgrave Macmillan, 2008).

[13] 杜幼康 [Du Youkang], "权力转移理论质疑——以新兴大国中印崛起为视角 [Questioning the Power Transition Theory: A Perspective of the Rise of Emerging Big Countries – China and India]," 《国际观察》 [*International Review*], no. 6 (2011), 32–39; 韦宗友 [Wei Zongyou], "新兴大国群体性崛起与全球治理改革 [The Rise of Emerging Powers as a Group and the Global Governance Reform]," 《国际论坛》 [*International Forum*] 13, no. 2 (March 2011), 8–14; 杨洁勉 [Yan Jiemian], "新兴大国群体在国际体系转型中的战略选择 [Emerging Powers Strategic Options on the Transitional International System]," 《世界经济与政治》 [*World Economics and Politics*], no. 6 (2008), 6–12; 赵干城 [Zhao Gancheng], "发展中大国兴起与国际体系 [The Rise of Large Developing Countries and the International System]," 《现代国际关系》 [*Contemporary International Relations*], no. 11 (2007), 35–39; 宋玉华 [Song

scholars' views on multilateral cooperation in global governance under the leadership of both the declining power – the United States – and the emerging power – China.[14] In addition, this observation is also consistent with those views grounded in defensive realism about China's peaceful rise in the international system.[15]

This optimism about China's and India's peaceful rise in the international system is supported, first of all, by the fact that China and India are willing to make compromises to the established great powers to ensure international security. As discussed in Chapter 4, concerning the controversial Iranian nuclear program, China and India have sided with the United States and other big powers to pressure Tehran to give up its nuclear weapons program, regardless of their heavy reliance on Iranian oil supplies.

Second, China and India are willing to cooperate with the established great powers to provide global public goods to the international community. In the energy sector, China and India have engaged in cooperation with institutions such as the IEA, initiated by the established great powers. In addition, China and India have begun to establish their own strategic petroleum reserves and to develop renewable energy, especially wind and solar power. In terms of climate change, China and India have agreed to undertake voluntary mitigation obligations to reduce their GHG emissions and to subject their

---

Yuhua] and 姚建农 [Yao Jiannong], "论新兴大国的崛起与现有大国的战略 [*The Rise of Emerging Countries and the Strategy of the Existing Great Powers*]," 《国际问题研究》 [*China International Studies*], no. 6 (2004), 50–54; Paul and Shankar, "Status Accommodation through Institutional Means."

[14] Karolina M. Milewicz and Duncan Snidal, "Cooperation by Treaty: The Role of Multilateral Powers," *International Organization* 70, no. 3 (Fall 2016), 841.

[15] There are two opposing views on the nature of China's rise: the defensive realists and the offensive realists. According to the former, China will rise peacefully: Charles L. Glaser, "A U.S.–China Grand Bargain? The Hard Choice between Military Competition and Accommodation," *International Security* 39, no. 4 (Spring 2015), 53; Michael D. Swaine, *America's Challenge: Engaging a Rising China in the Twenty-First Century* (Washington, D.C.: Carnegie Endowment for International Peace, 2011), 339. In stark contrast, according to the latter, China will not rise peacefully: John J. Mearsheimer, *The Tragedy of Great Power Politics*; John J. Mearsheimer, "China's Unpeaceful Rise," *Current History* 105, no. 690 (April 2006), 160–162; Aaron L. Friedberg, *A Contest for Supremacy: China, America, and the Struggle for Mastery in Asia* (New York: W.W. Norton, 2011).

actions to legally binding transparency. All of these efforts will benefit global energy security and climate protection.

Finally, although China and India have been increasingly rising in the international system, the capacity of both countries to shape the outcomes of international affairs, such as international climate change negotiations, has not been increased, but rather weakened compared with those of the established great powers. During the UNFCCC and the Kyoto Protocol negotiations, China and India played an important role in shaping some fundamental norms and principles, such as "common but differentiated responsibilities," reflecting the imbalance and differentiation between the commitments of developed and developing countries in the UNFCCC, and pushing for the exemption of developing countries from any mitigation obligations. However, these norms and principles were revised outright at COP-21 by the United States, together with the EU, meaning that China and India, as well as other developing countries, are now on an equal footing with developed countries to undertake actions to address climate change.

Index

Abdullah (King of Saudi Arabia), 135
ACU (Asian Clearing Union), 125, 136–137
Agarwal, Anil, 198, 211
air pollution, in India and China, 211, 270
al-Bashir, Omar, 138, 146, 150
  ICC arrest warrant for, 143
Amnesty International, 162–163
Annan, Kofi, 145
AOSIS (Alliance of Small Island States), 182, 198, 222, 281
APP (Asia Pacific Partnership on Clean Development and Climate Change), 190–191, 283
asymmetrical interdependence,
  and climate change diplomacy, 223–232
  and climate change policies, 266–269
  and energy diplomacy, 172–176
  faced by China and India, 63–76

Bali climate change summit (COP-13), 199, 221, 228, 230
bandwagoning in China's and India's climate change diplomacy, 282–290
BASIC group, 185–186, 281
Berlin Mandate, 197
Betsill, Michele M., 219
Bhushan, Chandra, 285
bilateral climate change arrangements of India and China, 192–196, 278, 284–286
Blair, Tony, 189
Bo Kong, 16, 169
Bonn climate summit (COP-5), 198
Brazil
  climate diplomacy of, 185
  energy security and climate change policies of, 301
BRIC countries (Brazil, Russia, India, and China), and global energy governance, 12
Brown, Gordon, 162
Buenos Aires climate summit (COP-4), 198
Bull, Hedley, 51
Bush, George W., 128–129, 132, 151, 161

Calder, Kent, 32–33
Cancun climate change summit (COP-16), 185, 199, 201, 206, 229–230, 284
car ownership
  in China, 42
  in India, 42
CDM (Clean Development Mechanism) projects, China's and India's participation in, 213–216, 231
CER (certified emission reduction) credits, 213
CERC (US–China Clean Energy Research Center), 228
China, 23–25
  asymmetrical interdependence faced by, 63–76, 172–175, 223–232, 266–269
  climate change diplomacy of, 11, 28–29, 179–180, 232–233, 280–290, 297
    bandwagoning in, 282–290
    dual-track diplomacy, 180–208
    two-level pressures on, 22, 208–232
  climate change policies of, 4–11, 26, 29–30, 209–211, 234–235, 295, 298

305

China (cont.)
  National Climate Change
    Programme, 227
  pressures from epistemic
    communities on, 235–238
  reactive, 238–248
  subordinated to energy security
    policies, 248–255
  two-level pressures on, 34–35,
    255–270, 297
  economic development/growth of, 6,
    23, 27, 39–48, 259
  energy diplomacy of, 27, 106–112,
    164–176, 280, 297
    with Iran, 112–138
    with Myanmar, 154–164
    with Sudan, 138–154, 296
  energy imports of, 42–47
  and asymmetrical interdependence,
    73–75
  energy insecurity of, 80
  energy security policies of, 4–6, 27,
    29, 79–85, 104–105, 107,
    293–294
    climate change policies
      subordinated to, 248–255
    dependency on United States in,
      70–73
    differences with India, 100–104
    proactive, 85–100
    two-level pressures on, 34–35, 297
  GHG emissions of, 3, 47–48, 238,
    255, 284
    limitations on, 189, 200–201,
      203–204, 218–219, 243
  and global energy and climate
    governance, 16–18, 30,
    278–294, 300–301
  in international system, 301–304
  nuclear non-proliferation policies
    of, 59
  planned economy of, 83
  political system of, 24
  status seeking by, 51–61, 170–172,
    217–223, 261–266
    superpower status, 51
  wealth seeking by, 39–48, 166–170,
    208–217, 256–261
China South–South Climate
  Cooperation Fund, 195

China's Energy Policy 2012, 82
climate change, definition of, 4
climate change diplomacy
  of China and India, 11, 28–29,
    179–180, 232–233, 280–290,
    297
  bandwagoning in, 282–290
  dual-track diplomacy, 180–208
  two-level pressures on, 22,
    208–232
climate change policies, 38–39
  of Brazil and South Africa, 301
  of India and China, 4–6, 9–11,
    29–30, 234–235, 295, 298
  and asymmetrical interdependence,
    266–269
  pressures from epistemic
    communities on, 235–238
  reactive, 238–248
  subordinated to energy security
    policies, 248–255
  two-level pressures on, 255–269,
    297
climate governance, global, 14–16,
  273–278, 298
  China's and India's role in, 16–18,
    30, 278–294, 300–301
Clinton, Hillary, 130
CNOOC (China National Offshore Oil
  Corporation), 107
  activities in Iran, 126, 173
CNPC (China National Petroleum
  Corporation), 106
  activities in Iran, 115, 126
  activities in Myanmar, 158
  activities in Sudan, 138, 143,
    296
  activities in United States, 173
CO_2 emissions, *see* GHG emissions
Cohen, Stephen, 53
Copenhagen climate change summit
  (COP-15), 185, 206, 230
  Accord of, 200, 204, 280
  failure of, 5

Darfur crisis, and Chinese energy
  diplomacy, 142–154
democratization of global climate
  governance, 15
Deora, Murli, 156

Index

developed countries, technology transfer obligations of, 223–224
developing countries
  exemption from limitations on GHG emissions for, 196–205
  proposals for GHG emission reductions by, 202–204, 220, 228
development plans
  of India and China, 83
    climate change in, 243, 247
    energy consumption targets in, 256–258
    energy security in, 83–88, 91–93, 107–109
diplomacy, *see* energy diplomacy; climate change diplomacy
Doha climate change summit (COP-18), 229, 231
domestic policies, *see* national politics/policies
Downer, Alexander, 163
Dubash, Navroz, 195, 211, 245
Durban climate change summit (COP-17), 185, 201, 229–230, 281

economic aid of China and India
  and climate change diplomacy, 187–188
  to Myanmar, 156
economic development/growth, *see also* wealth-seeking objective of states
  of India and China, 6, 27, 39–48, 259
  and climate change debate, 213
  and climate change policies, 256–259
  export dependency of, 67–70
economic systems of China and India, 101
economy, great-power status in, 57–58
Egeland, Jan, 152
EIA (US Energy Information Administration)
  on China's oil imports, 43, 112–138
  on India's NOCs, 167
emissions, *see* GHG emissions
The Emissions Gap Report 2015, 286

energy consumption of India and China, 94–95, 256–258
  targets in FYPs, 256–258
energy diplomacy, 109
  of China and India, 27, 106–112, 164–176, 280, 293–294
  with Iran, 112–138
  with Myanmar, 154–164
  with Sudan, 138–154, 296
energy efficiency in China and India, 94–98, 101–102
energy governance, global, 11–14, 273–278, 298
  China's and India's roles in, 16–18, 30, 278–294, 300–301
energy imports, *see also* oil, imports of China and India
  of India and China, 42–47
  and asymmetrical interdependence, 73–75
energy insecurity of India and China, 80
energy intensity in China and India, 94–98, 102, 219
energy security policies, 38–39
  of Brazil and South Africa, 301
  of India and China, 4–6, 27, 29, 79–85, 104–105, 107–109, 293–294
    climate change policies subordinated to, 248–255
    dependency on United States, 70–73
    differences, 100–104
    literature on, 7–9
    proactive, 85–100
    two-level pressures on, 297
Energy Technology Perspectives 2017 (IEA), 286
environmental colonialism, 212
epistemic communities, pressures on China's and India's climate change policies from, 235–238
equitable burden sharing in climate change approaches, 217–218
  Sino-Indian emphasis on, 180–181
ETS (emissions trading scheme)
  of China, 243–244
  of India, 247

EU
  anti-dumping disputes with China
    and India, 263–265
  bilateral agreements on climate
    change with China and India by,
    192–193
  and China's and India's climate
    diplomacy, 281
  climate change arrangements by,
    283–284
export dependency of China and India,
  67–70
extreme weather and climate events,
  impact in India and China of,
  260–261

Five-Year Plans (FYPs) of India and
  China, 83
  climate change in, 243, 247
  energy consumption targets in,
    256–258
  energy security in, 83–88, 91–93,
    107–109
fragmentation of global climate
  governance, 14, 287–288
free-riding by India and China in
  climate change negotiations,
  221–222
Friedberg, Aaron, 71
FTAs (free trade agreements) of India
  and China, 110
  with Myanmar, 156

G-77/China coalition, 180–182,
  186–187
  rejection of voluntary limitation of
    GHG emissions by, 197–199
G8/G7
  on climate change, Indian and
    Chinese participation in, 200
  and energy-climate debate,
    274–277
G8+5 Climate Change Dialogue,
  Chinese and Indian
  bandwagoning at, 189–190
G20
  on climate change, 192
  and global energy and climate
    governance, 13, 275, 277
Gates, Robert, 130

GDP (gross domestic product)
  Green, of China, 258–259
  of India and China, 40–41
  of United States, 63–64
  as wealth indicator of states, 38–50
Genocide Olympics Campaign,
  153–154
GHG emissions
  of India and China, 3, 24, 47–48,
    181–182, 238, 255, 284
  limitation obligations
    of developed countries, 213–216
    of India and China, 189, 203–204,
      218–219, 243
    of United States, 283–284
  negotiations on reduction of, Chinese
    and Indian approaches to,
    180–182, 196–205
Ghosh, Arunabha, 246
Gilpin, Robert, 31, 36–37
global energy and climate governance,
  11–16, 273–278, 298
  China's and India's roles in, 16–18,
    26, 30, 278–294, 300–301
global health governance, 301
Glosny, Michael A., 56
Goldthau, Andreas, 12
Gourevitch, Peter, 18
great-power status, *see also* superpower
  status
  of India and China, 51–61, 170–172,
    261, 263, 270
  markers of, 52–53
greenhouse gas emissions, *see* GHG
  emissions

HAL (Hindustan Aeronautics Limited),
  163
Harsanyi, John, 49
health governance, global, 301
Hood, Karl, 222
Hu Angang, 210, 242
Hu Jintao
  on climate change, 189, 251
  on Darfur crisis/Sudan, 138, 146,
    151–152
  on energy security, 80–81
  on Green GDP, 258–259
  on Iranian nuclear program, 113,
    124, 128–129, 131

Index

on US–China climate change cooperation, 228
human rights NGOs, pressures on India and China by, 151–152, 171–172, 176
Human Rights Watch, 152
Hurrell, Andrew, 197
Hydro Carbon Vision 2025 (India), 82
Hydro Power Policy (India), 89

IAEA (International Atomic Energy Agency), and Iranian nuclear program, 118–120, 122–124, 128–135
ICC (International Criminal Court), arrest warrant for al-Bashir, 143
ICCN, *see* UNFCCC, negotiations
identities, social, 49
IEA (International Energy Agency), 12–13
  China's and India's non-membership of, 280, 292–293
  on Chinese and Indian NOCs, 168
  on Chinese and Indian oil imports, 43, 166
  emergency system of, 291–292
  and global energy and climate governance, 275–277
  on nuclear energy, 91
  pressure on India and China by, 8, 237
  on world energy demand, 47
IMF (International Monetary Fund), voting quotas of India and China in, 57–58
INDCs (Intended Nationally Determined Contributions), 202–204, 281–282, 286–287
India, 23–25
  asymmetrical interdependence faced by, 63–76, 172–175, 223–232, 266–269
  climate change diplomacy of, 28–29, 179–180, 232–233, 280–290, 297
    bandwagoning in, 282–290
    dual-track diplomacy, 180–208
    two-level pressures on, 208–232
  climate change policies of, 4–11, 29–30, 211–212, 234–235, 295, 298

and foreign policies, 26
pressures from epistemic communities on, 235–238
reactive, 238–248
subordinated to energy security policies, 248–255
two-level pressures on, 22, 34–35, 255–270, 297
economic development/growth of, 6, 23, 27, 39–48, 259
energy diplomacy of, 27, 106–112, 164–176, 280, 297
  with Iran, 112–138
  with Myanmar, 154–164
  with Sudan, 138–154
energy imports of, 42–47
  and asymmetrical interdependence, 73–75
energy insecurity of, 80
energy security policies of, 4–6, 26–27, 29, 79–85, 104–105, 108–109, 293–294
  climate change policies subordinated to, 248–255
  differences with China, 100–104
  proactive, 85–100
  two-level pressures on, 34–35, 297
GHG emissions of, 3, 47–48, 255
  limitations on, 200–201, 203–204, 218–219
and global energy and climate change governance, 16–18, 30, 278–294, 300–301
in international system, 301–304
nuclear capabilities of, 55
  dependency on United States in, 71–73
political system of, 24
status seeking by, 51–61, 170–172, 217–223, 261–266
wealth seeking by, 39–48, 166–170, 208–217, 256–261
Indo-US civil nuclear agreement (2005), 131–133, 195
infrastructure of Iran, Chinese and Indian investments in, 117
inside-out approaches in IR theory, 19
institutional building for climate change, of India and China, 239–248

Integrated Energy Policy (India), 83
intellectual property rights, 225
interdependence, 266
international community
　great-power status recognition by, 60–61
　integration of China and India in, 290
International Federation for Human Rights, 161
international relations theory (IR theory)
　levels of analysis in, 18–23
　two-level pressures analytical framework in, 19, 21, 298–301
international system
　India and China in, 6, 11, 27–28, 301–304
　as IR level of analysis, 18, 20–21
IPCC (Intergovernmental Panel on Climate Change), 235–237
Iran, India's and China's energy diplomacy with, 4–6, 112–138, 170–171, 173–174, 297
Iran–Pakistan–India (IPI) natural gas pipeline, 116
IRENA (International Renewable Energy Agency), 13, 276–277
Israel
　pressure on India from, on Iranian energy ties, 135
　weapons sales to India, 174

Japan
　reactive policies of, 32–33
Jiang Yu, 119
Jiang Zemin, 107, 113, 154
Jinko Solar factory (China), fluoride leak from, 260
JODI (Joint Organizations Data Initiative), 275–276
Johnston, Alastair Iain, 50

Kalam, A. P. J. Abdul, 139, 155
Karti, Ahmed Ali, 149
Keohane, Robert
　on climate change governance, 14, 270, 274
　on interdependence, 62
　on IR theory, 19–21
　on issue areas, 4
　on rationality, 32
　on UNFCCC process, 189
　on wealth of states, 36–37
Kerry, John, 194
Khatami, Muhammad, 114
Krasner, Stephen, 20
Kyoto climate summit (COP-3), 197, 287–288
　Protocol of, 181, 196–205
　　CDM (Clean Development Mechanism) in, 213–216

Lantos, Tom, 132
Layne, Christopher, 66
levels-of-analysis debate in IR theory, 18–23
Leverett, Flynt, 45
Li Baodong, 149
Li Keqiang, 262
Lima climate change summit (COP-20), 202, 281–282
Liu Guijin, 148, 150
LMDC group, 185–186
loan/aid-for-oil deals, of India and China, 111
　with Sudan, 140–141
Lula da Silva, Luiz Inácio, 58

Ma Zhaoxu, 149
Mann, Thura Shwe, 155
Mearsheimer, John, 35, 37, 52
MEF (Major Economies Forum on Energy and Climate Change), 191
middle classes, in India and China, 42
military cooperation, use for energy diplomacy, 111, 117–118, 141–142, 157
military expenditure of United States, 64–65
minilateral forums for climate change, 14
Mistry, Dinshaw, 299
Modi, Narendra, 81, 124, 252
Morgenthau, Hans, 32, 35
Mukherjee, Pranab, 158
Mulford, David C., 133
multipolarity in global energy governance, 12

Myanmar, Chinese and Indian energy diplomacy with, 154–164, 171, 297

NAMAs (nationally appropriate mitigation actions), 199–200, 228
NAPCC (National Action Plan on Climate Change, India), 245–246, 253
Narrain, Sunita, 198, 211
National Electricity Policy (India), 89
national oil companies, *see* NOCs (national oil companies)
national politics/policies
  of China and India, 24
  on climate change, 209–212, 270
  and international system, 19–21, *see also* two-level pressures
  and China's and India's energy security and climate change policies, 23, 26
National Tariff Policy (India), 89
naval power
  as great-power status marker, 56–57
  of United States
    China's dependency on, 70–73
NCCP (National Climate Change Programme, China), 241, 253
negotiations, 220
Negroponte, John, 147
neoclassical realism, two-level analytical framework of, 22
neo-mercantilism, in Indian and Chinese energy policies, 7
Netherlands Environmental Assessment Agency, pressures on China by, 238
NGOs, human rights related, pressures on India and China by, 151–152, 171–172, 176
NOCs (national oil companies) of India and China, 4–7, 107–110
  activities in Iran, 115–117, 126–127
  activities in Myanmar, 158–159
  activities in Sudan, 138, 143–144
  facilitation of, 28, 175

and United States and Western allies, 137–138, 173–174
and wealth seeking, 166–170
nonlogical behavior, 34
norm-abiders, India and China as, 59–60
NPT (Nuclear Non-Proliferation Treaty), 55
  Chinese non-ratification of, 59
  Indian non-ratification of, 55, 72, 195
  on nuclear weapon states, 55
nuclear capabilities
  and great-power status, 54–56
  of India, 71–73
  of Iran, Chinese and Indian policies toward, 118–120, 122–124, 128–135, 170–171
  of United States, 65–67
nuclear energy development in China and India, 91–94
nuclear non-proliferation, Indian and Chinese policies on, 59–60
  and Iranian nuclear program, 118–120, 122–124, 128–135, 170–171
Nye, Joseph S., 37, 62, 65

Obama, Barack, 73, 129, 131, 134, 136, 184, 228
oil
  emergency systems for market disruptions, 291–292
  equity investments of China, 296
  imports of China, *see* loan/aid-for-oil deals
  imports of China and India, 43–44, 74, 99
    diversification of, 166–167
    from Iran, 115–117, 120–121, 124–125, 135–138
    from Sudan, 143–144
  industry
    of China, 106–108
    of India, 108–110
  SPRs (strategic petroleum reserves), of India and China, 98–100, 291–292
  supplies disruptions of, 45–47

Olympic Games
  in Beijing and Darfur crisis, 153–154
  hosting of, as great-power status marker, 58–59
ONGC (Oil and Natural Gas Corporation) Videsh Limited (OVL) (India), 108, 169
opprobrium costs, 50
outside-in approaches in IR theory, 20, 26

PACE (US–India Partnership to Advance Clean Energy), 195
Pareto, Vilfredo, 34
pariah states, Indian and Chinese relations with, 8, 27
  and energy diplomacy, 112, 167, 171–172, 297
Paris climate change summit (COP-21)
  Agreement of, 204, 207, 282, 284, 287–288
  success of, 5
patents
  for climate mitigation-related technologies, 225, 228
Patey, Luke A., 169
Paul, T. V., 55
Peters, Glen, 285
photovoltaic (PV) industry, in China, 259, 267–268
political systems, of India and China, 100
pollution problems, of India and China, 270
population size
  and great-power status, 51
  of India and China, 42
poverty in India and China, 41
proactive policies, 32–33
  of India and China
    on climate diplomacy, 232
    on energy diplomacy, 112–121, 138–144, 154–159, 164–176
    on energy security and climate change, 4–6, 23, 27, 75, 85–100
  in international relations, 298–299
Putin, Vladimir, 58
Putnam, Robert, 19, 21, 23, 299

Qin Gang, 238

Ramesh, Jairam, 183, 201, 220–221, 289
Rao, Nirupama, 120
rational choice theory, IR application of, 25, 34, 299–300
rationality, 25
  and self-interest of states, 31–32
reactive policies, 32–33
  of India and China
    on climate diplomacy, 222–223, 232
    on energy diplomacy, 121–138, 144–154, 159–160, 164–165
    on energy security and climate change, 6, 23, 27, 29, 238–248
  in international relations, 298–299
  and rational choice, 34
regime complex, of global climate change governance, 14
renewable energy development in India and China, 86–91, 102–103, 259–260, 263–265
  and asymmetrical interdependence, 266–269
Rose, Gideon, 22

safety of nuclear energy, 92
sanctions
  imposed on India and China by United States, 137–138
  imposed on Iran, 297
    Chinese and Indian policies toward, 119, 135–138
  imposed on Myanmar, 160
Saud al-Faisal (Prince), 130
Saudi Arabia, pressure on India and China from, on energy ties with Iran, 130, 133, 135, 174
Saunders, Phillip C., 56
Save Darfur Coalition, 151
Schweller, Randall L., 22, 52
security threats, *see also* energy security policies
  and Chinese and Indian energy dependencies, 45–47
Sein, Thein, 155
self-interest of states, rational pursuit of, 31–32
Sengupta, Sandeep, 197
Shwe, Than, 155

Singh, Manmohan, 81, 124, 134, 160, 190, 240, 262
Sino-Indian bilateral climate alliance, 182–185
Sinopec (China Petrochemical Corporation), 107
Sino-US bilateral climate cooperation, 278, 284–286
Smith, Adam, 37
SOEs (state-owned enterprises) in renewable energy industries (China), 88, 102
solar energy development
 in China, 88, 260
 in India, 90, 262, 268
solar panels, anti-dumping disputes on, 263–265
South Africa
 climate diplomacy of, 185
 energy security and climate change policies of, 301
South Sudan, secession from Sudan, 148–150
sports, great-power status seeking in, 58–59
SPRs (strategic petroleum reserves) of India and China, 98–100, 291–292
states, *see also* status seeking by states
 self-interest of, 31–32
status seeking by states, 49–53
 China and India, 51–61
  and climate change diplomacy, 217–223
  and climate change policies, 261–266
  and energy diplomacy, 170–172
Sudan, Chinese and Indian energy diplomacy with, 138–154, 171–172, 296–297
superpower status, 51, *see also* great-power status
 of China, 51
 of United States, 56, 63–67
sustainable development and climate change debate, 212
systems theory approaches in international relations, 20–21

Tajfel, Henri, 49
Tang Jiaxuan, 159

technology imports, China's and India's asymmetrical dependence on, 266–269
technology transfer and climate change diplomacy, 223–229
temperature, global rises and limitation efforts, 286–287
Ten-Thousand Enterprises Program (China), 101, 104–105
trade relations
 global governance of, 300
 of India and China
  with Myanmar, 156
  with Sudan, 139–141
  with United States, 172–173, 263–265
two-level pressures
 on China and India
  climate diplomacy, 208–232
  energy diplomacy, 166–176
  energy security and climate change policies, 22–23, 26, 34–35, 75, 255–270, 297
 in IR theory, 19, 21, 298–301

UN officials, pressure on China by, 152–153
UN peacekeeping forces in Darfur, 145
 Chinese promotion of, 146–148, 152–153
UN Security Council
 climate change debates in, 250–251
 permanent membership of
  of China, 24
  as great-power status marker, 52–53
  of India, 53–54, 73
 resolutions
  on Darfur crisis, 142–143, 145
  on Iranian nuclear program, 123, 129–131, 297
  on Myanmar, 157–158, 161
UNFCCC (United Nations Framework Convention on Climate Change)
 climate change definition of, 4
 and global energy and climate governance, 277–278, 298
 legitimacy of, 14
 negotiations, 183

UNFCCC (United Nations Framework Convention on Climate Change) (cont.)
  Chinese and Indian participation in, 9–10, 28, 180–189, 196–205, 217–223, 280–290, 304
  on MRV/transparency requirements, 205–208
  on technology transfer, 223–224
United States
  anti-dumping disputes with India and China, 263–265
  bilateral cooperation with India and China
    on climate change, 194–196, 278, 283–286
    on energy-related technology development, 228
  India's and China's asymmetrical dependence on, 67–73, 172–176
  pressures on India and China from, on Iranian energy ties, 127–138
  superpower status of, 56, 63–67

Vajpayee, Atal Bihari, 114
Van de Graaf, Thijs, 13
verification of GHG emission limitations, Chinese and Indian approaches to, 205–208
Victor, David, 14, 189, 214, 274
Vihma, Antto, 16
voluntary limitation of GHG emissions
  Chinese and Indian rejection of, 196–205
  by developing countries, 220, 228
  by United States, 283–284

Waltz, Kenneth
  on interdependence, 63
  on IR theory, levels of analysis in, 18–20
  on status of states, 51
  on superpower status, 50
  on survival of states, 36
Wang Guangya, 142, 146

Warsaw climate change summit (COP-18), 185, 202
wealth-seeking objective of states, 34–39
  of China and India, 39–48
  and climate change diplomacy, 208–217
  and climate change policies, 256–261
  and energy diplomacy, 166–170
weapons
  Chinese and Indian sales of
    to Iran, 117
    to Myanmar, 157, 162–163
    to Sudan, 141
  Israeli sales of, to India, 174
Wen Jiabao, 124, 146, 160, 206
Wendt, Alexander, 20–21
wind energy development
  in China, 88, 103, 263, 267
  in India, 103, 263
Witte, Jan Martin, 12
Woods, Ngaire, 57
World Bank, voting quotas of India and China in, 57–58
World Energy Outlook 2007 (IEA), 8, 237
World Energy Outlook 2014 (IEA), 91
World Energy Outlook 2015 (IEA), 168, 237
World Energy Outlook 2016 (IEA), 255
Wu Bangguo, 113

Xi Jinping, 81, 149, 251–252
Xiaoyu Pu, 52
Xie Zhenhua, 211, 267

Yanzhong Huang, 301
Young, Oran, 50

Zhai Jun, 147
Zhang Gaoli, 289
Zhou Wenzhong, 142
Zhuhai Zhenrong (company), 137
Zoellick, Robert, 128, 171